Nanostructure Science and Technology

Series Editor:
David J. Lockwood, FRSC
National Research Council of Canada
Ottawa, Ontario, Canada

More information about this series at http://www.springer.com/series/6331

Yi Ge • Songjun Li • Shenqi Wang
Richard Moore

Editors

Nanomedicine

Principles of Nanomedicine

Volume 1

 Springer

Editors
Yi Ge
Centre for Biomedical Engineering
Cranfield University
Bedfordshire, United Kingdom

Songjun Li
School of Materials Science and Engineering
Jiangsu University
Zhenjiang, China

Shenqi Wang
Advanced Biomaterials and Tissue
   Engineering Centre
College of Life Science and Technology
Huanzhong University of Science
   and Technology
Wuhan, China

Richard Moore
Biomimesis
Melton Mowbray
Leicestershire
United Kingdom

ISSN 1571-5744   ISSN 2197-7976 (electronic)
ISBN 978-1-4614-2139-9   ISBN 978-1-4614-2140-5 (eBook)
DOI 10.1007/978-1-4614-2140-5
Springer New York Heidelberg Dordrecht London

Library of Congress Control Number: 2014954585

Preface

This book is comprised of two complementary volumes: *Principles of Nanomedicine* (Volume 1) and *Perspectives of Nanomedicine* (Volume 2). The purpose of this arrangement is to provide a more comprehensive overview of a new and potentially revolutionary branch of healthcare – nanomedicine.

The European Technology Platform on Nanomedicine has defined nano-medicine as:

> the application of nanotechnology to achieve breakthroughs in healthcare. It exploits the improved and often novel physical, chemical and biological properties of materials at the nanometre scale. Nanomedicine has the potential to enable early detection and prevention, and to essentially improve diagnosis, treatment and follow-up of diseases.

Nanomedicine is one of the fastest-growing sub-disciplines of nanotechnology, which itself is described by the US National Nanotechnology Initiative as "science, engineering, and technology conducted at the nanoscale...". One nanometre (1 nm) is one billionth of a metre (10^{-9} m) and the nanoscale is usually referred to as being the range 1 to 100 nm. A further important consideration is that the term nanotechnology is often used for materials or surfaces that have been intentionally altered or manipulated at or around the nanoscale so as to provide useful new properties that come into play at this scale.

It is the ability to consistently measure and manipulate matter at the nanoscale that makes nanotechnology so potentially valuable in medicine. In biology, the basic building blocks of life are themselves nanostructures, and being able to interface nanoscale materials and structures with biological structures at this level gives rise to new opportunities to understand disease mechanisms and to intervene in them by exploiting the often novel characteristics of nanomaterials and nanoscale architecture.

Nanotechnology has the potential to make an impact on nearly all branches of medicine: from earlier and more accurate diagnosis, to the enhanced imaging of smaller and finer structures, to novel medical devices, to highly targeted and more effective drugs, to regenerative medicine and to more personalised models of medicine. Furthermore, nanomedicine has the potential to address currently unmet medical needs and growing challenges such as diseases associated with an ageing population and unhealthy lifestyles.

Nanomedicine is highly interdisciplinary in nature and brings together fields such as biology, materials science, engineering and information technology. For example, a typical biosensor used in medicine may comprise a biologically-based sensing component, a physical transducer to generate a measurable signal and an IT component to process the data generated.

The first volume, *Principles of Nanomedicine*, provides an overview of the key principles that underpin the development of nanomedicine and describes a range of applications that are emerging within this rapidly-growing new discipline. It consists of chapters that:

- Define nanomedicine and introduce its main sub-disciplines
- Describe "top-down" and "bottom-up" synthetic approaches
- Examine why nanomedicine can be described as both interdisciplinary and revolutionary
- Describe how life is a complex assembly of interacting nanoscale mechanisms and examine how nanobiology is driving new trends in medicine
- Explain the increasing role of nanotechnology in underpinning the diverse and rapidly-growing fields of tissue engineering and regenerative medicine
- Present examples of widely-used simulation and modelling methods that can contribute to the design of nanomaterials utilised in nanomedicine
- Describe a variety of medical nanomaterials including novel and biologically-inspired materials
- Describe multifunctional nanoparticles that can be used for theranostic and imaging applications
- Provide an overview of how nanotechnology can be used to design novel biosensors that provide powerful new diagnostic tools, for example for neurodegenerative diseases, cardiovascular diseases and cancers
- Examine the contribution nanotechnology can make to both improving medical devices and developing revolutionary, new advanced devices
- Examine the potential that remotely-controlled micro- and nano-robots may have for minimally-invasive procedures within the body and in microfluidic diagnostics
- Discuss the current challenges in pharmaceutical science associated with nanotechnology and how nanotechnology can be used in a variety of ways to address issues such as stability, solubility, efficacy, targeting to desired disease sites and overcoming physiological and anatomical barriers
- Explore issues surrounding toxicity and risk in nanomedicine, describe research on developing new risk assessment strategies, outline regulatory approaches and highlight areas for further risk research
- Examine the key ethical principles that underpin the implementation of nanomedicine and other emerging medical and related technologies

Nanomedicine will only fulfil its potential if it can be brought successfully and safely to the market and to the clinic. The second volume, *Perspectives of Nanomedicine*, therefore builds upon those principles outlined in Volume 1 with a series of perspectives on how the commercialisation and clinical implementation of nanomedicine can be facilitated and the challenges that need to be addressed to accomplish these objectives.

Many expectations have been placed on the medical applications of nanotechnology. Chapter 13 examines these expectations, considers the phenomenal pace of research and development, and seeks to differentiate between the hyperbole accorded to some themes of research and the reality of applying the results of this research within a clinical setting.

Nanomedicine cannot expect to have a significant market and clinical impact unless it can be taken up and developed commercially. Chapter 14 explores the business dimensions of developing nanomedical products, including finance and cost issues, business models and strategies, and potentially disruptive effects on existing business models and product markets.

Nanotechnology continues to develop at an astonishing rate and much of this progress has the potential to make a positive impact on medicine. However concerns, such as the potential toxicity of some nanomaterials, remain. Chapter 15 examines the lessons, both positive and negative, that can be learned from this rapid evolution and from on-going research in nanotechnology in order to drive nanomedicine further in the right direction.

Nanotechnology is one of a number of emerging and enabling technologies that are likely to revolutionise the way healthcare is delivered. Chapter 16 explores this intersection of technologies and discusses some of the changes that are likely.

Nanotechnology holds much promise for overcoming challenges in drug design, development and delivery. A number of nano-based pharmaceuticals have already been approved with a great number more in product development pipelines. Chapter 17 explores the need for future regulatory systems that will be able to keep pace with rapid developments in nanopharmaceuticals and that can deliver effective and safe products to the healthcare systems.

Although widely applicable for the diagnosis and treatment of many diseases, one of the fields in which research in nanomedicine has progressed the furthest is in the diagnosis and treatment of cancer. Chapter 18 describes how operating at the nanoscale allows for an impressive level of diversity in approaches and capabilities that enable a range of nanoparticles to address an equally diverse range of targets and provide an integrated, personalized approach to diagnosis and therapy, especially in cancer disease management.

With a rapidly growing number of nanomedical products in research and development, it is useful to review those steps and processes a potential nanomedical product will need to pass through before it reaches the market or clinic. Chapter 19 reviews these processes and other challenges and provides insights into the data and evidence that developers will need to accumulate to achieve the objectives of market entry and clinical adoption.

In these two volumes of the book, each chapter is accompanied by an extensive list of references that will help direct the reader towards additional sources of information.

The editors would sincerely like to thank Springer Science+Business Media LLC, and especially Dr. Kenneth Howell and Ms. Abira Sengupta, for everything that made the publication of this book possible.

Last but not least, welcome to the magic world of nanomedicine. The next new era of nanomedicine is coming!

Bedfordshire, UK	Yi Ge
Zhenjiang, China	Songjun Li
Wuhan, China	Shenqi Wang
Melton Mowbray, UK	Richard Moore

Contents

Chapter 1
Nanomedicine: Revolutionary Interdiscipline

Ferdia Bates

1.1 Introduction

Nanomedicine can be defined as the application of nanotechnology to medicine. One may be forgiven for thinking that, given the fact that it is defined as a subsection of nanotechnology, its boundaries are finite; this assumption is, however, deceptive. The application of the field to medicine has opened up an entirely new horizon of applications and possibilities. Application of this technology to the physiological system has also provided an opportunity to study these biological systems on an altogether more intimate level which has allowed scientists to replicate and refine the established mechanisms in order to apply them in the form of innovative and highly beneficial new technology. A second assumption arising from this definition, especially given the title of this work, is that the cause for referring to nanomedicine as an interdiscipline stems from it being an amalgamation of medicine with an emerging technological field. This, again, is not the case. Delving into this exciting and emerging discipline can be, at times, similar to plunging into a Carroll-esque rabbit-hole with regard to the capabilities of materials that, in their bulk forms, border on the mundane.

Approaching the research and development of nanomedicine chronologically, the technology as a concept can be traced back to the lecture given by the infamous Richard Feynman at Caltech in 1959 titled "There's plenty of room at the bottom". In this talk, Feynman outlined the potential that the harnessing nanoscopic materials could yield [1]. Using the directions given by Feynman, the logical train of thought would lead the inquisitive researcher directly from macro to nano or, from top down. This method of research and development was, and still is, an extremely fruitful method for synthesis and research of nanotechnology; indeed, the products of

F. Bates (✉)
Universitat Autònoma de Barcelona, Barcelona, Spain
e-mail: ferdia.bates@mariecurie.cat

© Springer Science+Business Media New York 2014
Y. Ge et al. (eds.), *Nanomedicine*, Nanostructure Science and Technology,
DOI 10.1007/978-1-4614-2140-5_1

top-down synthesis of nanotechnology are an integral part of modern life. At this point in time, one would be hard pressed not to find an electronic device with did not contain a large amount of nanoscopic circuitry.

This being said, however, for the true revolution, one must look up rather than down. Bottom up synthesis of nanotechnology may not be immediately intuitive for the layman observer but it has proven to be the most fruitful in terms of the yield of technology. Indeed, the nanomedicine that is used in the creation of new medical techniques is predominantly created using the bottom-up techniques and thus, it provides the most justification for the use of the word 'revolution' [2].

Nanomedicine, to reiterate, is far from a finite field. It encompasses a great scope of technology all of which is at the forefront of research and development. For this paper, the sub headings of nanobiotechnology, nanotechnology and nanobiomimetics are used in the definition of nanotechnology as an interdiscipline [3–6]. What follows in a brief discussion and explanation as to why nanomedicine has captured the attention of researcher, physicians and laymen alike.

1.2   Enhancement of Existing Technology: The Top-Down Revolution

When discussing nanotechnology, particularly if one does not have previous experience in the area, the logical thought process would be to start at the macroscopic, or bulk modulus, and to reduce the dimensions until nanoscopic proportions have been achieved. Indeed, this is how much of the nanotechnology currently on the market is synthesised. Lithography, in various forms is one of the most widely spread forms of this top down method, indeed, the computer or mobile phone that is in the vicinity or on the person of the reader is sure to have numerous circuit boards which were printed utilising a lithographic process. Techniques such as photolithography have been an integral part of the fabrication process of microchips in the computing industry for 50 years [7], with the dimensions of the circuits on the chips getting smaller and smaller in accordance with Moore's law, which predicts that computing circuits should double in power or half in size every 18 months [8]. It may surprise the reader as to the degree to which this form of nanotechnology is incorporated into everyday life. There are, in fact, several types of this technique in practice today, all of which work at the nano level. One might very well question the use of the word 'Revolution' in the context of technology of this age however, the advancement and refinement of this technology must also be acknowledged; at this point, techniques such as photolithography have been refined to low nanoscale proportions [9]. Indeed, it is this refinement that directly facilitates the increase in computing power that this generation has become so accustomed to.

Nanomaterials have also been used to replace their bulk counterparts. In this manner, though nanomaterials are being used, they are merely being exploited as smaller versions of their bulk counterparts rather than as an entirely new material such as will be discussed below (see Sect. 1.5.2). Utilising the example of carbon nanofibres being used to replace carbon fibre as the providers of tensile strength in composite materials or the addition of nanoparticles as strength enhancement agents, a notable and highly useful increase in material strength though it still only serves the same purpose as it did before in the sense that only its existing properties have been enhanced rather than new ones created [10, 11].

The 'top-down' aspect of nanomedicine is perhaps neglected in the face of the more glamorous 'bottom-up' technologies but it nonetheless plays a crucial role in the development of the field albeit, using nanotechnology's role in increasing computer processing power as an example, a back ground one.

1.3 From the Bottom-Up, Introduction of New Technology and the True Revolution

It is important to make the distinction between this section of nanomedicine and the previous one. While the products yielded from top down synthesis are highly innovative, costs for the synthesis process are high and yields are low. Bottom-up synthesis, on the other hand, relies on the chemical synthesis of nanomaterials through various processes. The product achieved is heterogeneous, pure, and greatly more cost effective than its top-down predecessor.

This synthesis procedure has allowed for production of high quality particles and nanostructures on a scale that has allowed the wide spread development of the field [12, 13]. It is also through this synthesis that the smallest particles can be synthesised, those with the most cutting edge functions such those with the ability to transcend the blood brain barrier (see Sect. 1.5.2), a capability that has one of the farthest reaching implications in terms of transmuting new areas of research into curative techniques and thus halting ailments which until now have existed with meagre attenuation from existing clinical treatments, in particular, neurodegenerative conditions.

'Bottom-up' is somewhat of an umbrella term in the sense that it incorporates a great deal of techniques and can refer to both precipitation of a polymer [14] or equally the agglomeration of gold atoms to form nanoparticles [15]. The binding factor between these synthesis processes is that both particles were begot from a mixture of fluids in which they grew to form a colloidal suspension. As was said above, the key advantage to this technique is the ability to produce monodisperse particles of extremely small size.

1.4 Interdisciplinary Medicine, Why the Fuss?

Collaboration and amalgamation of ideas and applications within the various traditional disciplines of medicine has always occurred. An obvious example of this is the prescription of a single medication for several different ailments in a manner such as to 'harness' side effects observed when characterising the drug during trials; some well-known examples of these being aspirin [16, 17], duloxetine [18, 19] and Viagra [20, 21]. Likewise, the marriage of disciplines to birth an entirely new sub-specialty in response to demand is by no means uncommon; Palliative care, Medical Ethics and Biomedical Engineering to name a few. Nanomedicine, however, must be classed as something entirely different for the predominant reason that it uses such a term as interdiscipline in a far more intensive manner.

1.5 Technologies of Nanomedicine

The term 'nanomedicine' was coined to not only to describe the application of several separate technologies to medicine, but also to encompass the hybridisation of these technologies within the application of medicine. The three main subsections of nanomedicine, as shown in Fig. 1.1, can be given as nanotechnology, nanobiotechnology and nanobiomimetics. These three particular groupings were chosen because of their easily definable boundaries. Indeed these three research streams were all well established as individual fields before they were ever combined under the universal banner of nanomedicine. Thus, the term 'nanomedicine' can be used to describe any one of these sections or, more importantly, any one combination of these three sections. The word of "combination" is the key component in

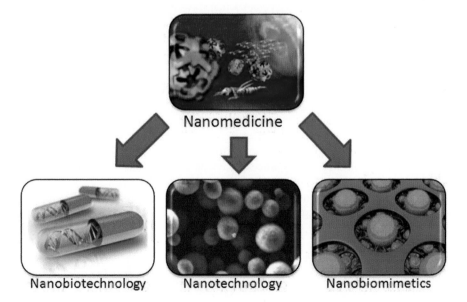

Fig. 1.1 The three main subdivisions of Nanomedicine (Copyright-free images)

the description of nanomedicine particularly when using such implicit adverbs as 'revolutionary'. Indeed, it can be seen from the brief definitions given in Sects. 1.2 and 1.3. The majority of the advantageous treatments and techniques that are described as nanomedicine come from the hybridisation of two separate components. For instance, the delivery of a therapeutic nanoparticle to a site by incorporating a specific biological or non-biological ligand on to its surface; or vice versa, the delivery of a therapeutic compound by attaching it to a magnetic nanoparticle for example.

1.5.1   Nanobiotechnology

To start with nanobiotechnology, it can be defined as the global title given to the intersection of nanotechnology and biology, in other words, it is the exploitation of nanobiological entities for the application to medicine [22]. This can be taken as the oldest form of nanomedicine to be practiced, as demonstrated by the publication dates of the landmark papers cited. One of the first examples for the application of such technologies can be found almost 40 year ago with the discovery of monoclonal antibodies [23]. The degree of integration of this technology into daily modern life may come as a surprise to the layman. Indeed, such common house hold items as the pregnancy test rely on Enzyme-linked Immunosorbent Assay (ELISA) tests which, in turn, rely on enzyme sequences of nanoscopic dimensions; it is to these enzymes that the modern test owes its accuracy [24]. ELISA tests are also used extensively as diagnostic tools within the clinical setting for the determination of disease through the quantification of proteins or antigens with, in the case of the diagnosis of HIV, accuracies up to 98 % [25, 26]. Nanobiotechnology also holds a special prevalence in therapeutics. Techniques such as gene [27] and protein [27, 28] therapy all rely on extracted biological sequences. Intracellular delivery techniques such as viral capitation [29] also owe their existence to nanobiotechnology. Such biological entities also provide the mechanism for many active therapeutic delivery protocols using the phenomenon of antibody-antigen binding by attaching the relevant ligand sequence to the surface of the delivery vessel thus creating a biological recognition element. The age of the field by no means implies that it is a relic of the past; indeed, nanobiotechnology has kept pace with its sister streams of nanomedicine in terms of innovation and development. Prototypical examples of this can be found in the development of nanobodies [30]. Nanobodies are approximately one order of magnitude smaller than their parent molecules, conventional monoclonal antibodies, and thus have greater stability as well as a reduced chance of causing an immunogenic response in vivo. This is advantageous for the therapy because, with this reduction in immunogenicity, comes improved residence time and biodistribution thus improving overall efficacy [31]. Synthesis of these nanobodies is achieved by isolating and replicating the particular site on the antibody where the desired interaction takes place.

1.5.2   Nanotechnology

This brings the discussion neatly onto the second facet of nanomedicine, nanotechnology, which is probably the subsection most commonly associated with nanomedicine. This term is used, in this case, in the more specific sense of all non-biological technologies on the nanoscale. Nanoparticles are the prototype example of this and can be used in a sweeping reference to any nanoscopic rod, tube, cube, sphere etcetera, in the sense they are completely alien to the physiological system and thus have completely unique behaviours; this is opposed to biotechnology discussed above in which existing properties are harnessed and exploited. Such nanomaterials have proven to be highly versatile in both their properties and their applications.

What gives nanotechnology its celebrity is, perhaps, the fact that all nanomaterials are of identical composition to their bulk moduli; though they exhibit properties which are more attributable to quantum mechanics than conventional Newtonian behaviour. This is caused by nanomaterials' extremely high surface area to volume ratio, demonstrated in Fig. 1.2, which increases exponentially with respect to decrease in particle size. With this increase in surface area also comes an increase in surface atoms, which are less rigidly bound atoms that consequently are more disposed to reaction with their environment.

The extent of this decrease in size also allows nanoparticles to reach physiological crannies which were inaccessible to conventional molecules. The 'crannies' of interest for therapeutics, are the blood-brain [32, 33] and blood-testis [34, 35] barrier both of which are notoriously difficult to bypass and thus delivery of treatments to these two vital organs have remained a challenge to modern science. The small size of nanoparticles allows them to navigate the tight junctions which have been impermeable to so many other treatments.

This phenomenon has been harnessed to create some of the most notable innovations in nanomedicine [36].

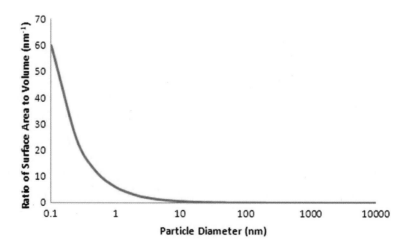

Fig. 1.2 The size effect: the relationship between particle size and surface area to volume ratio for a theoretical spherical particle

1.5.3   The Revolution of Interdiscipline

Active delivery of therapeutics is perhaps one of the most widely known techniques associated with nanomedicine. This is done by attaching a therapeutic to a vector which can then be directed to the targeted site. An example of this is the use of magnets to guide iron oxide nanoparticle-conjugated therapeutics to the target site, most commonly; this is a cancerous tumour. The therapeutic can be, and this is where the revolution begins to manifest, either a conventional chemotherapeutic, a gene or protein as was discussed above, or the particle itself [37–40]. In all cases, increased specificity and efficacy are achieved through the use of nanomaterials.

Conventional therapies can also be enhanced by combining them with nanotechnology; contemporary chemotherapeutics can receive a great decrease in observed side effects through the delivery of the drug to the target site through encapsulation within liposomes. Liposomes are bi-lipid membranes capable of encapsulating several varieties of molecules including hydrophobic, hydrophilic and crystalized drugs as well as therapeutic gene sequences. The liposomes' advantage lies in their ability to be used as triggered release vessels for molecules thus creating a prodrug derivative of the conventional chemical. They are also highly useful for 'disguising' a molecule to improve biodistribution and residence time as well as allowing hydrophobic molecules to disburse in aqueous systems, something highly advantageous in the physiological environment [41–43]. Indeed, in the quest for increased efficacy of treatment, several different components can be combined in order to create a hybrid treatment designed for a very specific target site, thus guarding against non-specific damage to the surrounding tissue. In one example, [44] a magnetisable element has been added to a thermally sensitive liposome containing chemotherapeutics with a temperature dependent trigger; in this strategy, even when the liposome is manually directed to the target site, release will only take place if the thermal parameter, caused by tumour-associated inflammation, is met. Even at this point, the chemotherapeutic still requires the thermal element as well, all of which is designed to defend the patient against the life threatening and debilitating side effects that one was obliged to endure in the past.

To elaborate on the use of nanoparticles themselves as therapeutics, in the case of magnetic particles, they hold a significant advantage over conventional treatment, as do many nanomedications, of being initially inert, or rather, a 'prodrug', this implies that the particle will remain non-interactive until such a time as is appropriate for it to execute its function which, in the case of the magnetic particles, is done via an oscillating magnetic field thus concentrating the effect to the target area and minimising non-specific damage to the surrounding tissue [39]. Of course there are several different nanomaterials that can be used as either detection, delivery or therapeutic agents, some of the most common materials used are heavy atom-elements such as gold [45] and platinum [46] or specific structures of elements such as carbon nanotubes [47]. Though there is an Aladdin's cave of applications of nanotechnology within this field, as this work is not intended to be a review of such, it will not be dwelled upon.

Non-biological nanotechnology has also offered safer solutions to innovative techniques produced by nanobiotechnology. Again, to give but a single example of this, the use of viral delivery vectors in gene therapy, as discussed above, can be quite dangerous given the handling of infectious viruses required to replace the viral payload with the therapeutic; the use of anionic polymer nanoparticles has been proposed as an effective alternative delivery vector [48, 49].

Self-assembly of complex nanostructures has also been permitted through the interdisciplinary collaborations within nanomedicine. Study of DNA, RNA or protein-mediated self-assembly of biological structures has led to another revolutionary line of research, namely, the exploitation of this biological mechanism to facilitate the design and synthesis of complex nanostructures [50]. Again, this innovative new technique stems from the research lines of surface functionalization of nanoparticles, more specifically, the conjugation of a ligand to the surface of a nanoparticle so as to add an antigen-antibody recognition site to the particle. It is a short step to then remove the target site from its host and instead attach it to a different particle thus creating the blue print for the orientation and organisation of the particles and consequently removing the need for costly and time consuming manipulation [51, 52]. Indeed, the potential for this application is truly great; nanoscaffolds for the repair or construction of proteins or nanoparticle arrays for tissue engineering applications [53] are already being proposed and designed.

Thus the unique abilities that the combination of these two already highly innovative subsections should be becoming apparent; the enhancement and optimisation of treatments that has been observed because of such collaborations should, in itself, be ample cause to bestow the title of 'revolutionary' on the field of nanomedicine. There is, however, a third subsection which has come, perhaps as the final destination to the train of thought that amalgamated nanobiotechnology with nanotechnology. This is nanobiomimetics.

1.5.4 Nanobiomimetics

Nanobiomimetics can be defined as the design and synthesis of nanomaterials using the structures and mechanisms of biological systems. It is the logical progression from the previous two subsections of nanomedicine described in the previous sections. Biomimetics, in fact, is a discipline that stretches back over half a century. It is however the mimicry of biological structures at the nanoscale that makes this facet noteworthy [54]. Some of nature's most intriguing mechanisms are confined to the nanoscale, this newfound ability to mimic and refine or improve them holds many significant advantages and thoroughly demonstrates nanomedicine's worthiness for the title of a 'revolutionary interdiscipline', or at the very least acts as a demonstration of how nanomedicine has truly come full circle, the full revolution if you will. To reiterate, the intent of this piece is not to provide a review of the state of the art, but rather to globally demonstrate the innovation of the field. The prototypical example of nanobiomimicry can be taken as the development of aptamers and molecularly imprinted polymers or 'MIPs'. In the case of MIPs, it must be added that the use of the term 'nano-mimicry' comes from the ability of the polymer to form imprint sites on a scale of singular molecules on the angstrom (0.1 nm) level. These imprints are contained within pores within the structure of the MIP while the polymer itself can be manipulated to form both particles as well as layers and films on surfaces. Within the molecular imprinting of particles, two distinct strategies can be defined; that of bulk polymerisation, where a polymer 'brick' is allowed to form and is then broken to yield imprinted monoliths, and precipitation polymerisation where nano and macro particles are grown in a solvent suspension. These can be named the top-down and bottom-up synthesis protocols within the field of MIP particles. The applicability of MIPs to the field of nanobiomimetics is due to the increasing volumes of literature being published reporting imprinted nanoparticles via this 'bottom-up' strategy of particle formation [55, 56].

MIPs seek to serve the same purpose as monoclonal antibodies did in the 1970s in the sense that recognition of a molecular target or antigen can be achieved by imprinting a template molecule into the surface of a polymer. This imprint renders the polymer a 'synthetic antibody', or rather, synthetic nanobody, as only the specific binding sites remain [57, 58]. MIPs and nano-MIPs are vastly cheaper than biological ligands as well as having superior stability and extremely high selectivity and specificity [59]. Indeed, this imprinting technique is far from limited to the recognition of biological ligands, MIPs can be imprinted with pharmaceutical compounds for example, thus making them a highly versatile tool in both the synthesis process as well as delivery, distribution and controlled release of a drug [60, 61].

Aptamers can be defined as artificial oligonucleotides or peptide sequences [62] which can be chemically synthesised to form a recognition element for anything from amino acids and proteins to drug molecules or other miscellaneous molecules. The versatility of these ligands is complimented by their inherent lack of immunogenicity in vivo. Having been first reported in the nineties, aptasensors, which are sensors utilising immobilised aptamers as their detection mechanism, have proven to be superior to biological antibodies through superior chemical stability as well as cost effectiveness [63]. Their versatility has also led to their wide spread use as sensors in both laboratory and clinical settings for diagnostic and discriminatory applications [64, 65].

1.6 Conclusions and Future Outlook

Nanomedicine is an innovative and emerging field with roots which can be traced back through several decades. Though nanotechnology is interspersed throughout, and is inseparable from modern life, the true capabilities of nanomedicine are still being realised. Top-down synthesis of nanomaterials, which is expensive and low-yielding, has given way to bottom-up synthesis which has proven to be a vastly more cost effective and efficient way to synthesise products as well as providing results with a much higher degree of precision.

Nanomedicine is an umbrella term used to encompass all applications of nano-technology to the medical field. The 'bottom-up' nanotechnologies can be divided into three further subcategories, namely nanobiotechnology, nanotechnology and nanobiomimetics. The interdiscipline of nanomedicine comes with the combination of these categories, which are themselves well defined fields, to form innovative novel multidimensional therapies designed to enable the discrimination between the targeted site and the surrounding healthy tissue. What makes these combinational therapies exceptional, and nanomedicine truly a revolutionary interdiscipline, is the unlimited number of dimensions a single treatment can have and thus, a so-called 'intelligent' medicine can be developed which will only execute its purpose once the target site has been reached and identified. The spirit of this interdiscipline and 'combinational' ethos can be found embodied in the discipline of nanobiomimetics whereby the biological systems themselves are studied and replicated in order to achieve a hybrid mechanism which has advantages over both the original biological system and the synthetic alternative.

All of these factors allow for nanomedicine to garner with the title of a 'revolutionary interdiscipline'.

References

1. Feynman RP (1960) There's plenty of room at the bottom. Eng Sci 23:22–36
2. Sabatier PA (1986) Top-down and bottom-up approaches to implementation research: a critical analysis and suggested synthesis. J Public Policy 6:21–48
3. Brzicová T, Feliu N, Fadeel B (2014) Research highlights: highlights from the last year in nanomedicine. Nanomedicine 9(1):17–20
4. Bogue R (2013) Inspired by nature at the nanoscale. Sens Rev 33:19–24
5. Bhushan B (2010) Springer handbook of nanotechnology. Springer, Berlin
6. Whitesides GM (2003) The 'right' size in nanobiotechnology. Nat Biotechnol 21:1161–1165
7. Noyce RN (1977) Microelectronics. Sci Am 237:63–69
8. Brock DC (2006) Understanding Moore's law: four decades of Innovation. CHF Publications, Philadelphia
9. Shi Z, Kochergin V, Wang F (2009) 193 nm superlens imaging structure for 20 nm lithography node. Opt Express 17:11309–11314
10. Veedu VP, Cao AY, Li XS, Ma KG, Soldano C, Kar S, Ajayan PM, Ghasemi-Nejhad MN (2006) Multifunctional composites using reinforced laminae with carbon-nanotube forests. Nat Mater 5:457–462
11. Quaresimin M, Varley RJ (2008) Understanding the effect of nano-modifier addition upon the properties of fibre reinforced laminates. Compos Sci Technol 68:718–726
12. Lee SW, Chang W-J, Bashir R, Koo Y-M (2007) "Bottom-up" approach for implementing nano/microstructure using biological and chemical interactions. Biotechnol Bioprocess Eng 12:185–199
13. Wang YL, Xia YN (2004) Bottom-up and top-down approaches to the synthesis of monodis-persed spherical colloids of low melting-point metals. Nano Lett 4:2047–2050
14. Haupt K, Mosbach K (2000) Molecularly imprinted polymers and their use in biomimetic sensors. Chem Rev 100:2495–2504
15. Zhang H, Fung KH, Hartmann J, Chan CT, Wang DY (2008) Controlled chainlike agglomeration of charged gold nanoparticles via a deliberate interaction balance. J Phys Chem C 112:16830–16839

16. Krein SL, Vijan S, Pogach LM, Hogan MM, Kerr EA (2002) Aspirin use and counseling about aspirin among patients with diabetes. Diabetes Care 25:965–970
17. Chan AT, Fuchs CS, Ogino S (2009) Aspirin use and survival after diagnosis of colorectal cancer. Gastroenterology 136:A55
18. Elmissiry M, Mahdy A, Ghoniem G (2011) Treatment of female stress urinary incontinence: what women find acceptable and the impact of clinical and urodynamic evaluation on their final choice. Scandanavian J Urol Nephrol 45:326–331
19. Frampton JE, Plosker GL (2007) Duloxetine: a review of its use in the treatment of major depressive disorder. CNS Drugs 21:581–609
20. Cheitlin MD, Hutter AM Jr, Brindis RG, Ganz P, Kaul S, Russell RO Jr, Zusman RM (1999) Use of sildenafil (Viagra) in patients with cardiovascular disease. Technology and Practice Executive Committee. Circulation 99:168–177
21. Lalej-Bennis D, Sellam R, Selam JL, Slama G (2000) How to prescribing Viagra in practice.... Diabetes Metab 26:416–420
22. Dordick JS, Lee KH (2014) Editorial overview: nanobiotechnology. Curr Opin Biotechnol 28:iv–v
23. Schwaber J, Cohen EP (1973) Human x mouse somatic cell hybrid clone secreting immuno-globulins of both parental types. Nature 244:444–447
24. Bastian LA, Nanda K, Hasselblad V, Simel DL (1998) Diagnostic efficiency of home pregnancy test kits. A meta-analysis. Arch Fam Med 7:465–469
25. Kannangai R, Ramalingam S, Prakash KJ, Abraham OC, George R, Castillo RC, Schwartz DH, Jesudason MV, Sridharan G (2001) A peptide enzyme linked immunosorbent assay (ELISA) for the detection of human immunodeficiency virus type-2 (HIV-2) antibodies: an evaluation on polymerase chain reaction (PCR) confirmed samples. J Clin Virol 22:41–46
26. Cordes RJ, Ryan ME (1995) Pitfalls in HIV testing. Application and limitations of current tests. Postgrad Med 98:177–180
27. Schwarze SR, Ho A, Vocero-Akbani A, Dowdy SF (1999) In vivo protein transduction: delivery of a biologically active protein into the mouse. Science 285:1569–1572
28. Jo D, Liu DY, Yao S, Collins RD, Hawiger J (2005) Intracellular protein therapy with SOCS3 inhibits inflammation and apoptosis. Nat Med 11:892–898
29. Yamanaka R (2004) Alphavirus vectors for cancer gene therapy (Review). Int J Oncol 24:919–923
30. Cortez-Retamozo V, Backmann N, Senter PD, Wernery U, De Baetselier P, Muyldermans S, Revets H (2004) Efficient cancer therapy with a nanobody-based conjugate. Cancer Res 64:2853–2857
31. Goel A, Colcher D, Baranowska-Kortylewicz J, Augustine S, Booth BJM, Pavlinkova G, Batra SK (2000) Genetically engineered tetravalent single-chain Fv of the pancarcinoma monoclonal antibody CC49: improved biodistribution and potential for therapeutic application. Cancer Res 60:6964–6971
32. Kreuter J, Shamenkov D, Petrov V, Ramge P, Cychutek K, Koch-Brandt C, Alyautdin R (2002) Apolipoprotein-mediated transport of nanoparticle-bound drugs across the blood-brain barrier. J Drug Target 10:317–325
33. Modi G, Pillay V, Choonara YE, Ndesendo VMK, Du Toit LC, Naidoo D (2009) Nanotechnological applications for the treatment of neurodegenerative disorders. Prog Neurobiol 88:272–285
34. Liu G, Swierczewska M, Lee S, Chen X (2010) Functional nanoparticles for molecular imaging guided gene delivery. Nano Today 5:524–539
35. Chen Y, Xue Z, Zheng D, Xia K, Zhao Y, Liu T, Long Z, Xia J (2003) Sodium chloride modified silica nanoparticles as a non-viral vector with a high efficiency of DNA transfer into cells. Curr Gene Ther 3:273–279
36. Krpetić Z, Anguissola S, Garry D, Kelly PM, Dawson KA (2014) Nanomaterials: impact on cells and cell organelles. In: Nanomaterial. Springer, Netherlands, pp 135–156
37. Dobson J (2006) Magnetic nanoparticles for drug delivery. Drug Dev Res 67:55–60

38. Mcbain SC, Yiu HHP, Dobson J (2008) Magnetic nanoparticles for gene and drug delivery. Int J Nanomedicine 3:169–180
39. Kim DH, Kim KN, Kim KM, Lee YK (2009) Targeting to carcinoma cells with chitosan- and starch-coated magnetic nanoparticles for magnetic hyperthermia. J Biomed Mater Res A 88A:1–11
40. Tani J, Faustineand Sufian JT (2011) Updates on current advances in gene therapy. West Indian Med J 60:188–194
41. Kito A, Yoshida J, Kageyama N (1987) Basic studies on chemotherapy of brain tumors by means of liposomes: affinity of sulfatide-inserted liposomes to human glioma cells. Brain Nerve 39:783–788
42. Kikuchi H (2004) Application of stealth liposomes to cancer chemotherapy and gene therapy. Biotherapy 18:353–360
43. Fortier C, Durocher Y, De Crescenzo G (2014) Surface modification of nonviral nanocarriers for enhanced gene delivery. Nanomedicine 9(1):135–151
44. Pradhan P, Giri J, Rieken F, Koch C, Mykhaylyk O, Doblinger M, Banerjee R, Bahadur D, Plank C (2010) Targeted temperature sensitive magnetic liposomes for thermo-chemotherapy. J Control Release 142:108–121
45. Giljohann DA, Seferos DS, Daniel WL, Massich MD, Patel PC, Mirkin CA (2010) Gold nanoparticles for biology and medicine. Angew Chem-Int Ed 49:3280–3294
46. Salata O (2004) Applications of nanoparticles in biology and medicine. J Nanobiotechnol 2:3
47. Liu Z, Tabakman S, Welsher K, Dai H (2009) Carbon nanotubes in biology and medicine: in vitro and in vivo detection, imaging and drug delivery. Nano Res 2:85–120
48. Xiong MP, Bae Y, Fukushima S, Forrest ML, Nishiyama N, Kataoka K, Kwon GS (2007) pH-responsive multi-PEGylated dual cationic nanoparticles enable charge modulations for safe gene delivery. ChemMedChem 2:1321–1327
49. Wu GY, Wu CH (1987) Receptor-mediated in vitro gene transformation by a soluble DNA carrier system. J Biol Chem 262:4429–4432
50. Hirst AR, Escuder B, Miravet JF, Smith DK (2008) High-tech applications of self-assembling supramolecular nanostructured gel-phase materials: from regenerative medicine to electronic devices. Angew Chem-Int Ed 47:8002–8018
51. Chhabra R, Sharma J, Liu Y, Rinker S, Yan H (2010) DNA self-assembly for nanomedicine. Adv Drug Deliv Rev 62:617–625
52. Yan H, Yin P, Park SH, Li HY, Feng LP, Guan XJ, Liu DG, Reif JH, Labean TH (2004) Self-assembled DNA structures for nanoconstruction, DNA-based molecular electronics: international symposium on DNA-based molecular electronics, 725:43–52
53. Hosseinkhani H, Hosseinkhani M, Kobayashi H (2006) Design of tissue-engineered nanoscaffold through self-assembly of peptide amphiphile. J Bioact Compat Polym 21:277–296
54. Kumar CSSR (2010) Biomometic and bioinspired nanomaterials. Wiley, Weinheim
55. Li S, Ge Y, Piletsky SA, Lunec J (2012) Molecularly imprinted sensors: overview and applications. Elsevier, Boston
56. Haupt K (2003) Peer reviewed: molecularly imprinted polymers: the next generation. Anal Chem 75:376 A–383 A
57. Bossi A, Bonini F, Turner APF, Piletsky SA (2007) Molecularly imprinted polymers for the recognition of proteins: the state of the art. Biosens Bioelectron 22:1131–1137
58. Bui BTS, Haupt K (2010) Molecularly imprinted polymers: synthetic receptors in bioanalysis. Anal Bioanal Chem 398:2481–2492
59. Ruigrok VJ, Levisson M, Eppink MH, Smidt H, Van Der Oost J (2011) Alternative affinity tools: more attractive than antibodies? Biochem J 436:1–13
60. Lulinski P (2011) Molecularly imprinted polymers in pharmaceutical sciences. Part II. Applications in pharmaceutical analysis. Polimery 56:3–10
61. Lulinski P (2010) Molecularly imprinted polymers in pharmaceutical sciences. Part I. The principles of molecular imprinting. Applications in drug synthesis and drug delivery systems. Polimery 55:799–805

62. O'sullivan CK (2002) Aptasensors–the future of biosensing? Anal Bioanal Chem 372:44–48
63. Lim YC, Kouzani AZ, Duan W (2010) Aptasensors: a review. J Biomed Nanotechnol 6:93–105
64. Giovannoli C, Baggiani C, Anfossi L, Giraudi G (2008) Aptamers and molecularly imprinted polymers as artificial biomimetic receptors in affinity capillary electrophoresis and electro-chromatography. Electrophoresis 29:3349–3365
65. Abelow AE, Schepelina O, White RJ, Vallee-Belisle A, Plaxco KW, Zharov I (2010) Biomimetic glass nanopores employing aptamer gates responsive to a small molecule. Chem Commun (Camb) 46:7984–7986

Chapter 2
Nanobiology in Medicine

Hariprasad Thangavel

2.1 Introduction to Nanobiology

The term 'nano' is a Greek word for 'dwarf', meaning one billionth. It was first used by N. Taniguchi in 1974. A quest for nano began from the noble lecture, 'There is plenty of room at the bottom' presented by the Noble Laureate, Professor Richard Feynman in 1959. In the 1980s, K. Eric Drexler popularized the word 'nanotechnology'. His idea was to build machines on the molecular scale [1]. The principle behind this technology is engineering and manufacturing structures, devices, and systems that have novel properties at the atomic and molecular level. Later, scientists and researchers successfully employed nanotechnology to explore the boundaries of biomedical sciences. They made it possible to use biological processes to construct biocompatible nanostructures. There are several approaches to construct nanostructures and they are top-down (miniaturization) approach, bottom-up (building from atoms and molecules) approach and functional (building materials with desired functionality) approach [2].

Biological studies focused at extremely minuscule to molecular levels are termed as Nanobiology. Most of the fundamental biological functions take place at the level of molecular machineries that have a size range of less than 100 nm. Figure 2.1 demonstrates a good size comparison of biological structures in the nanometric scale [3]. The emergence of nanobiology made opportunities to better understand the functions of these molecular machineries with the help of scanning probe microscopy, modern optical techniques, and micro-manipulating techniques.

H. Thangavel (✉)
Department of Chemistry and Chemical Technologies, University of Calabria,
Rende (CS), Italy
e-mail: hariprasad.thangavel@unical.it

© Springer Science+Business Media New York 2014
Y. Ge et al. (eds.), *Nanomedicine*, Nanostructure Science and Technology,
DOI 10.1007/978-1-4614-2140-5_2

Fig. 2.1 Relative sizes of
different biological structures
in nanometric scale [3]

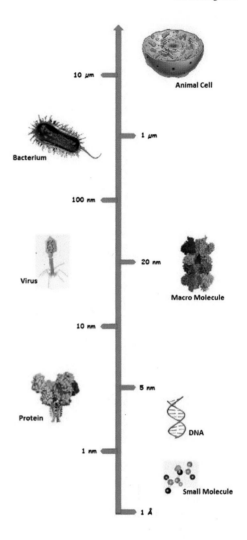

Nanobiology research interests can be roughly grouped into two basic categories: nanotechnologies applied to biological systems, and development of biologically-inspired nanotechnologies. The main reason for this categorization is to differentiate between the sources of inspiration. However, nanobiology research are mainly focussed at nanobiological structures and systems, biomimetics, nanomedicine, nanoscale biology, and nanointerfacial biology [4]. The tools, techniques, and technologies derived from nanobiology are applied to medical field directly and indirectly, giving rise to a new ground of medicine termed Nanomedicine. Nanobiology by itself is a blend of various different disciplines not limited to physics, chemistry, biology, computation, and engineering [4]. The different themes of nanobiology can be well understood from Fig. 2.2.

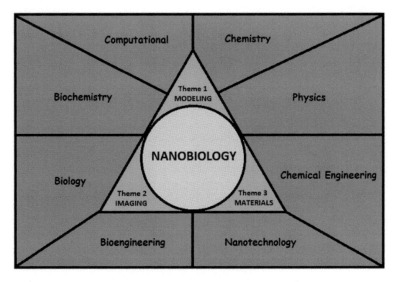

Fig. 2.2 Nanobiology map showing three different themes [4]

Nanobiology can make significant contributions to medicine by achieving success in early detection and diagnostic strategies, and in treatment and prevention of fatal diseases. Based on enhanced efficacy, nanoparticles of polymers, metals or ceramics can conflict life challenging conditions like cancer, AIDS, heart and brain disorders. They can invade bacteria, virus, parasites, etc. Nanoparticles are generally characterized based on their size distribution, shape, surface area, surface reactivity, surface charge, chemical composition, and aggregation state. Factors such as specificity, solubility, stability, biocompatibility, biodegradability, and pharmacokinetics are to be considered predominantly while using nanoparticles in medicine. The purpose of employing nanobiology in medicine is to diagnose diseases early and accurately; and to treat them effectively without any side effects. It has proven potential to cure human cancer by specifically killing the targeted cancer cells leaving the surrounding healthy tissues harmless. Nano-drug delivery system can provide new insight for the treatment of tuberculosis (TB). DNA-based nanotechnology is a new arrival in molecular medicine [5]. Nano-drug therapy sounds promising compared to traditional therapies [6]. It is strongly believed that nanobiology has all the probability to benefit mankind by improving the quality of life, without any doubts.

2.2 Nanobiology and Human Biology

In terms of medical application, the main thing to consider about nanomaterials is their sustainability in the human body. These small materials exhibiting unique surface properties at nanoscale size could be toxic, causing adverse ill effects.

Table 2.1 FDA-approved gold based nanomaterials used in therapeutics [8]

Product/ brand name	Component/ active ingredient	Delivery route	Target	Company	Current status
Verigene	Gold	In vitro diagnostics	Genetics	Nanosphere	FDA-approved
Aurimmune	Colloidal gold nanoparticle coupled to TNF-α and PEG-Thiol (~27 nm)	Intravenous	Solid tumor	Cyt-Immune sciences	Phase-II
Auroshell	Gold coated silica nanoparticles (~150 nm)	Intravenous	Solid tumor	Nanospectra biosciences	Phase-I
Combidex (Ferumoxtran-10)	Iron oxide nanoparticles (17–20 nm)	Intravenous	Tumor imaging	Advance magnetics	NDA filed

The nanomaterials used on purpose should be able to dissolve inside the body without leaving any side effects when they are no longer needed. Nanomaterials have an unknown behavioural property when compared with bulk materials. For example, nanomaterials made of inert element like gold become extremely active at nanometric dimensions [7] and it is one of the most used nanomaterial in medicine. A few FDA-approved gold based nanomaterials which are currently used in therapeutics are listed in Table 2.1. In a research conducted by Han et al., functionalized gold nanoparticles proved to be highly attractive for drug delivery because of their distinctive dimensions, surface tenability, and controlled drug release [9].

The nanoparticles enter human body through four major routes: nasal, oral, dermal and intravenous. Upon entry, they can be distributed throughout the body including brain, heart, lungs, gut, liver, spleen, kidney, and skin. Inside the body, nanoparticles behaviour can be disturbed or altered by factors such as hydrophobic, hydrophilic, lipophobic, lipophilic, active catalysis or passive catalysis. Nanoparticles enter the cell by one of the following four mechanisms: passive diffusion, facilitated diffusion, active transport, and endocytosis. The passive diffusion is achieved by electrochemical or concentration gradient driven mechanism. Positively charged small particles of approximately 20 nm size undergo passive diffusion. Facilitated diffusion is for small particles of size 10–30 nm and these particles get internalized via selective membrane protein channels and concentration gradient driven mechanism. The active transport and endocytosis are meant for larger particles of size ranging from 50 to 500 nm. The former is facilitated by transport protein and are energy dependent, against concentration gradient whereas the later is collective term for energy dependent internalization of substances, forming vesicles.

Several pathways of nanoparticle endocytosis include, phagocytosis, Clathrin-mediated endocytosis, Caveolae-mediated endocytosis, macropinocytosis, other Clathrin- and Caveolae-independent endocytosis, transcytosis (occurs in epithelial cells in blood brain barrier). The possible fates of internalized nanoparticles in the

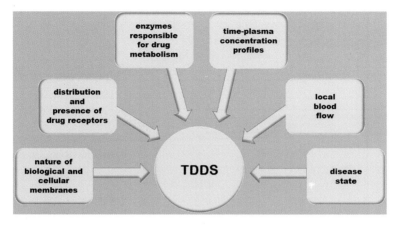

Fig. 2.3 Parameters to be considered for designing effective targeted drug delivery system (TDDS) [8]

cells are: enzymatic degradation or destruction by pH effect (acidic), exocytosis or transcytosis, transportation to intracellular locations escaping from endolysosomal compartment. The internalization of nanoparticles in the cells can cause molecular irregularities and incompatibilities resulting in fatal disorders ranging from interstitial fibrosis in respiratory system to acute coronary syndrome in vascular system. Because of some unresolved complications, not all the nanoparticles are readily used for medical applications. Only very few nanoparticles such as liposomes, dendrimers, organic polymers, quantum dots are tested to be safe in medicine and are found to be promising for in vivo imaging, in vitro diagnostics, and drug delivery. A few important parameters that are to be considered while designing an effective targeted drug delivery system are shown in Fig. 2.3.

2.3 Nanomaterials for Medicine

Materials with structural elements having dimensions in the range of 1–100 nm are termed as nanomaterials. Nanomaterials is a common term applicable for nanoscale materials, nanophase materials, and nanostructured materials. These three materials are different from one another; nanoscale materials, where the material itself fall under nanoscale regime; nanophase materials, these are hybrid materials having nanoscale phase or component; nanostructured materials, here the material structure will have nanoscale size or features [10]. Nanomaterials exhibit unique characteristic properties when compared to their bulk states. These improved properties along with large surface areas made them ideal for use in medical field. Nanomaterials play vital role in medical diagnosis and therapeutics ranging from fluorescent imaging to site-specific targeted drug delivery. More recently, Zhang et al. discussed how the nanomaterials could be designed based on their interactions with biological systems to

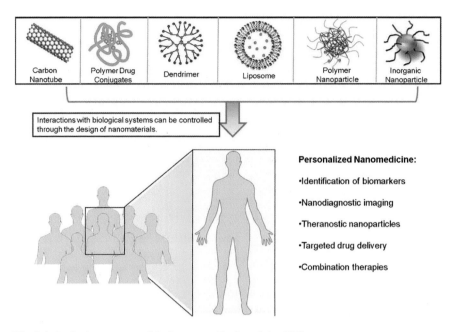

Fig. 2.4 Designing nanomaterials for personalized medicine [11]

meet their role in medical diagnostic, imaging and therapeutics [11]. The graphical view of the above context is depicted in Fig. 2.4. Nanomaterials of biological interest are mainly from two distinct groups: organic nanoparticles such as dendrimer, polymer; and inorganic nanoparticles such as gold, silver, iron, titanium dioxide. The other miscellaneous groups include carbon-based nanoparticles, organic and inorganic hybrids, liposomes, and protein and peptide-based nanoparticles.

Some examples of commonly used nanoparticles in medicine: multifunctional nanoparticles (best suited for intravenous delivery), lipid and polymer nanoparticles (trigger strong immune response), gold and magnetic nanoparticles (promising response in targeted drug delivery), virus based nanoparticles (acts as nanocarriers), dry powder aerosol (inhalable nanoparticles for lung specific treatment), smart nanomaterials (tunable by external stimuli).

Gold nanoparticles can be used for detection purposes. The gold nanoparticle with a DNA linker can result in a large size complex particle. A characteristic colour transition from pink to gray can be seen with naked eye. In medical diagnostics, gold nanoparticles can be used for laboratory analysis. Recently, a study on the development of a gold nanoparticle amplified agglutination system for weak blood group determination has been reported by Wiwanitkit et al. [12, 13]. In fact, gold nanoparticles found its entry in medical filed way back in 1800s. It was used as an important ingredient in many classical Chinese medicinal remedies. Jung et al. treated site-specific coupling of protein G to DNA oligonucleotide with a biolinker for efficient antibody immobilization [14]. The study reported that antibody targeting on glass slides could be immobilized using this linker system without modifying or spotting antibodies, and the protein G-DNA conjugate brought a simple but effective method to label DNA-functionalized gold nanoparticles with target antibodies. Jung et al. concluded that the DNA-linked protein G construct introduced in this study offered a useful method to manage antibody immobilization in many immunoassay systems [14].

Silver nanoparticles is another commonly used nanomaterial for medical research which is proven to have promising antibacterial properties. Silver nanoparticles changes its colour with size. Silver particle having a size smaller than 10 nm appears to be golden yellow in colour. The colour further changes into red and black when the particle size increase. In 2006, Panacek et al. reported that the size of silver particles have a significant impact on their antibacterial activity. The study revealed that a very low concentration of silver gave antibacterial characteristic and thus the bactericidal property was found to be size-dependent of the silver particles [15]. In another quoted work, Gogoi et al. reported that above a certain concentration, silver nanoparticles were found to reduce the sizes of treated bacteria in addition to their characteristic bactericidal activity [16]. Pal et al. identified that silver nanoparticles underwent a shape-dependent interaction with *Escherichia coli* in their work on determining the antibacterial properties of differently shaped silver nanoparticles against the gram-negative bacteria both in liquid broth and on agar plates [17]. Also, an increase in the antibacterial activities of penicillin G, amoxicillin, erythromycin, clindamycin, and vancomycin in the presence of silver nanoparticles was reported by Shahverdi et al. [18]. As like gold nanoparticle, silver nanoparticles can also be used in the DNA linker system.

Superparamagnetic iron oxide nanoparticles are widely used in molecular and cellular imaging [19]. The major advantages of using iron oxide-based nanomaterials are their nontoxic property and biocompatibility. In a study reported by Thorek et al. superparamagnetic iron oxide nanoparticles demonstrated their utility as a novel tool for enhancing magnetic resonance contrast, allowing researchers to monitor physiological and molecular changes, in addition to previously monitored anatomical changes [20]. Hu et al. fabricated magnetic sponge-like hydrogels called ferrosponges by using an *in-situ* synthesis of magnetic iron nanoparticles in the presence of various concentrations of gelatin [21]. By using these unusual magnetic sensitive properties of the ferrosponges, a new drug delivery system can be designed to use in medicine. Titanium dioxide is another inorganic oxide nanoparticle which is recently found to be useful in medical application. Nano TiO_2 is studied to have good antibacterial properties as like silver nanoparticles. An effective organic degradation process was documented by Peralta-Hernandez et al. in their work on investigating the photo catalytic properties of nanostructured TiO_2 – carbon films obtained by means of electrophoretic deposition [22].

Viral nanoparticles (VNPs) and Virus-like particles (VLPs) are naturally occurring bionanomaterials [23]. Both of them are very promising candidates for developing 'smart' devices which can be used in several medical applications ranging from tissue-specific imaging to targeted-drug delivery. The major advantages of using viral particles are their exceptional stability and biocompatibility. To add a few more, they are monodisperse, programmable, multifunctional, and easy to produce on large scale [24]. Cohen et al. investigated the use of aptamer-labeled MS2 bacteriophage capsids for targeted in vitro photodynamic therapy [25]. In their study, they demonstrated a unique virus-based loading strategy for efficient targeted delivery of photoactive compounds for site-specific photodynamic cancer therapy using biologically derived nanomaterials [25]. More recently in a similar study, Zeng et al. affixed folic acid (FA) as targeting moiety on the rigid Cucumber Mosaic Virus capsid and stuffed the interior cavity of CMV with substantial load of doxorubicin (Dox) to design a controlled drug delivery system for cancer therapy [26].

2.4 Nanobiology: Applications in Medicine

Nanobiology can help medicine in many aspects by providing a new range of tools and techniques that can be applied in early detection, disease diagnosis, non invasive treatment, and also in disease prevention. Some of the applications of nanomaterials to medicine include: fluorescent biological labels for imaging, drug and gene delivery platform, detection of pathogens and proteins, probing of DNA structure, tissue engineering and regeneration, tumour destruction via heating, separation and purification of biological molecules, magnetic resonance imaging (MRI) contrast enhancement, and phagokinetic studies [27].

2.4.1 Diagnostic Applications of Nanoparticles

In medicine, nanoimaging and nanovisualization contributes a lot more in recent diagnostics. The use of nanoparticles in imaging can provide new insights to medical diagnostics. Nanomaterials and nanotechnology combined with novel devices have the potential to address emerging challenges in medical field. The application of nanomaterials for imaging and visualization will offer fast, sensitive, and cost effective solutions for the modern clinical laboratory. Rotomskis et al. reported that the properties of nanomaterials such as quantum effect and surface area effect could improve the sensitivity of biological detection and imaging at least by 10–100 folds [28]. Basically, the nanodiagnosis can be classified into naked eye diagnostic system, immunological diagnostic system, and molecular diagnostic system. The protein precipitation systems to qualitatively test the protein in body fluids are the best example for naked eye detection system. The colour changing property of nanomaterials due to aggregation of particles can be applied to diagnostic tests in medicine [29]. Wiwanitkit et al. proposed gold nanoparticle as an alternative tool for the detection of microalbuminuria [12, 13]. Normal urine triggers the precipitation of gold nanoparticle solution resulting in a gray coloured mixture, while urine samples with protein do not. In addition to protein detection, the naked eye detection system using nanoparticles can also be applied to detect small substances such as hormones in body fluid. Gold nanoparticle solution can be used for the detection of human choriogonadotropin (hCG) in urine, which is a basic diagnostic test to confirm pregnancy in females. The cost of gold nanoparticle is cheaper than the urine strip test. In another study, Bauer et al. aimed at the direct detection of sub-molecular layers of DNA with naked eye, based on the understanding of absorption property of metal nanoparticles [30]. The study focused on the nanolayer coated metallized-PET-chip setup and on the synthesis of DNA nanoparticle conjugates suitable for resonance amplified absorption -point of care tests and the applied usage of those particles in the direct visualization of DNA-DNA binding events [30].

Immunodiagnosis by nanoparticle is the new stage of immunological test in medicine. Over the last decade, the immunological diagnostic systems using nanoparticles achieved considerable momentum in the field of medicine. Immunological diagnostic systems employ nanoparticle labeling. The best example for nano-enabled immunological diagnosis is luminescent quantum dot in immunoassay. Quantum dots are emerging as a new class of biological labels with properties and applied usages that are not available with traditional organic dyes and fluorescent proteins [31]. Zhu et al. reported quantum dots as a new immunoflorescent detection system for *Cryptosporidium parvum* and *Giardia lamblia* [32]. The study concluded that this new fluorescence system exhibited superior photostability, gave 1.5–9 fold higher signal-to-noise ratios than traditional organic dyes in detecting *C. parvum*, and allowed couple-colour detection for *C. parvum* and *G. lamblia* [33]. In addition to quantum dots, superparamagnetic nanoparticles are also used in immunoassays. Kuma et al. reported their development of a liquid phase immunoassay system using magnetic nanoparticles [34]. They developed a highly sensitive immunoassay system using Fe_3O_4 magnetic nanoparticles [34]. Other than these, nanoparticles like silicon di oxide, europium-doped lanthanum fluoride, europium-doped gadolinium oxide, cadmium telluride are also used in nanodiagnosis [35–38]. In another work, Hwang et al. proposed the use of gold nanoparticle-based immunochromatographic test for identification of *Staphylococcus aureus* from clinical specimens [39]. Hwang et al. reported that this detection method was fast, easy to perform, and had a long shelf life at room temperature [39]. More recently, Jiang et al. reported a single step synthesis method to produce water soluble Ag_2S quantum dots for in vivo fluorescence imaging [40]. This work proposed potential cadmium-free (and lead-free) quantum dots for nanodiagnostics and in vivo imaging [40].

Finally, molecular diagnosis is rapidly advancing with the help of nanotechnology. There is a steady progress in the use of electrochemical biosensors for DNA analysis, over the past few years. In 2005, Shen et al. studied polymerase chain reaction (PCR) of nanoparticle-bound primers [41]. They concluded that with either one or two primers respectively bound to the nanoparticle surface, PCR could proceed completely under optimized conditions, having been subjected to certain rules [41]. Kalogianni et al. reported a simple and inexpensive assay that allowed visual detection and demonstration of the PCR-amplified sequences by hybridization within minutes [42]. According to their study, the nanoparticles bound to the target DNA through hybridization, and the hybrids were captured by immobilized streptavidin at the test zone of the strip, producing a characteristic red line [42]. In another quoted study, the use of plasmonics-based nanoprobes that acted as molecular sentinels for DNA diagnostics were demonstrated. The plasmonics nanoprobe was composed of a metal nanoparticle and a stem-loop DNA molecule tagged with a Raman label, and the nanoprobe utilized the specificity and selectivity of the DNA hairpin probe sequence to detect a specific target DNA sequence of interest [43]. The study completely demonstrated the specificity and selectivity of the plasmonics nanoprobes to detect PCR amplicons of the HIV gene [43].

2.4.2 Therapeutic Applications of Nanoparticles

In recent days, the drug formulation and development becomes easier with the advancement of nanopharmacology. The application of nanotechnology in pharmacology is aimed at finding out novel pharmacological molecular entities; targeted site-specific drug delivery within the body; and providing personalized treatment to reduce side effects and increase drug effectiveness [44]. A graphical view on the beneficial effects of targeted drug delivery system (TDDS) are illustrated in Fig. 2.5 [8].

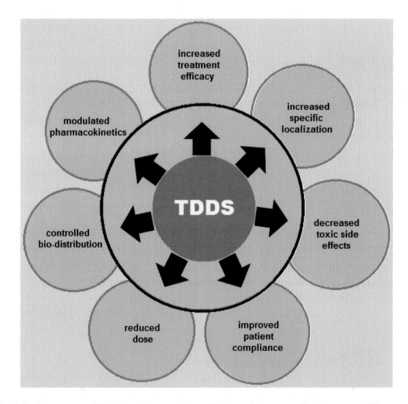

Fig. 2.5 Advantages of TDDS to improve drug efficiency for personalized treatment [8]

Table 2.2 Nanoparticles currently used in therapeutics

Type of nanoparticles	Material used	Application
Polymeric nanoparticles	Biodegradable polymers	Controlled and targeted drug delivery
Quantum dots	CdSe-CdS core-shell	Targeting and imaging agent
Nanopores	Aerogel, which is produced by sol–gel chemistry	Controlled release drug carriers
Nanowires or carbon nanotubes	Metals, semiconductors or carbon	Gene and DNA delivery
Nanoshells coated with gold	Dielectric (typically gold sulphide or silica) core and a metal (gold) shell	Tumor targeting
Liposomes	Phospholipid vesicles	Controlled and targeted drug delivery
Ceramic nanoparticles	Silica, alumina, titania	Drug targeting and biomolecules delivery
Polymeric micelles	Amphiphilic block co-polymers	Systemic and controlled delivery of water-insoluble drugs

Nanoparticles can easily cross any of the biological barriers in human body as their size falls in the nanometer range. A list of nanoparticles and their applications in medical therapeutics are given in Table 2.2. When compared with classical drug administrations, nanosystems provide more advantages with greater therapeutic effect. Nanotherapy has all the potential to deliver treatment for fatal diseases like human cancer, HIV-AIDS, etc., Nanoparticles act as vector in drug delivery systems. The drug can be directly attached (entrapped, coated, or functionalized) to the nanoparticles. After drug discovery, a large number of compounds failed to prove high solubility, stability and bioavailability and are dropped. Nanotechnology-based drug formulation could meet out all the requirements to sustain as a good drug agent [45]. Different nanoparticles such as micelles, liposomes, solid lipid nanoparticles, polymeric nanoparticles, pegylated nanostructures, nanocrystals, nanobodies, cyclodextrin, dendrimers, and metallic nanoparticles are used for drug delivery in medicine. By using these nanoparticles in drug delivery, a novel platform to deliver drug targeted to a specific tissue with a controlled release rate can be developed. Nanoparticles and liposomes (with or without pegylation), dendrimers and micelles are candidate carriers for tumor-specific drug delivery [46]. In medical therapeutics, nanoparticles are applied to surgical medicines, dentistry, gene therapy, stem cell therapy, tissue engineering and regeneration.

Nanosurgery is a new concept in surgical medicine. But, nanosurgery for human beings has rarely been achieved. There are some advents in eye surgery. Femtosecond laser pulses, emitted from lasers working in the near-infrared, based on multiphoton effects allowing both imaging and laser effects to be generated which are in the submicron range and which do not cause collateral damage are available [47]. Kohli et al. reported membrane surgery and nanosurgical cell isolation using high-intensity femtosecond laser pulses [48]. They demonstrated the applicability of using ultra short laser pulses for performing surgery on live mammalian cells [48]. In 2007, another study reported on corneal multiphoton microscopy and intratissue optical nanosurgery by nanojoule femtosecond near-infrared pulsed lasers [49]. In this study, multiphoton microscopy including multiphoton autofluorescence imaging and second-harmonic generation was used as a new diagnostic tool to perform tissue nonlinear optical tomography with submicron resolution [49]. Nanoneurosurgery is another interesting emerging nanosurgery. In addition, nanocoated tools can also be useful for surgery and orthopaedics. Although bone is a very diverse tissue providing different functions within the body, recently identified nanobiomaterials shown promising solution to orthopaedic problems [50]. Chris Arts et al. reported the use of a bioresorbable nano-crystalline hydroxyapatite paste in acetabular bone impaction grafting [51]. In this work, destructive lever-out tests and in vivo animal tests were performed with various combinations of materials. Chris Arts et al. concluded that TCP-HA granules with a nano-crystalline hydroxyapatite paste could be a valuable addition when TCP-HA ceramic granules are being used for acetabular bone impaction grafting procedures [51].

Nanotechnology has a significant involvement in dentistry materials. He and Swain used a nano-based indentation system to determine the indentation stress-strain response of two kinds of dental ceramics, one kind of dental alloy and healthy enamel [52]. They reported that strong and tough materials with primarily elastic response, such as toughened ceramics, were required to enable dental crown/bridges to have long-term reliability [52]. In another study, Fu et al. studied effects of dental bleaching on micro- and nano- morphological alterations of the enamel surface [53]. The study concluded that the thickness of the enamel smear layer was significantly reduced due to the bleaching process [53]. Hairul Nizam et al. performed a nanoindentation study of human premolars subjected to bleaching agent [54]. According to their study, the exact mechanism by which hydrogen peroxide impacts the dentin and enamel had yet to be completely elucidated; however, it was observed to have an undermining effect on the nanomechanical properties of teeth [54]. Lee et al. reported the changes of optical properties of dental nano-filled resin composites after curing and thermocycling [33]. The objective of their study was to access the colour changes after curing, polishing, and thermocycling of a nano-filled resin composite. Lee et al. concluded that changes in colour and translucency after curing, polishing, and thermocycling varied by the shade group [33].

Gene therapy is the advanced therapeutic concept at present. It involves gene manipulation and transfer. Gene therapy implies local or systemic administration of a nucleic acid construct that can prevent, treat and even cure diseases by changing the expression of genes that are responsible for the pathological condition. This therapy is the hope for treating presently incurable diseases. Viruses are used in most of the clinical experiments today; however, they do have significant drawbacks. Therefore, non-viral vectors based on lipids, hydro-soluble polycations, other non-condensing polymers and nano or microparticles/capsules have been proposed [55]. Both biodegradable and non-biodegradable inorganic particles can be completely fabricated on the nanoscale with the attributes of binding DNA, internalizing across the plasma membrane and finally releasing it in the cytoplasm for final expression of protein [56]. In addition to the classical intravenous injection system, polymer-based nanoparticle technologies for oral gene therapy is in continuous development [57]. Pan et al. revealed that polyamidoamine dendrimers-modified magnetic nanoparticles might be a good gene delivery system and have potential applied usages in cancer therapy, molecular imaging and diagnosis [58]. In another study, Yamada et al. noted that hepatitis B surface antigen-L (HBsAg-L) particles were able to deliver payloads with high specificity to human hepatocytes. The study indicated that the L particle was a suitable cell- and tissue-specific gene/medicine transfer vector [59].

Stem cell therapy and tissue engineering are few other advanced therapeutic concepts. Stem cell therapy is mainly applied for the treatment of congenital defects and malignancies, whereas tissue engineering is targeted at regenerative medicine. Murugan and Ramakrishna reported that electrospinning was a straightforward, cost-effective technique that could be applied to the fabrication of nano-featured scaffolds suitable for tissue engineering, and it offered usefulness over conventional scaffold methodologies [60]. Chen et al. performed a meta-analysis of applied usage of electrostatic spinning technology in nano-structured polymer scaffold [61]. The study concluded that the nano-structured polymer scaffold could support the cell adhesion, proliferation, site, and differentiation, and this kind of scaffold had a considerable value in the tissue engineering field [61]. In another study, Huang et al. reported their development of nano-sized hydroxyapatite-reinforced composites for tissue engineering scaffolds [62]. In a more recent study, Corradetti et al. illustrated the use of affinity targeted biodegradable nanoparticles to mediate paracrine stimulation as an alternate to withstand the growth and pluripotency of mouse embryonic stem cells [63]. This approach will extremely contribute to the scalable manufacture of stem cells and the clinical delivery of new advanced cellular therapies for regenerative medicine [63].

2.5 Conclusions and Future Outlook

Nanobiology signifies the merger of biological research with nanotechnologies and it is a multidisciplinary field where a wide range of applications such as using bionanomaterials in engineering or engineered nanomaterials in biology and medicine are studied with diverse viewpoints. It confers innumerable welfare measures on humanity like advancement in medicine, nanoparticle coated zidovudine as in AIDS therapy, treating tumour of precise points, storage of hydrogen fuel and drug delivery material. The use of nanomaterials in surgical medicine could help in the development of many nanosurgical tools and nanoprosthetic devices. The concept of nanosurgery is to minimize blood loss, inter- and post-operative complications and to decrease the period of hospitalization for the patient. The advances in nanobiology could help the researchers to create future "SMART MATERIALS" that play an important role in diagnostics and in efficient drug delivery. Nanotechnology will also help in the formation of brand new molecular systems which might perfectly resemble existing living systems. Though nanobiology has several achievements and promises to add to its credit, it also requires our much appreciated patience. One should always realise that nanobiology can be a promising avenue by which medicine can advance, instead of expecting it to revolutionise medicine.

Moreover, researchers are also working to discover the potential long-term toxicity of nanoparticles, and their metabolic and degradative mechanisms. It is necessary to understand the fate of the drug once delivered to the nucleus and other sensitive organelles. In order to avoid public distrust and address the risks and potential hazards of emerging nanotechnologies: methods and tools should be developed to identify and characterize nanomaterials in biological matrices; international guidance should be developed on the effective exposure control; well-characterized stable benchmark and reference materials should be developed and used for toxicology studies; required data should be collected on human exposure, biomonitoring, and health outcomes that might be related to exposure; education and training should be given to researchers, manufacturers, and users of nanomaterials regarding the safe development and use of nanomaterials; existing regulations should be evaluated and new regulations should be developed if necessary; international guidance should be developed and shared on the best available practices for working with nanomaterials.

The in depth journey through nanobiology will lead one to end up in picobiology. Also, in nature there are many things exist below 1 nm level. Many single molecules are in the picolevel. In volume sense, many substances such as vitamins exist in the picolevel. The science focusing on objects that fall in the range of 1–100 pm is believed to be the next for nanoscience. But, the main question is when we can reach picotechnology.

References

1. Tolochko N (2009) History of nanotechnology. In: Valeri N, Chunli B, Sae-Chul K (eds) Nanoscience and nanotechnologies. Encyclopaedia of Life Support Systems (EOLSS), Developed under the auspices of the UNESCO, Eolss Publishers, Oxford. http://www.eolss.net
2. Teo B, Sun A (2006) From top-down to bottom-up to hybrid nanotechnologies: road to nanodevices. J Clust Sci 17(4):529–540
3. Yanamala N, Kagan V, Shvedova A (2013) Molecular modelling in structural nano-toxicology: interactions of nano-particles with nano-machinery of cells. Adv Drug Deliv Rev 65(15): 2070–2077
4. Kroll A (2012) Nanobiology-convergence of disciplines inspires great applications. Cell Mol Life Sci 69(3):335–336
5. Mohri K, Nishikawa M, Takahashi Y, Takakura Y (2014) DNA nanotechnology-based development of delivery systems for bioactive compounds. Eur J Pharm Sci 58:26–33
6. Yatoo MI, Saxena A, Malik MH, Kumar MK, Dimri U (2014) Nanotechnology based drug delivery at cellular level: a review. J Anim Sci Adv 4(2):705–709
7. Handy R, Shaw B (2007) Toxic effects of nanoparticles and nanomaterials: implications for public health, risk assessment and the public perception of nanotechnology. Health Risk Soc 9(2):125–144

8. Kumar A, Zhang X, Liang X (2013) Gold nanoparticles: emerging paradigm for targeted drug delivery system. Biotechnol Adv 31(5):593–606
9. Han G, Ghosh P, Rotello V (2007) Functionalized gold nanoparticles for drug delivery. Nanomedicine 2(1):113–123
10. Mark J, Vesselin N, Yeoheung Y (eds) (2009) Nanomedicine design of particles, sensors, motors, implants, robots and devices, 1st edn. Artech House, Boston/London
11. Zhang X, Xu X, Bertrand N et al (2011) Interactions of nanomaterials and biological systems: implications to personalized nanomedicine. Adv Drug Deliv Rev 64(13):1363–1384
12. Wiwanitkit V, Seereemaspun A, Rojanathanes R (2007) Increasing the agglutination reaction in slide test for weak B blood group by gold nanoparticle solution: the first world report. J Immunol Methods 328(1–2):201–203
13. Wiwanitkit V, Seereemaspun A, Rojanathanes R (2007) Gold nanoparticle as an alternative tool for urine microalbumin test: the first world report. Ren Fail 29(8):1047–1048
14. Jung Y, Lee J, Jung H et al (2007) Self-directed and self-oriented immobilization of antibody by protein G-DNA conjugate. Anal Chem 79(17):6534–6541
15. Panacek A, Kvitek L, Prucek R et al (2006) Silver colloid nanoparticles: synthesis, characterization, and their antibacterial activity. J Phys Chem 110(33):16248–16253
16. Gogoi S, Gopinath P, Paul A et al (2006) Green fluorescent protein-expressing *Escherichia coli* as a model system for investigating the antimicrobial activities of silver nanoparticles. Langmuir 22(22):9322–9328
17. Pal S, Tak Y, Song J (2007) Does the antibacterial activity of silver nanoparticles depend on the shape of the nanoparticle? A study of the Gram-negative bacterium *Escherichia coli*. Appl Environ Microbiol 73(6):1712–1720
18. Shahverdi A, Fakhimi A, Shahverdi H et al (2007) Synthesis and effect of silver nanoparticles on the antibacterial activity of different antibiotics against *Staphylococcus aureus* and *Escherichia coli*. Nanomedicine 3(2):168–171
19. Corot C, Robert P, Idee J et al (2006) Recent advances in iron oxide nanocrystal technology for medical imaging. Adv Med Delivery 58(14):1471–1504
20. Thorek D, Chen A, Czupryna J et al (2006) Superparamagnetic iron oxide nanoparticle probes for molecular imaging. Ann Biomed Eng 34(1):23–38
21. Hu S, Liu T, Liu D et al (2007) Nano-ferrosponges for controlled medicine release. J Control Release 121(3):181–189
22. Peralta-Hernandez J, Manriquez J, Meas-Vong Y et al (2007) Photocatalytic properties of nanostructured TiO_2-carbon films obtained by means of electrophoretic deposition. J Hazard Mater 147(1–2):588–593
23. Yildiz I, Shukla S, Steinmetz N (2011) Applications of viral nanoparticles in medicine. Curr Opin Biotechnol 22(6):901–908
24. Steinmetz N, Manchester M (eds) (2011) Viral nanoparticles: tools for material science and biomedicine, 1st edn. Pan Stanford Publishing, Singapore
25. Cohen B, Bergkvist M (2013) Targeted in vitro photodynamic therapy via aptamer-labeled, porphyrin-loaded virus capsids. J Photochem Photobiol B: Biol 121:67–74
26. Zeng Q, Wen H, Wen Q et al (2013) Cucumber mosaic virus as drug delivery vehicle for doxorubicin. Biomaterials 34(19):4632–4642
27. Salata O (2004) Applications of nanoparticles in biology and medicine. J Nanobiotechnol 2(3):1–6
28. Rotomskis R, Streckyte G, Karabanovas V (2006) Nanoparticles in diagnostics and therapy: towards nanomedicine. Medicina (Kaunas) 42(7):542–558
29. Service R (2005) Nanotechnology: color-changing nanoparticles offer a golden ruler for molecules. Science 308(5725):1099
30. Bauer M, Haglmuller J, Pittner F et al (2006) Direct optical visualization of DNA-DNA interaction by nanoparticle-capture on resonant PET-films. J Nanosci Nanotechnol 6(12):3671–3676
31. Smith A, Nie S (2004) Chemical analysis and cellular imaging with quantum dots. Analyst 129(8):672–677

32. Zhu L, Ang S, Liu W (2004) Quantum dots as a new immunofluorescent detection system for *Cryptosporidium parvum* and *Giardia lamblia*. Appl Environ Microbiol 70(1):597–598

33. Lee Y, Lim B, Rhee S et al (2004) Changes of optical properties of dental nano-filled resin composites after curing and thermocycling. J Biomed Mater Res Part B Appl Biomater 71(1):16–21

34. Kuma H, Tsukamoto A, Saitoh K et al (2007) Development of liquid phase immunoassay system using magnetic nanoparticles. Rinsho Byori 55(4):351–357

35. Jia-Ming L, Hui Z, Aihong W et al (2004) Determination of human IgG by solid-substrate room-temperature phosphorescence immunoassay based on an antibody labelled with nanoparticles containing Rhodamine 6G luminescent molecules. Anal Bioanal Chem 380(4):632–636

36. Pi D, Wang F, Fan X et al (2005) Luminescence property of Eu^{3+} doped LaF_3 nanoparticles. Spectrochim Acta A Mol Biomol Spectrosc 61(11–12):2455–2459

37. Nichkova M, Dosev D, Perron R et al (2006) Eu^{3+} doped Gd_2O_3 nanoparticles as reporters for optical detection and visualization of antibodies patterned by microcontact printing. Anal Bioanal Chem 384(3):631–637

38. Li X, Wang L, Zhou C et al (2007) Preliminary studies of applied usage of CdTe nanocrystals and dextran-Fe_3O_4 magnetic nanoparticles in sandwich immunoassay. Clin Chim Acta 378(1–2):168–174

39. Hwang S (2006) Gold nanoparticle-based immunochromatographic test for identification of *Staphylococcus aureus* from clinical specimens. Clin Chim Acta 373(1–2):139–143

40. Jiang P, Zhu C, Zhang Z et al (2012) Water-soluble AG_2S quantum dots for near-infrared fluorescence imaging in vivo. Biomaterials 33(20):5130–5135

41. Shen H, Hu M, Wang Y et al (2005) Polymerase chain reaction of nanoparticle-bound primers. Biophys Chem 115(1):63–66

42. Kalogianni D, Goura S, Aletras A et al (2007) Dry reagent dipstick test combined with 23S rRNA PCR for molecular diagnosis of bacterial infection in arthroplasty. Anal Biochem 361(2):169–175

43. Wabuyele M, Vo-Dinh T (2005) Detection of human immunodeficiency virus type 1 DNA sequence using plasmonics nanoprobes. Anal Chem 77(23):7810–7815

44. O'Malley P (2010) Nanopharmacology: for the future-think small. Clin Nurse Spec 24(3):123–124

45. Marcato P, Duran N (2008) New aspects of nanopharmaceutical delivery system. J Nanosci Nanotechnol 8(5):1–14

46. Alexis F, Rhee J, Richie J et al (2008) New frontiers in nanotechnology for cancer treatment. Urol Oncol: Semin Ori Invest 26(1):74–85

47. Fankhauser F, Niederer P, Kwasniewska S et al (2004) Supernormal vision, high-resolution retinal imaging, multiphoton imaging and nanosurgery of the cornea – a review. Technol Health Care 12(6):443–453

48. Kohli V, Elezzabi A, Acker J (2005) Cell nanosurgery using ultrashort (femtosecond) laser pulses: applied usage to membrane surgery and cell isolation. Lasers Surg Med 37(3):227–230

49. Wang B, Halbhuber K (2006) Corneal multiphoton microscopy and intratissue optical nanosurgery by nanojoule femtosecond near-infrared pulsed lasers. Ann Anat 188(5):395–409

50. Balasundaram G, Webster T (2007) An overview of nanopolymers for orthopaedic applied usages. Macromol Biosci 7(5):635–642

51. Chris J, Verdonschot N, Schreurs B et al (2006) The use of bioresorbable nano-crystalline hydroxyapatite paste in acetabular bone impaction grafting. Biomaterials 27(7):1110–1118

52. He L, Swain M (2007) Nanoindentation derived stress-strain properties of dental materials. Dent Mater 23(7):814–821

53. Fu B, Hoth-Hanning W, Hanning M (2007) Effects of dental bleaching on micro- and nano-morphological alterations of the enamel surface. Am J Dent 20(1):35–40

54. Hairul Nizam B, Lim C, Chng K et al (2005) Nanoindentation study of human premolars subjected to bleaching agent. J Biomech 38(11):2204–2211

55. Piskin E, Dincer S, Turk M (2004) Gene delivery: intelligent but just at the beginning. J Biomater Sci Polym Ed 15(9):1181–1202

56. Chowdhury E, Akaike T (2005) Bio-functional inorganic materials: an attractive branch of gene-based nano-medicine delivery for 21st century. Curr Gene Ther 5(6):669–676
57. Bhavsar M, Amiji M (2007) Polymeric nano- and microparticle technologies for oral gene delivery. Expert Opin Med Delivery 4(3):197–213
58. Pan B, Cui D, Sheng Y et al (2007) Dendrimer-modified magnetic nanoparticles enhance efficiency of gene delivery system. Cancer Res 67(17):8156–8163
59. Yamada T, Ueda M, Seno M et al (2004) Novel tissue and cell type-specific gene/medicine delivery system using surface engineered hepatitis B virus nano-particles. Curr Drug Targets Infect Disord 4(2):163–167
60. Murugan R, Ramakrishna S (2006) Nano-featured scaffolds for tissue engineering: a review of spinning methodologies. Tissue Eng 12(3):435–447
61. Chen D, Li M, Fang Q (2007) Applied usage of electrostatic spinning technology in nano-structured polymer scaffold. Chin J Reparative Reconstr Surg 21(4):411–415
62. Huang J, Lin Y, Fu X et al (2007) Development of nano-sized hydroxyapatite reinforced composites for tissue engineering scaffolds. J Mater Sci Mater Med 18(11):2151–2157
63. Corradetti B, Freile P, Pells S et al (2012) Paracrine signalling events in embryonic stem cell renewal mediated by affinity targeted nanoparticles. Biomaterials 33(28):6634–6643

Chapter 3
Tissue Engineering *In Vivo* with Nanotechnology

Erik Taylor, Dave A. Stout, George Aninwene II, and Thomas J. Webster

3.1 *In vivo* Nanotechnology Tissue Engineering Studies

The main goal of tissue engineering is to replace diseased and/or damaged tissues with biological substitutes or biocompatible materials that can restore and maintain normal tissue or biological functions. There have been major advances in the areas of cell and organ transplantation, as well as advances in materials science and engineering which have aided in the continuing development of tissue engineering and regenerative medicine [1–7]. But with all of these advancements, there still is a need for better materials and techniques to facilitate tissue growth. One field that can promote this advancement is nanotechnology, which focuses on gathering simple nanosized materials to form complex structures. In a sense, nanotechnology involves utilizing materials which possess at least one physical dimension between 1 and 100 nm to construct structures, devices, and systems that have novel properties [7] – in fact, many biological components, like DNA, involve some aspect of nano-dimensionality [7]. With the knowledge of the nano-dimensionality of nature, it has

E. Taylor • G. Aninwene II
Department of Chemical Engineering, Northeastern University,
360 Huntington Avenue, Boston, MA 02115, USA

D.A. Stout
Center for Biomedical Engineering, Brown University, Providence, RI 02917, USA

T.J. Webster (✉)
Department of Chemical Engineering, Northeastern University,
360 Huntington Avenue, Boston, MA 02115, USA

School of Engineering, Northeastern University,
313 Snell Engineering Building, Boston, MA 02115, USA
e-mail: Th.Webster@NEU.edu

© Springer Science+Business Media New York 2014
Y. Ge et al. (eds.), *Nanomedicine*, Nanostructure Science and Technology,
DOI 10.1007/978-1-4614-2140-5_3

Fig. 3.1 A schematic for the use of bio-inspired tissue engineering techniques with the basic understanding of materials and the use of nanotechnology comprise the field of Nanomedicine. Nanomedicine has logically given rise to high interest in using nanomaterials for tissue engineering applications since these materials have a significant impact towards inhibiting infections, decreasing inflammation, and promoting tissue growth

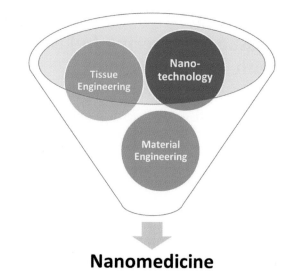

logically given rise to the interest in using nanomaterials for tissue engineering with these materials having the potential to significantly impact tissue engineering – a new field known as nanomedicine (Fig. 3.1).

Numerous papers, articles, books and research has gone into the utilization, development, analysis, implications and understanding of nanotechnology *in vitro* tissue engineering experiments with the understanding that future research will explore its *in vivo* associations [7–11]. To see where nanotechnology has made an impact from *in vitro* to *in vivo*, we present below some of the most promising tissue engineering *in vivo* experiments. Since nanotechnology has had a large impact *in vivo* on many aspects of medicine, we have delineated studies into particular organs/cells of interest and materials used.

3.1.1 Bone Studies

When looking at natural bone one can see a complex system with intricate hierarchical structures assembled through orderly deposition of nanometer sized hydroxyapatite minerals of low crystallinity within a type I collagenous matrix [12]. Below are *in vivo* experiments divided by the type of nanotechnology material investigated.

3.1.1.1 Nanomaterial-Coating

To mimic the surface of bone, one such practice is to coat implant surfaces. This will allow for the implant to be porous and allow for tissue ingrowth into the implant leading to direct anchorage to host bone – known as biological fixation [13, 14].

One such way of doing this is by soaking the material in an aqueous solution containing major inorganic components present in the body (such as HCO_3^-, Ca^{2+}, HPO_4^{2-}, and Mg^{2+} ions) and then adding, using plasma spray techniques, a material coating [15].

A study using nano-hydroxyapatite (HA) as the coating material, ingrowth chambers implanted into the lateral metaphysics of the distal femur of skeletally mature large coonhound dogs for a period of 8 weeks found that one is capable of promoting bone ingrowth and apposition, affecting the behavior of osteoblasts (bone forming cells) [15]. It was shown that more bone formation was generated in the channel lines with the apatite coating than one with non-coating titanium surfaces [15].

3.1.1.2 Injectable Nanomaterials

Injectable hydrogels have been used to create nanocomposites for tissue engineering as a means to deliver drugs and as a model for extracellular matrices due to their ability to have good biocompatibility, biodegradability and intrinsic cellular properties [16–18].

For example, an *in vivo* project involving 16 male Sprague Dawley® (SD) rats (320–350 g) were used to determine the degradation of an injectable bone regeneration nanocomposite which was comprised of nano-hydroxyapatite/collagen particles in alginate hydrogel carrier [19]. Alginate is a well-known natural polysaccharide compound of 1,4-linked β-D-mannuronate and 1,4-linked α-L-guluronate found in numerous biomedical applications due to its good biocompatibility, low toxicity and inexpensive cost [20–22]. It was shown that after 24 weeks, the injectable bone composite was almost fully degraded and was replaced by connective tissue as well as no fibrous capsule and acute inflammatory reactions were found [19].

Hosseinkhani et al. also showed that using a self-assembled injectable nanohydrogel comprised of a peptide-amphiphile and bone morphogenetic protein-2 (BMP-2) induced bone formation in Fischer male rats, aged 6 weeks [23]. It was demonstrated that where the injection site occurred, they found significant homogenous ectropic bone formation compared to a non-injected site.

3.1.1.3 Nanofiber Materials

Nanofibers are usually defined as fibers with one dimension ≤ 100 nm made of many different materials [24]. In the tissue engineering realm, the appeal of nanofibers is their structural similarity to native extra cellular matrix (ECM) proteins – this network creates a dynamic, three-dimensional nano and microenvironment in which cells are maintained. Signals are transmitted between the cell nucleus and the ECM enabling communication for cell adhesion, migration, growth, differentiation, programmed cell death, modulation of cytokines and growth factor activity, and activation of intracellular signaling [25, 26].

One such *in vivo* example is the use of silk fibroin nanofibers for its attractive properties: biocompatibility, oxygen and water vapor permeability, and biodegradability [27]. Using 30 adult male New Zealand white rabbits, two 8-mm-diameter defects were created, one on each side of the midline where a silk fibroin nanofiber was implanted. After 12 weeks of implantation, *in vivo* examination of the implanted silk fibroin nanofibrous membranes showed a negligible inflammatory response including occasional lymphocytes, eosinophils, foreign-body giant cells, and the subsequent formation of a relatively thick fibrous capsule, thus providing an alternative nanomaterial for bone growth [28]. Also, mimicking the natural nanofibers of the ECM can decrease the inflammation response since inflammatory cells would not recognize it as a foreign material.

3.1.1.4 Nanocomposite Materials

One of the major paradigms in bone tissue engineering is the scaffold, which functions as a structural support and transport vehicle for cells and bioactive molecules necessary for the formation of new bone [29, 30]. The ideal scaffold would need to have many properties that would support adequate bone tissue growth: mechanical properties equal to that of bone, degrade upon bone tissue growth, biocompatibility, pore interconnectivity, porosity, and promote bone tissue ingrowth.

One such way to overcome this obstacle is the use of nano reinforced composites. This would support strength and reinforcement as well as porosity and durability [31]. One such example is the use of ultra-short single walled nanotubes (SWNTs) for reinforcing synthetic polymeric scaffold materials. Reinforced ultra-short SWNT composites were implanted in rabbit femoral condyles and in subcutaneous pockets [32]. Results indicated that the reinforced composites showed favorable hard and soft tissue responses as well at 12-weeks the composites had a threefold greater bone tissue ingrowth and a reduced inflammatory cell density and increased tissue organization compared to their non-reinforced scaffold counterparts [32] – thus providing evidence that nanotechnology can strengthen bone tissue engineering scaffolds to promote bone growth.

Also, when embedding nanofibers into a scaffold one could increase a scaffold's biocompatibility properties and bone formation ability. For example, Osathanon et al. showed using a mouse calvarial defect system that by creating nanofibrous fibrin scaffolds and adding nanocrystalline hydroxyapatite one could promote bone formation around the defected area [33]. Also, they showed when adding micro/nano pore sizes and embedding rhBMP-2 into the nano-based scaffold, one could increase bone formation even more.

Another way to use nanocomposite materials is to use amorphous tricalcium phosphate (TCP) embedded in poly-lactic-co-glycolic acid (PLGA). Amorphous TCP has been shown to be very soluble and reactive due to a high enthalpy of formation [34] and bioactivity [35]. TCP/PLGA nanocomposites were transplanted into New Zealand white rabbits on the nasal and cranial bone and results showed newly formed bone after 4 weeks of implantation which was significantly increased when compared to non-nanocomposite reinforced PLGA scaffolds [36].

3.1.2   Neural Studies

Due to the ever increasing complexity of the central nervous system, many challenges have surfaced (such as vascularization, junction connection, cell interactions and toxicity (Fig. 3.2)) which have hindered the development of nanomaterial tissue engineering applications. Due to this difficult challenge, most studies have been aimed at the regeneration of an injured spinal cord and peripheral nerves.

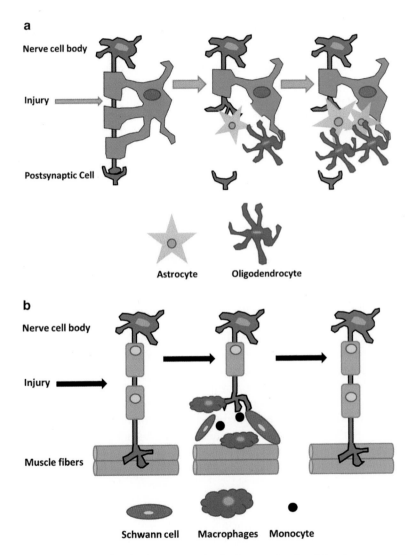

Fig. 3.2 Schematics of injured nerve regeneration in the central and peripheral nervous systems. (**a**) Central nervous system recovery process with glial scar tissue formation and (**b**) peripheral nervous system recovery process involving the activity of Schwann cells, macrophages, and monocytes (Image are adapted and redrawn from previous work) [37, 38]

3.1.2.1   Nanofiber Materials

A leading material for mimicking ECM properties for neural tissue is nanofibers
due to their material characteristics and size. Within neural tissue engineering,
nanofibers are usually broken up into two sub-categories (self-assembling and electrospun) depending on the materials synthesis. Below are *in vivo* experiments
divided by the nanotechnology materials used.

3.1.2.1.1   Self-assembling Nanofibers

A self-assembling nanofiber is composed of a peptide molecule that can instinc-
tively aggregate from an aqueous solution into a stable nanofiber network due to
non-covalent interactions in the presence of a salt solution or by changing the
pH. These molecules are formed from a hydrophobic tail and a hydrophilic peptide
head, which will self-assemble into a cylindrical nanofiber in which the hydrophobic tail will form the core of the fiber and the hydrophilic peptide head will be present at the surface of the fiber [39]. Also, the peptide terminus can be designed to
incorporate specific functional ligands that can enhance the surface of the nanofiber
scaffold: one can add bone marrow homing proteins [40], insulin growth factors
[41] or many other cell enhancing systems [40].

In vivo research has demonstrated the benefits of using self-assembled nanofibers in neural tissue engineering parameters. RADA16-I (reassembly of an ionic
self-complementary peptide RADARADARADARADA that forms a well-defined
nanofiber scaffold, the 16-residue peptide forms stable β-sheet structure and undergoes molecular self-assembly into nanofibers) has been shown to repair a disor-
dered optic track and return practical vision [42], help in the reconstruction of the
lost tissue in acutely injured brains [43] and bridge the injured spinal cord site of
rats after transplantation [44].

Upon adding an IKVAV laminin epitope to the peptide-amphiphile self-
assembling nanofibers, then injecting these molecules into a mouse spinal cord
lesion, reduced astrogliosis resulted – an irregular surge in the number of astrocytes
due to various types of injury caused by physical, chemical, and pathological trauma
[45] causing neural cell death, and increased Schwann cells at the site of injury [46].
Furthermore, this technique promoted the regeneration of both sensory and motor
fibers through the lesion site and indicated greater behavioral improvements via the
use of the IKVAV laminin epitope.

3.1.2.1.2   Electrospun Nanofibers

Electrospinning is a process by which an electric field is used to cause a viscous
polymer solution to overcome surface tension and create a charged jet that travels
through air [47]. The solvents evaporate from the liquid jet as it travels through the
air, leaving a charged fiber that can be manipulated by an electric field. This method
has been shown to be able to produce nano-fibers can be used in a wide array of

fields, from filters, sensors, textiles, and clothing, to drug delivery systems, medical devices and tissue grafts [48]. It has been shown that electrospun nano-fibers mats that can have similar surface morphology and architecture to cell extracellular membranes found in the body [49], thus they are good candidates for tissue grafts.

In vivo studies have shown the possible uses of electrospun nanofibers in the neural tissue engineering. Electrospun materials made out of chitosan [50], layering systems [51], poly-caprolactone (PCL)/poly-lactic-co-glycolic acid (PLGA) blends [52] and a copolymer of PCL and ethyl ethylene phosphate with encapsulated glial cell-derived neurotrophic factors (GDNF) [52] have all been used in a rat model of sciatic nerve transection with positive results ranging from the promotion of the ingrowth of connective elements to increasing neural cell processes.

Also, Rochkind et al. showed when an electrospun tubular scaffold containing bundles of parallel nanofibers made of a biodegradable dextran sulfate-gelatin co-precipitate seeded with human nasal olfactory mucosa cells or human embryonic spinal cord cells was implanted for the regeneration of a rat spinal cord, the nano-material promoted the partial recovery of function in one to two limbs just 3 months after implantation [53].

3.1.2.1.3 Carbon Based Nanofibers

Despite concern over toxicity, carbon based nanomaterials have attracted wide neural medical attention due to their material characteristics such as conductivity, high strength and biologically inspired size. Webster and colleagues imbedded neural stem cells (Wistar rats, isolated for later implantation from 1 to 3 day old neonate) into carbon nanofibers to see if neural capabilities in rats could be reestablished after an induced stroke.

In vivo preliminary results demonstrated that unfunctionalized carbon nanofibers/nanotubes can: (1) conduct electricity when implanted into damaged, non-conductive, regions of the brain; (2) mimic the nanometer features of key proteins found in neural tissue (such as laminin) to promote neuron functions and decrease glial scar tissue formation; (3) anchor stem cells to promote their differentiation into neurons when implanted into damaged regions of the brain; (4) remain non-toxic and non-migratory when used at low concentrations in the brain; and (5) most importantly, return motor skills to stroke-induced rats at least three times faster than injecting stem cells alone.

Furthermore, it was shown that only the neural stem cells impregnated with carbon nanofibers differentiated into neurons (Figs. 3.3 and 3.4). This may be due to: (1) high electrical conductivity of carbon nanofibers, (2) high surface reactivity of carbon nanofibers which according to *in vitro* studies increased the adsorption of laminin from serum, and/or (3) size similarities to laminin (70 nm cruciform protein). Moreover, this could be because the carbon nanofibers allowed for a favorable substrate for the neural stem cells to adhere to, thus, keeping them in the local environment (i.e., anchoring them) so that they could be stimulated to differentiate into neurons by the conductive carbon nanofibers.

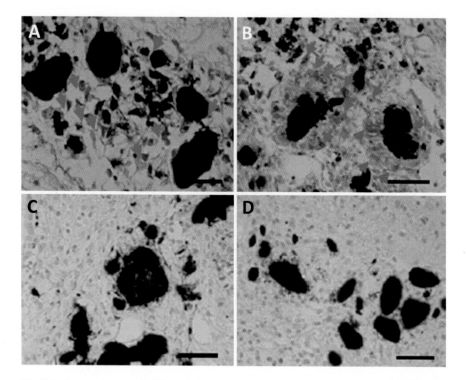

Fig. 3.3 Increased stem cell differentiation into neurons after 3 weeks implantation of hydropho-bic carbon nanotubes and stem cells into stroke damaged neural tissue of rats. Immunostaining of stroke damage brain sections where stem cells and carbon nanotubes (*black*) were implanted. (**a**) Nestin = marker for stem cells; (**b**) MAP2 = marker for neurons; (**c**) GFAP = marker for astrocytes; and (**d**) CD11b = marker for meningeal cells. All samples were double stained with BrdU. *Dark brown* (indicated by *pink arrows*) show positively double stained neurons with no scar tissue form-ing cells (astrocytes and meningeal cells) surrounding carbon nanotubes. Bars = 50 μm. Similar results shown up to 8 weeks [54]

3.1.3   *Wound Repair Studies*

Wound repair or healing is a process in which the skin, or other organ-tissue, repairs itself after injury [55] both unintentional (falling off a bike and scraping one's hand) or intentional (surgery). In normal skin, the epidermis and dermis exists in an equi-librium state, forming a defensive barrier against the external, or outside, environ-ment. Once the barrier is broken, the physiological process of wound "healing" begins. This process is expressed in four sequential and overlapping phases: (1) hemostasis, (2) inflammatory, (3) proliferative and (4) remodeling [56]. Through the advent of nanotechnology, nanomaterials have been able to help in the wound repair process by affecting all of these overlapping phases as described latter.

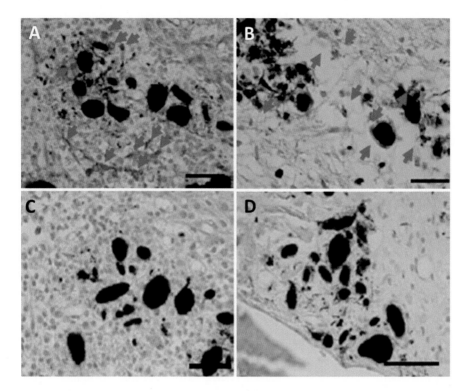

Fig. 3.4 Increased stem cell differentiation into neurons after 3 weeks implantation of hydrophilic carbon nanotubes and stem cells into stroke damaged neural tissue of rats. Immunostaining of stroke damage brain sections where stem cells and carbon nanotubes (*black*) were implanted. (**a**) Nestin = marker for stem cells; (**b**) MAP2 = marker for neurons; (**c**) GFAP = marker for astrocytes; and (**d**) CD11b = marker for meningeal cells. All samples were double stained with BrdU. *Dark brown* (indicated by *pink arrows*) show positively double stained neurons with no scar tissue forming cells surrounding carbon nanotubes. Bars = 50 μm. Similar results shown up to 8 weeks [54]

3.1.3.1   Nanocomposites

Nanocomposites have been used in wound repair to help in the promotion of tissue growth after surgery. An *in vivo* study using a BMP-2 coupled nanosilver-PLGA composite (BMP-2/NS/PLGA) showed that one could help in bone graft therapy. Metallic nanosilver (20–40 nm) combined with PLGA composites and BMP-2 were implanted into 16–18 week old SD rats with infected femoral defects [57]. After 12 weeks, inspection of the bone graft and wound, silver composite grafts were able to heal the femoral defect implanted with BMP-2 coupled with 2 % nanosilver particles, whereas the PLGA BMP-2 failed to heal the defect, thus provided evidence that nanomaterials can help in wound healing.

Nanocomposite *in vivo* studies have also demonstrated wound healing in gastrointestinal tissue. For example, gastrointestinal perforation is one of the major causes of bacterial peritonitis that usually leads to severe sepsis [58]. Postoperative anastomotic breakdown, which is one of the major complications after gastrointestinal operations, and can also cause bacterial peritonitis [59]. Therapeutic treatment by suture repair of a perforated/leaked lesion is crucial for wound repair, but not always foolproof. One way of helping this predicament is the use of nanotechnology. Fujie et al. showed by creating and transplanting a nanocomposite/sheet made of polysaccharides with a thickness of tens of nanometers into a rat's punctured murine cecum after 2 weeks, one could promote the growth fibroblasts on the muscular surface and completely seal the hole [60]. This was also shown with nanosheets composed of Tetracycline(TC) sandwiched between a poly(vinylacetate) (PVAc) layer and polysaccharide nanosheets. It was shown on mice that the growth of fibroblasts around the site of injury, lipocytesor fibroblasts specifically, bridged the tissue-defect site without any associated inflammatory reactions, resulting in an almost complete regeneration of the mucosal defect [61].

3.1.3.2 Nanofiber Materials

Again, nanofiber materials are usually defined as fibers with at least one dimension ≤ 100 nm and can be made of many different materials [24]. In the tissue engineering realm, the appeal of nanofibers is their structural similarity to native extra cellular matrix (ECM) – this network creates a dynamic, three-dimensional microenvironment in which cells are maintained.

In wound healing, nanofibers have aided medicine by creating sutures (a stitch used to hold tissue together) made of nano-systems. One such *in vivo* trial loaded cefotaxime sodium as a model drug into poly(L-lactic acid) (PLLA) using two different electrospinning methods – blend and coaxial electrospinning, which were collected on a grounded rotating edge-sharpened disk collector with a diameter of 280 mm and a maximal angular velocity of 1500 rpm to collect the aligned fiber to obtain fiber diameters between 238 and 965 nm [62]. Finally, the nanofiber sutures were fabricated using a mini type braiding machine, followed by dipping into a chitosan solution for 0.5 h and drying at ambient conditions to obtain a 0.35 ± 0.05 mm material, which is comparable to the diameter to most commercial sutures [62]. Using SD rats, the researchers implanted four different material sutures (blend PLLA-cefotaxime sodium, core-sheath PLLA-cefotaxime sodium, commercial silk and commercial PLLA sutures) and looked at tissue morphology around the wound site. After 3 days of implantation, it was found that conglutination and edema phenomena occurred more frequently in the wound area of the silk and the core-sheath suture groups, whereas the PLLA and the blend suture groups resulted in comparatively smooth anastomosis. These results show that by adding a "nano" approach to wound healing one could promote the process of wound healing.

Another *in vivo* study showed the use of nanofiber sutures to decrease skin dehiscence in rats after 1 week of implantation. This was done by creating materials with biodegradable poly(ester urethane) urea (PEUU) and poly(lactic-co-glycolic) acid (PLGA), where PLGA was loaded with antibiotic tetracycline hydrochloride and electrospun – where the fiber diameters for the PLGA tetracycline hydrochloride materials were between 100 and 200 nm and the PEUU fibers were around 390 ± 120 nm [63]. Materials were then combined to create a nanofiber composite with a weight percentage of 10 % tetracyline added; known as PEUU-PLGA-tet-10 and implanted into a contaminated abdominal wall from adult female Lewis rats. After 1 week of implantation, the rats were sacrificed and using digital image processes of images of the suture, it was shown that skin dehiscence decreased by 21 % using the PEUU-PLGA-tet-10 versus plain PEUU and abscess formation was not observed in the PEUU-PLGA-tet-10 nanofiber composites, whereas moderate to severe abscess formation was observed in the plain PEUU system.

3.1.3.3 Nanoparticles

Nanoparticles are usually defined as particles with sizes between about 1 and 100 nm that show properties not found in bulk samples [64]. Within the wound healing realm, nanoparticles have shown promising *in vivo* results.

Greenhalgh and Turos studied the use of antibiotic-conjugated polyacrylate nanoparticles on rats for enhanced wound healing on dermal abrasions. The study demonstrated favorable activity of the nanoparticle channeled drug system for systemic or topical application in a murine model. It was also shown that both routes were promising, and when topically applied, the emulsion enhanced wound healing by an average of 3–5 days [65].

Silver nanoparticles have been used on an array of applications ranging from wound dressings [1, 66, 67] to household products like socks and deodorants [68] or paints [69] due to their antimicrobial activity. One such *in vivo* study by Tian et al. showed that silver nanoparticles accelerate wound healing and achieve superior cosmetic outcomes [67]. It was shown in thermal injury mice model, the deep partial-thickness wounds normally healed after 35.4 ± 1.29 days, but in animals treated with silver nanoparticles, healing took place in 26.5 ± 0.93 days. Also, when comparing the appearance of healed wounds it was found that wounds in the silver nanoparticle group showed the most resemblance to normal skin, with less hypertrophic scarring and nearly normal hair growth on the wound surface.

Furthermore, wound healing was also investigated in diabetic mice. In this model, excised wounds treated with silver nanoparticles completely healed in 16 ± 0.41 days after injury, whereas mice in the control group (no nanoparticles) required 18.5 ± 0.65 days [67]. In the nondiabetic littermates, silver nanoparticles still accelerated wound healing relative to the control group.

Using silver nanoparticles and S-nitrosoglutathione (GSNO) to treat burn wounds, Melo et al. also showed promotional wound healing [70]. Shaved dorsum rats were exposed to intense heat (90 °C) and burned, then the nanoparticle compounds were topically administered immediately and up to 28 days after the burn injury, four times a day. Looking at toxicity, results showed no significant difference in level of urea, creatinine, aminotransferases, and hematological parameters in the control burn groups and treated burn group – indicating no substance induced toxic effects in the kidneys and/or the liver.

3.1.4   Other Tissue Engineering Applications

Due to nanotechnology's vast implications for tissue engineering, *in vivo* experiments with nanomaterials are being conducted on many different organs and cell types. Below are a few examples of *in vivo* experiments where nanotechnology is being utilized.

3.1.4.1   Vascular System

For vascular applications, nanotechnology has fit very well due to the fact that nanomaterials are able to support cellular attachment for endothelial or other cells on the graft surface to avoid platelet adhesion and thrombus formation as well as resist shear stress and the pressure of the blood stream. *In vivo* experiments have shown that one of the suitable purposes of nano-scaffolds for vascular grafts is aligned nanofibers that achieve the structural properties of the endothelium with elongated vascular endothelial cells [71]. For example, aligned Poly-L-Lactide Acid (PLLA) nanofibrous scaffolds used as an artery bypass with or without mesenchymal stem cells allowed the infiltration of vascular cells and matrix remodeling while the grafts with mesenchymal stem cells showed anti-thromogenic properties [71].

3.1.4.2   Cardiac System

Stem cell therapy for heart attack (myocardial infarction) and chronic ischemic heart disease has grown significantly in the past decade, but the benefits of cell therapy have still not been clearly shown. A new approach that has gained attention is the use of nanotechnology, via nanofibers to enable the local intramyocardial release of stem cells, growth factors and/or chemokines [72]. Using male SD rats, Segers et al. were able to modify a SDF-1 chemokine that was resistant to matrix metalloproteinase-2 and exopeptidase cleavage tethered to self-assembling nanofibers RADA16-II which promoted the chemotactic recruitment of stem cells and improved cardiac functions after a myocardial infarction.

Another such example used insulin growth factor IGF-1, a cardiomyocyte growth and differentiation factor, which was bound to biotinylated peptide nanofibers in a sandwich method [41]. *In vivo* results showed that IFG-1 bound to biotinylated peptide nanofibers, when injected into the myocardium, enabled the local delivery of IGF-1 and, in grouping with transplanted cardiomyocytes, reduced apoptosis and increased the growth of the transplanted cells.

3.1.4.3 Visual System

Research has shown that limbal stem cells are a stem cell population responsible for renewing and regenerating the corneal epithelium. Just like the cardiac system, the use of nanotechnology, via nanofibers, can be used as a transportation vessel to enable the local release of limbal stem cells and growth factors [73]. A recent study, using rabbits, implanted an electrospun nanofiber scaffold to transfer corneal epithelial or oral mucosal cell sheets which improved corneal healing and inhibited the local inflammatory reaction [74].

3.2 Nanotechnology for Improved *In Vivo* Antimicrobial Response

Medical device infections can be frequent and costly depending on the device location and the duration of use. Yet the benefits from these devices outweigh this low probability detriment and therefore continue to be used clinically. For example, peripheral or central intravenous catheters (CVCs) resulting in bloodstream infections (BSI) occur in about 4–5 out of every 1,000 CVC devices inserted [75, 76] with an attributed cost per infection estimated at US$34,508–$56,000 [77, 78], and the annual cost of caring for patients with CVC-associated BSIs ranges from $296 million to $2.3 billion [79]. However, CVCs are necessary for the delivery of fluids and medication or for monitoring patient health (such as through the drawing of blood or monitoring of blood pressure).

In addition to transcutaneous extracorporeal devices or other medical devices that are constantly exposed to the nonsterile environment outside the body, implanted tissue replacement devices are also susceptible to infections, resulting in implant failure. For example, prosthetic joint replacements are permanently implanted to alleviate pain, promote mobility, and improve the quality of life, but such implantations also suffer from the risk of infection, which occurs in about 1–1.5 % of all total hip and knee arthroplasties (THAs and TKAs, respectively) in the USA [80]. Although the chance of infection is rare in these procedures, the problem is significant, as periprosthetic implant infections, which are also known as septic failures and cost about US$70,000 per episode, are the most common cause of revision surgery in all TKAs (25 %), the third most common cause in all THAs (15 %), and the most common reason for removal of all TKAs (79 %) and THAs (74 %) [80–82]. Prosthesis device infections are some of the most striking medical device infections due to the widespread use of prosthesis devices, but other implanted medical devices, such as intrauterine devices, mechanical heart valves, pacemakers, tympanostomy tubes, and voice prostheses, can similarly suffer from infection, and could benefit from new techniques to stop infections [83].

Towards the goal of infection prevention, *in vivo* studies are also being conducted to examine the implications of nanotechnology in response to microbial challenges. This growing direction in tissue engineering focuses on improved antimicrobial response in biomaterials, simulating the immune response of the body which is compromised by the placement of current engineered materials. As many of these *in vivo* studies focus on tissues, this section will also be broken down by the tissue investigated and nanomaterial used.

3.2.1 Wound Repair

3.2.1.1 Antimicrobial Nanofeatured Biomaterials

An overwhelming number of studies focus on improving the wound healing response of materials *in vivo* (Table 3.1). One such method incorporates antimicrobials or antibiotics into nanofeatured tissue engineering devices. In one such work, the authors developed braided drug-loaded nanofibers for suturing and wound repair [62]. Structural applications of electrospun nanofibers were explored for use in sutures. Loading of cefotaxime sodium (CFX-Na) as a model drug into poly(L-lactic acid) (PLLA) was also carried out using two different electrospinning methods, i.e. blend and coaxial electrospinning, which were collected on a grounded rotating edge-sharpened disc collector with a diameter of 280 mm and a maximal angular velocity of 1,500 rpm to collect the aligned fibers to obtain fibers with diameters of between 238 and 965 nm having an average size of 667 nm. Post processing of the fibers using a combination of twisting and hot-stretching at 50 °C with a tension ratio of 50 % to release internal tension and improve the size stability of the materials. Sutures were fabricated using a minitype braiding machine, followed by dipping into a chitosan solution for half an hour, and finally drying at ambient conditions to obtain 0.35 ± 0.05 mm material comparable in diameter to most commercial 2–0 sutures.

Table 3.1 Major players in wound healing (Adapted from [84])

Cell type	Source	Timing and behavior
Keratinocytes	Epidermal wound edges and cut appendage stumps	Migration commences after a brief lag phase
Fibroblasts/ myofibroblasts	Connective tissue wound edges and fibrocytes from circulation	Invasion to form wound granulation tissue commences early, but transformation into myofibroblasts is later
Endothelial cells	Nearby blood vessels	Vasculature near to wound site becomes activated rapidly to allow diapedesis, but sprouting is later
Platelets	Spill from damaged blood vessels	Immediately at the site of tissue damage
Mast cells	Small numbers resident in tissue; others by migration through the pores in blood vessels, (diapedesis) from adjacent vessels	Appear to have an early role in regulating the later inflammatory response by neutrophils
Neutrophils	Diapedesis from adjacent vessels	Earliest of the leukocytes to derive from the blood
Macrophages	Some tissue-resident cells but majority derive from blood-borne monocytes	Secondary influx from the blood after neutrophils have killed the immediate foreign-organism invaders

Antibacterial efficacy against *Staphylococcus aureus* (*S. aureus*; AATCC 6538) and *Escherichia coli* (*E. coli*; AATCC 8099) was carried out using a zone of inhibition study [62]. Zones of inhibition diameters of 28.32–36.29 mm were found with an *S. aureus* challenge using blend sutures and core-sheath sutures respectively, versus 25.33 and 26.12 mm for the same materials challenged with *E. coli*. The minimum inhibitory concentration (MIC) for all strains of CFX-Na-sensitive bacteria was 1.4 and 0.15 µg/ml for *E. coli* and *S. aureus*, respectively [85].

The next goal of the sutures was to determine if the materials could keep a wound closed, the most important function of a suture [62]. For this, four types of sutures were implanted including blended PLAA-CFX-Na sutures, core-sheath PLLA. CFX-Na sutures, commercial silk sutures, and commercial PLLA sutures. Silk is the most commonly used surgical suture material, and can be used in a variety of suturing and legating procedures, but is known to be non-absorbable, meaning the material does not lose tensile strength after 60 days of implantation *in vivo* as caused by material breakdown, causes a foreign body response, a type of chronic immune response to implanted biomaterials, and is more susceptible to infection [86]. Therefore, controlled absorbable sutures are desirable if the process is well controlled, especially if prevention of infection is possible.

For an *in vivo* study, the researchers looked at inflammation, tissue morphology (biocompatibility), and bacterial response [62]. After 3 days of implantation, it was found that conglutination and edema phenomena occurred more frequently in the wound area of the silk and the core-sheath suture groups, whereas the PLLA and the blend suture groups results in relatively smooth anastomosis. These results indicated that the commercial PLLA suture group and the blend suture groups had better performance than the silk group and the core-sheath suture group. This could be explained by the burst release of antibiotics by the blended suture group, reducing inflammation caused by infection, whereas blend sutures released antibiotics more slowly (as observed *in vitro*). Other studies have also shown that PLLA sutures could cause less of an inflammatory reaction compared with the silk sutures [87]. After 4 weeks of implantation, the wounds healed completely with all suture groups.

Previous studies have indicated that low inflammatory reactions should be characterized by an increased ratio of fibroblast cells over inflammatory cells, or by reduced number of macrophages [88]. The biocompatibility study found silk sutures caused a greater foreign body reaction, with macrophages, whereas blend and core-shell structures had better performance in inflammation reduction [62]. This could be explained by fewer lymphocytes around the blend and core-sheath sutures as compared to the commercial silk or PLLA sutures. Secondly, more fibroblasts were found in the blend and core-sheath sutures, which is an indication of favorable tissue recovery.

In the final study by these researchers, the four sutures were evaluated for bacterial contamination [62]. After the wound was re-opened, a muscle sample was taken from the animals, homogenized, and 100 uL was taken and diluted in physiological saline. Then, 100 ul of the diluted suspension was transferred onto an LB plate and subsequently cultured in an incubator at 37 °C for 12 h. The number of bacterial colonies was calculated to obtain the bacterial count, and it was express as the number of bacteria present in 1 g of the musculature. Each experiment was completed with five replicated, and the results were expressed as an average.

Interestingly, this study found that the musculature was asepsis prior to suture implantation, and 3 days after implantation the quantity of bacterial colonies became the largest [62]. As time increased, the number of colonies decreased and after 28 days, few bacteria were observed due to the wound being gradually restored. The authors observed that commercial silk suture group had a serious bacterial infection, while the drug-loaded sutures had preferable performance. Quantification by the authors showed that the actual difference between bacterial colony counts was about 26,000 for the silk, 23,000 for the PLLA groups, 19,000 for the core-sheath, and 17,000 for the blend sutures respectively after 3 days. Although the differences between the sutures became more modest as time proceeded to 7, 14, 21, and finally to 28 days, bacterial contamination was consistently lower. These results were also consistent with the lower inflammatory response observed with the nano-fabricated blend and core-sheath sutures as compared to silk and PLLA commercial products.

3.2.1.2   Antimicrobial Nanoparticles Improve Wound Repair

Nanoparticles are also being used as an injectable or topical to improve tissue responses and wound healing. In one set of *in vivo* studies, polyacrylate nanoparticle emulsions were used for topical wound healing or systemic applications [65]. This study found that the antibiotic-conjugated polyacrylate nanoparticles had activity against MRSA *in vitro* and no cytotoxicity against human dermal cells. The water-based emulsion is capable of solubilizing lipophilic antibiotics for systemic administration, and has shown to protect antibiotics from hydrolytic cleavage by penicillinases, thus improving the activity of penicillin against MRSA. This study demonstrated the favorable activity in murine models of the drug for systemic application by intraperitoneal injection or topical application in a wound abrasion model. *In vivo*, both routes were favorable, showing no signs of inflammation, such as redness or abnormal cytokine release, and when topically applied, the emulsion enhanced wound healing by an average 3–5 days. These nanoparticles might therefore afford promising opportunities for treating both skin and systemic infections. It has also been found that topical delivery of silver nanoparticles promotes wound healing [67]. The beneficial effects of silver nanoparticles on wound healing were investigated in an animal model, and it was found that rapid healing and improved cosmetic appearance occur in a dose-dependent manner. Silver nanoparticles also show antimicrobial properties resulting in a reduction of wound inflammation and modulation of cytokines, a potential explanation of these beneficial outcomes in tissue regeneration.

3.2.1.3   Biomaterials Coupled with Antimicrobial Nanoparticles

Nanoparticles are also being used as a component in biomaterials to improve the antimicrobial response. In one such work, *in vivo* wound healing and antibacterial performances of electrospun nanofiber membranes, materials for wound healing and preventing infection, were fabricated [89]. This study looked at PVA, PCL, PAN, and other polymers using an electrospinning technique, and the wound-healing response was examined *in vivo* using female SD rats. Either wool protein or silver nanoparticles were incorporated into nanofibers. It was found that wound healing performance was mainly influenced by the porosity, air permeability, and surface wettability of the nanofiber membranes. Interestingly, fiber diameter and antibacterial activity had little effect on wound healing efficiency. Instead, nanofeatured mats were predicted to inhibit exogenous bacterial infiltration via a sieve effect, whereby the multi-layered structure featured nanometer pore sizes smaller than bacteria. Simultaneously, hydrophilic and nano-porous materials effectively removed excess fluids from the wound, which the authors concluded may further reduce the chances of infection. Yet, a limitation to this study is that the authors did not model a contaminated wound site, such as could be caused during accidental injury or surgical intervention.

Waterborne polyurethane-silver nanocomposites have been used to fabricate biomaterials for preventing infection in catheters & subcutaneous wounds [90]. Using polyurethane (PU) with nanosilver (~5 nm size), the material could be loaded up to 30 ppm with a good dispersion, as confirmed by transmission electron microscopy, whereas at higher concentrations the nanosilver aggregated (Fig. 3.5). Important previous observations using nano-gold and nano-silver contributed to the design of this study through determination that the nanoparticle component improved material properties such as microphase separation on the surface of H_{12}MDI-based PU for enhanced the biostability and biocompatibility *in vitro* and *in vivo*, with nanosilver at 30 ppm being more effective [91]. In conjunction with similar materials property improvements also found in this study, the nano-components enhanced fibroblast attachment and endothelial cell response, while allowing a reduction in monocyte and platelet activation, relative to PU alone or nanocomposites with other silver concentrations, and biocompatibility was confirmed in a subcutaneous rat model. The adhesion of *Bacillus subtilis*, *E. coli*, or Ag-resistant *E. coli* on PU-Ag nanocomposites was significantly lower at all concentrations of nanosilver tested, in addition to bacteriostatic ability, whereas PU did not. Commercial catheters were also coated with PU-nanosilver at 30 pmm, and these were inserted into rat jugular veins for evaluation. The results indicated milder inflammation after 3 months, with the wall thickness of the veins being smaller than that with the commercial catheters or the pure PU-coated catheters. Finally, it was concluded that the PU-nanosilver could reduce vein occlusion and improve blood compatibility through an overall better anti-fouling response, and should also be considered as a cardiovascular biomaterial.

3.2.2 Bone Defect and Fracture Repair

A concern about bone tissue infection, termed osteomyelitis, has prompted researchers to incorporate antimicrobial aspects into bone tissue engineering systems. In one such study, the use of BMP-2 coupled with nanosilver-PLGA composite grafts to

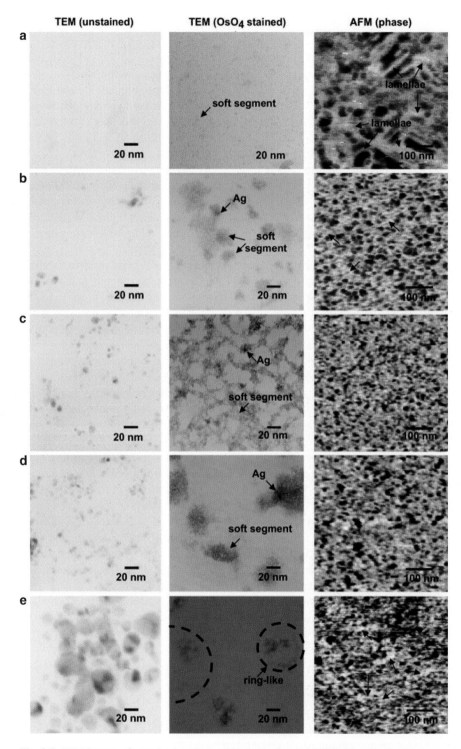

Fig. 3.5 TEM images of unstained or OsO$_4$ stained samples and AFM phase images for (**a**) the pure PU as well as the PU–Ag nanocomposites containing (**b**) 15 ppm, (**c**) 30 ppm, (**d**) 50 ppm or (**e**) 75 ppm, of nano Ag. *Arrows* in AFM phase images indicate the presence of lamellae [90]

induce bone repair in grossly infected segmental defects used materials for treating infection and healing bone defects [57]. The authors of this study used metallic nanosilver (20–40 nm) combined with PLGA composite grafts. To make grafts, the authors mixed PLGA with various weight percentages of nanosilver (with respect to PLGA), and poured the mixture over a bed of sieved sugar particles (200–300 μm) to generate a paste. This process was used to generate porosity in the structure after removal of the sugar particles via soaking in water. The paste was stacked in a Teflon mold to generate cylindrical grafts, dried for 12 h, lyophilized for 4 h, and washed with distilled water to leach out the sugar, and finally sterilized with 70 % ethanol and further dried. Bone morphogenic protein 2 (BMP-2), an osteoinductive protein used to repair many bone fractures and bone defects, was also injected into some grafts to prepare BMP-2/NS/PLGA bone grafts. The authors used a vancomycin and methicillin resistant clinical strain of *S. aureus*, Mu50, and cultured the microorganisms in the presence of grafts with up to 2 % nanosilver. The authors found that with an *in vitro* challenge of 10^7 CFU/graft, 0.1 % nanosilver resulted in only a slight decrease of growth, a 0.5 % nanosilver graft slowed the emergence of exponential phase by 13 h, and weight percentages of 1 % or higher completely inhibited growth. Using a higher inoculate of 10^8 CFU/graft, 0.1 % and 0.5 % nanosilver had no significant effect, 1 % and 1.5 % slowed the emergence of exponential phase by 10 and 15 h respectively, and 2 % completely inhibited growth. Using this information, the authors decided to use an *in vivo* rat femoral segmental defect model, and with PLGA grafts of 0 %, 1 %, or 2 % nanosilver implanted into SD rats along with an injection of 10^8 CFU (results of this study shown in Fig. 3.6). The authors found that the grafts did not inhibit adherence, proliferation, alkaline phosphatase activity, or mineralization of on growth MC3T3-E1 pre-osteoblasts compared to PLGA controls, or the osteoinductivity of BMP-2. The silver composite grafts were able to heal a femoral defect implanted with the BMP-2 coupled with 2 % nanosilver particles-PLGA were the BMP-2 in PLGA grafts without silver failed to heal the defects.

A very innovative method utilizing nanotechnology to improve antimicrobial response simultaneous to bone tissue engineering is the combination of immune cell regulating functions into a scaffold material. In one study using this concept, multilayer polypeptide nanoscale coatings incorporating IL-12 was used for the prevention of biomedical device-associated infections and fracture repair [92]. This study incorporated IL-12 (an immunoregulatory cytokine) into nanoscale coatings using electrostatic layer-by-layer self-assembly. Specifically, IL-12 activates natural killer cells and macrophages, leading to phagocytosis of bacteria. The coatings were made of albumin, PLL, and PLGA, which was coated onto materials such as quartz, stainless steel, and titanium. When the coating was applied to stainless steel K-wires, the authors found that IL-12 was released *in vitro*, with most of the coating being released within 10 days. Coated wires were prepared and implanted into an open femur fracture model using SD rats to test for the *in vivo* coating stability. For this, six wires were prepared with three implanted and three maintained as controls. An *in vivo* bacterial challenge was also carried out, whereby upon fracturing of the femur, rats were injected with 100 μL suspension of 10^2 CFU/0.1 ml *S. aureus* directly into the site. The fractures were left open for 1 h, and then fixed using an intra-medullary

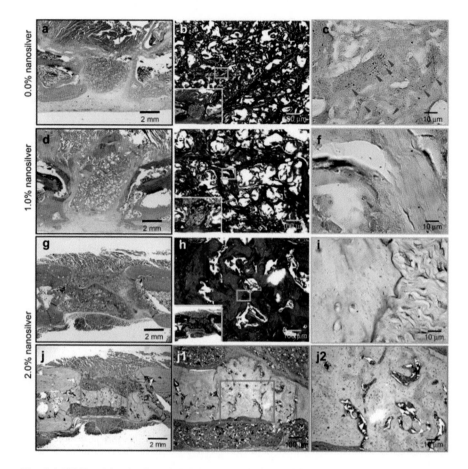

Fig. 3.6 H&E staining (**a**, **d**, and **g**), Masson's trichrome staining (**b**, **e**, and **h**), Taylor modified Brown and Brenn gram stain (**c**, **f**, and **i**) and immunostaining of OCN (**j**, **j1**, and **j2**) of 10^8 CFU *S. aureus* Mu50 contaminated rat femoral segmental defects implanted with 0.0 % (**a–c**), 1.0 % (**d–f**), and 2.0 % (**g–j2**) nanosilver-PLGA bone grafts coupled with 30 μg/ml BMP-2 at 12 weeks post implantation, respectively. Almost no bone regenerated in BMP-2/0.0 %-NS/PLGA (control BMP-2 coupled control PLGA) implanted groups (**a**, and **b**) with obvious continued bacterial contamination (**c**, *red arrows*). Less bone regenerated in the defect area of BMP-2/1.0 %-NS/PLGA implanted groups (**d** and **e**), while only limited bacterial colonies were observed (**f**, *red arrow*). BMP-2/2.0 %-NS/PLGA grafts promoted significantly greater bone formation to form a mineralized bony bridge between the two defect ends (**g**, **h**, and **j**) by eliminating bacteria in the defect area (**i**), Higher magnification figures show active bone regeneration around the mineralized bridge and in the marrow-like cavities in the bridge (**j1**, and **j2**) [57]

stainless steel K-wire. After quantitative colony counts, infections were defined as >2–5 bacterial colonies per plate (or 200–500 CFU per gram of tissue). Using this method, the authors found that IL-12 coated stainless steel K-wires with 10.6 ng dose or higher, as obtained by manipulating the concentration of the IL-12 loading solution and the number of coating nanolayers, had a lower rate of infection (20 % with 10.6 ng IL-12 versus 90 % for control).

3.2.3   Other Tissue Engineering Applications

Understanding of nanotechnology as a tool to improve antimicrobial response is growing as more nanomaterials are being developed and tested. Yet, few *in vivo* studies of nanomaterials have stepped outside of wound repair and bone repair towards improved bacteria inhibition on tissue engineering devices. Some exceptions are gastrointestinal and abdominal tissue repair which have been investigated and will be described below.

3.2.3.1   Gastrointestinal Tissue Repair

Nanotechnology has been used in the production of antibiotic-loaded nanosheets for the treatment of gastrointestinal tissue defects [60]. In a series of studies, ultrathin polymer films (nanosheets) composed of polysaccharides were used as a wound dressing in gastrointestinal tissue defects (See Fig. 3.7 for details about preparation of and tissue repair with nanosheets). To prepare these materials, the authors used a layer-by-layer assembly method using biocompatible and biodegradable chitosan and sodium alginate, whereby single layers, each approximately 29 nm, were prepared via electrostatic interactions by spin-coating on SiO_2 substrates [93]. Because the nanosheets are too thin to lift and use surgically alone, a water soluble polyvinyl alcohol (PVA) with a thickness of about 70 μm was incorporated, which could be dissolved upon placement at the site of injury by the simple addition of saline. These nanosheets showed favorable properties as a tissue engineering device sealing gastrointestinal or other tissues, acting as a physical barrier to bacteria, while attaining equal performance in sealing to sutures, with a less invasive procedure, and without postoperative adhesions [60, 61, 93]. Despite the sealing and healing effects, it was found that bacteria could penetrate through the wound dressing because of the ultrathin structures [60, 61]. To reduce bacterial penetration, tetracycline was incorporated between a polyvinyl-acetate (PVAc) layer and polysaccharide nanosheets [60].

The tetracycline was released for 6 h under physiological conditions. *In vivo* studies showed that overlapping therapy significantly increased mouse survival rate after cecal puncture, suppressing intraperitoneal bacterial count and leukocyte count, therefore showing the antibacterial and anti-inflammatory properties of the wound covering, and providing an efficient way for surgeons to repair tissues in various scenarios.

3.2.3.2   Abdominal Tissue Repair

Abdominal cavity coverage and wound repair is being investigated via elastic, biodegradable polyurethane/poly(lactide-co-glycolide) fibrous sheets with controlled antibiotic release via two-stream electrospinning materials were fabricated [63]. Materials made were a biodegradable poly(ester urethane) urea (PEUU) and a poly(lactide-co-glycolide) (PLGA), where PLGA was loaded with antibiotic

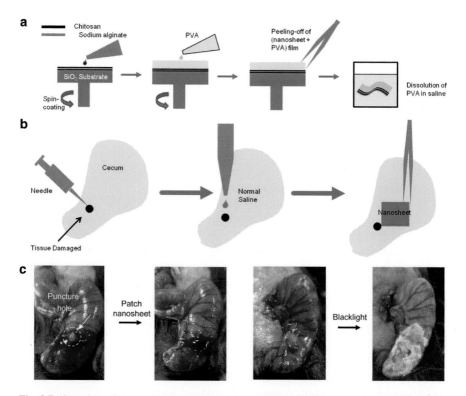

Fig. 3.7 Overview of the methods required for tissue repair using biocompatible and biodegradable nanosheets. (**a**) Nanosheets composed of chitosan and sodium alginate were prepared by layer-by-layer assembly on a SiO₂ substrate, a thicker 70 μm polyvinyl alcohol (PVA) layer allows handling with tweezers, and dissolution of the PVA layer takes place in normal saline. (**b**) After tissue damage caused by a needle puncture into the cecum, the defect site is prepared with normal saline, and the nanosheet is simply applied to the site of damage resulting in PVA dissolution and sealing. (**c**) Macroscopic images of tissue repair after treatment with antibiotic loaded nanosheets containing tetracycline (*TC*), where the TC nanosheets are viewed under a blacklight in the final image (the nanosheet is visible due to fluorescence of TC), thus preventing penetration of bacteria (Image in (**a**) and (**b**) are adapted and redrawn from previous work [60, 93])

tetracycline hydrochloride (PLGA-tet). Using electrospinning, it was possible to create a uniform blend of PEUU and PLGA-tet fibers with nanofeatures. The fiber diameters for the PLGA-tet materials were between 100 and 200 nm, and varied without an obvious trend dependent on the concentration of tetracycline added whereas PEUU fibers were larger with sizes of 390±120 nm. The release profile was controlled, with an initial burst release of 3 h for PLGA or up to 96 h for PEUU-PLGA-tet, followed then by a slower release with 55–90 % remaining after 14 days for the PLGA-tet scaffolds and 0–65 % for the PEUU/PLGA-tet scaffolds remaining after the same period.

The samples were then incubated and growth inhibition of *E. coli* was tested for 24 h after prior incubation for 0, 3, and 7 days in PBS [63]. In this test, the authors measured the "antibacterial diameters" using a method similar to zone of inhibition and were able to conclude that after incubation for 3 or 7 days, antibacterial activity tended to decrease, where only PEUU-PLGA-tet 10 (the weight percentage of tetracycline added was 10 %) was found to have a significantly larger antibacterial diameter than other composites. Using adult female Lewis rats, fibrous sheets were implanted into a contaminated abdominal wall model. This was achieved by first exposing the abdominal cavity through the creation of a 3 cm long, full-thickness incision approximately 2 cm inferior to the xiphoid process, followed by lavage of the peritoneal cavity, and finally injection of a rat stool slurry (0.25 ml) consisting of 1 g rat stool homogenized in 20 ml of normal saline. Using 7–0 polypropylene with over and over sutures, a 2.5×0.5 cm piece of either PEUU or PEUU/PLGA tet-20 sheet was interposed within the incision space (n=5 for each sample), and closed with 4–0 Vicryl interrupted suture. The animals were sacrificed after 1 week implantation period. The implant was quantitatively assessed using digital image processing of images of the suture line for wound dehiscence. Following opening of the sutures, the implant sites were then qualitatively scored for the extent of abscess formation, or the amount of pus with resulting swelling and inflammation. Abscess formation was also not observed in the PEUU-PLGA tet-20 composite, whereas moderate to severe abscess formation was observed in the PEUU control sample. Using similar methods, it was found that skin dehiscence, or breaking open of the wound, was also decreased by 21 %, using PEUU/PLGA tet-20 versus PEUU.

3.3 Immune Response to Nanomaterials

While there may be several reasons why an implant may fail, inflammation of the implant area is both a major cause, and major indicator of implant failure. According to one study, in their set of 282 cases of failed stainless steel orthopedic implants, 10.6 % exhibited inflammation of the surrounding tissues [94]. Inflammation rates may change depending on the material implanted, location of implantation, and the function of the implanted device; however, inflammation is still a major issue to consider when developing a new implantable material or wound treatment. Implanted materials that cause excess inflammation lead to significant damage of the surrounding tissues and result in the loosening or total failure of the implanted material.

3.3.1 Inflammation Pathway

3.3.1.1 Foreign Body Response Summery

Although the Egyptians recognized inflammation by 1650 BC, it was not until 25 AD that the Roman Cornelius Celsius first defined the process by the observable features of redness, swelling, heat and pain that present with inflamed tissues on the body surface [95]. Inflammation is considered a normal part of healing, and is closely linked to the activation of the immune system [95]. There are predictable stages to the inflammation/foreign body response that follows any injury or implantation. The first step is nonspecific protein absorption and adhesion of cells, such as monocytes, leukocytes and platelets [19]. This is followed by the formation of giant cells and cytokine release [19]. Neutrophils arrive on the scene very early after any tissue damage, and in an attempt to kill any microbes they may flood the damaged area with free radicals that kill many otherwise-healthy host cells as well as the target infectious agents [84]. This becomes a major issue in chronic wound situations, and may lead to chronic inflammation and scarring. In the case of implanted materials the implanted device may be encapsulated by fibrous tissue [19]. The materials that are considered to be biocompatible reduce the inflammation process by either avoiding the initiation of the foreign body response, by avoiding or reducing the initiation of one or more steps in the foreign body response, or by acceleration the process to reduce the inflammation period.

3.3.1.2   Inflammation in the Central Nervous System

In cases of injury to the central nerves system injury there is a rapid response from the astrocytes in the injured tissue, which may result in the formation of a gial scar, also known as an astrogliosis [96]. The formation of a gial scar may result in the interference of neural repair and the impedance of axon regeneration and extension [96]. Thus, in cases of materials to be implanted or used to treat injuries in the central nerves system it is important to use materials or techniques that will minimize inflammation and thus minimize scar formation.

3.3.1.3   Injectable Scaffolds

A good nerve conduit must be biocompatible and exhibit extremely low immunogenic, cytotoxic, and immunogenic responses. Studies by Chuna et al. have illustrated that synthetic nano-fibers made of self-assembling peptide sequences hold promise in the creation of fibrous biodegradable nerve scaffolds [97]. Studies have shown that a scaffold fibers made of self-assembling natural L-amino acids, RADA16 (Ac-RADARADARADARADA-COHN2), can be functionalized with case specific functional sequences to promote healing and avoid triggering detectable immune response or inflammatory reactions [39].

The peptide sequence RADARADARADARADA (RADA16-I) has also been shown to effectively self-assemble into self-assemble into high-order interwoven nanofiber scaffold hydrogels with extremely high water contents [98]. Meng et al. preformed trials where they created controlled second degree burn wounds on mouse models [98]. Application of this self-assembling liquid dressing to a burn site was able to reduce inflammation, accelerate wound closure, and promote "vigorous" "healthy" healing when used as a wound dressing to treat second degree burns in a rat model.

3.3.1.4 Different Inflammation Cascades

Inflammation after injury or implantation is an important consideration. A biomaterial or treatment must be chosen which will minimize the scarring of the site, yet restore the function of the regenerating tissue. The following section briefly discusses some examples of nanomaterials that have been shown to reduce inflammation *in vivo*.

Using composite biomaterials such as bone-like nano-hydroxyapatite/collagen/ poly(lactic acid) (nHAC/PLA) may provide a means for the repair of bone defects that does not require the use of bone allografts, and reduces the amount of inflammation over the course of the healing process [99]. By using materials that can actively support the attachment and re-growth of desired tissues, without triggering immune response, inflammation after an implant surgery can be minimized.

Additionally, it has been shown that silver nanoparticles are able to reduce inflammation through cytokine modulation [70]. Work by Ding et al. illustrated that incorporating nanosilver into scaffolds may be able to inhibit inflammation by reducing the invasion of inflammatory cells after surgery [100]. A study by Melo et al. showed that silver nanoparticles were able to accelerate healing of burn related wounds by modulating local and systemic inflammatory response following burn injury [70]. This indicates that silver nanoparticles in wound repair may be able to reduce post injury inflammation and scar formation, thus leading to accelerated tissue re-growth. As opposed to other implantable materials that just attempt to minimize the activation of the inflammation pathways, it seems that silver nanoparticles actively reduce the inflammatory response mechanisms.

3.3.2 Future of Nanomaterials Inflammation

As new biological nanomaterials are created and implemented, care needs to be taken to ensure that their implantation does not lead to unwanted inflammation at the site of implantation or to systemic inflammation after implantation. Future studied will need to focus on not only suppressing the inflammation following implantation, but also controlling and guiding cellular response, and protein interactions to result is faster, healthy tissue regeneration and shorter recovery times. Broader *in vivo* studies will need to be conducted focusing solely on the inflammation response of a wide array of nanomaterials implanted into a wide array of body systems.

3.4 Conclusions and Future Perspectives

Tissue engineering with nanotechnology is no longer in its infancy; with the growing number *in vivo* studies, we are beginning to understand its benefits. Specifically, nanotechnology offers biomimetic features to scaffolds, such as structural similarities to

ECM, and the ability to form constructs through the process of self-assembly towards improved tissue engineering strategies in bone, neural, wound repair, and through-out the body. More recently, researchers and doctors have switched focus towards the impact of nanotechnology on other important clinical problems, such as infections and excess inflammation caused by tissue engineering with conventional procedures (using micrometer or larger sized materials). Here, it has been found that nanotechnology can provide some very innovative approaches towards simultaneously controlling bacterial growth through loading of drugs (such as nanoparticles), preventing the penetration of (micron-sized) bacteria through nano-featured materials, and controlling the inflammatory response. With the growing applications of nanomedicine throughout the body, future studies must also consider the long-term impacts of implanted nanomaterials. Specifically, toxicity of nanomaterials is still not well understood, and researchers using *in vivo* models will benefit. In summary, the future of *in vivo* nanomedicine is very bright.

Acknowledgements The authors would like to thank the following organizations for funding: Hermann Foundation, Veterans Affairs Pre-doctoral Fellowship, the National Science Foundation Graduate Research Fellowship Program, (NSF #92412) and the NSF Graduate K-12 program.

References

1. Fong J, Wood F, Fowler B (2005) A silver coated dressing reduces the incidence of early burn wound cellulitis and associated costs of inpatient treatment: comparative patient care audits. Burns 31:562–567
2. Guilak F, Butler DL, Goldstein SA, Baaijens F (2014) Biomechanics and mechanobiology in functional tissue engineering. J Biomech 47(9):1933–1940
3. Harrison BS, Atala A (2007) Carbon nanotube applications for tissue engineering. Biomaterials 28:344–353
4. Kummer KM, Taylor E, Webster TJ (2013) Biological applications of anodized TiO_2 nanostructures: a review from orthopedic to stent applications. Nanosci Nanotechnol Lett 4:483–493
5. Martin I (2014) Regenerative medicine applications in organ transplantation. Regen Med 9(3):267–268
6. Mironov V, Kasyanov V, Markwald RR (2008) Nanotechnology in vascular tissue engineering: from nanoscaffolding towards rapid vessel biofabrication. Trends Biotechnol 26:338–344
7. Silva GA (2006) Neuroscience nanotechnology: progress, opportunities and challenges. Nat Rev Neurosci 7:65–74
8. Gadegaard N, Martines E, Riehle MO, Seunarine K, Wilkinson CDW (2006) Applications of nano-patterning to tissue engineering. Microelectron Eng 83:1577–1581
9. Murugan R, Ramakrishna S (2006) Nano-featured scaffolds for tissue engineering: a review of spinning methodologies. Tissue Eng 12:435–447
10. Peter XM (2008) Biomimetic materials for tissue engineering. Adv Drug Deliv Rev 60:184–198
11. Saw SH, Wang K, Yong T, Ramakrishna S (2007) Polymeric nanofibers in tissue engineering. Nanotechnologies for the life sciences. Wiley-VCH Verlag GmbH & Co. KGaA, Weinheim
12. Rho J-Y, Kuhn-Spearing L, Zioupos P (1998) Mechanical properties and the hierarchical structure of bone. Med Eng Phys 20:92–102
13. Engh CA, Bobyn JD, Glassman AH (1987) Porous-coated hip replacement. The factors governing bone ingrowth, stress shielding, and clinical results. J Bone Joint Sur Br 69-B:45–55

14. Franchi M, Fini M, Martini D, Orsini E, Leonardi L, Ruggeri A et al (2005) Biological fixation of endosseous implants. Bone Struct Health Dis 36:665–671
15. Li P (2003) Biomimetic nano-apatite coating capable of promoting bone ingrowth. J Biomed Mater Res Part A 66A:79–85
16. Cosgriff-Hernandez E, Mikos A (2008) New biomaterials as scaffolds for tissue engineering. Springer, Netherlands, pp 2345–2347
17. Kretlow JD, Young S, Klouda L, Wong M, Mikos AG (2009) Injectable biomaterials for regenerating complex craniofacial tissues. Adv Mater 21:3368–3393
18. Slaughter BV, Khurshid SS, Fisher OZ, Khademhosseini A, Peppas NA (2009) Hydrogels in regenerative medicine. Adv Mater 21:3307–3329
19. Tan R, Feng Q, She Z, Wang M, Jin H, Li J et al (2010) In vitro and in vivo degradation of an injectable bone repair composite. Polym Degrad Stab 95:1736–1742
20. Augst AD, Kong HJ, Mooney DJ (2006) Alginate hydrogels as biomaterials. Macromol Biosci 6:623–633
21. Kikuchi A, Okano T (2002) Pulsatile drug release control using hydrogels. Adv Drug Deliv Rev 54:53–77
22. Tønnesen HH, Karlsen J (2002) Alginate in drug delivery systems. Drug Dev Ind Pharm 28:621–630
23. Hosseinkhani H, Hosseinkhani M, Khademhosseini A, Kobayashi H (2007) Bone regeneration through controlled release of bone morphogenetic protein-2 from 3-D tissue engineered nano-scaffold. J Control Release 117:380–386
24. Ramakrishna S, Fujihara K, Teo W-E, Lim T-C, Ma Z (2005) An introduction to electrospinning and nanofibers. World Scientific, Singapore
25. Li X, Yang Y, Fan Y, Feng Q, Cui FZ, Watari F (2014) Biocomposites reinforced by fibers or tubes as scaffolds for tissue engineering or regenerative medicine. J Biomed Mater Res Part A 102(5):1580–1594
26. Susan L et al (2006) Biomimetic electrospun nanofibers for tissue regeneration. Biomed Mater 1:R45
27. Santin M, Motta A, Freddi G, Cannas M (1999) In vitro evaluation of the inflammatory potential of the silk fibroin. J Biomed Mater Res 46:382–389
28. Kim K-H, Jeong L, Park H-N, Shin S-Y, Park W-H, Lee S-C et al (2005) Biological efficacy of silk fibroin nanofiber membranes for guided bone regeneration. J Biotechnol 120:327–339
29. Ma PX (2004) Scaffolds for tissue fabrication. Mater Today 7:30–40
30. Mistry AS, Mikos AG (2005) Tissue engineering strategies for bone regeneration. In: Yannas IV (ed) Regenerative medicine II. Springer, Berlin/Heidelberg, p 129
31. Baler K, Ball JP, Cankova Z, Hoshi RA, Ameer GA, Allen JB (2014) Advanced nanocomposites for bone regeneration. Biomater Sci. doi:10.1039/c4bm00133h
32. Sitharaman B, Shi X, Walboomers XF, Liao H, Cuijpers V, Wilson LJ et al (2008) In vivo biocompatibility of ultra-short single-walled carbon nanotube/biodegradable polymer nanocomposites for bone tissue engineering. Bone 43:362–370
33. Osathanon T, Linnes ML, Rajachar RM, Ratner BD, Somerman MJ, Giachelli CM (2008) Microporous nanofibrous fibrin-based scaffolds for bone tissue engineering. Biomaterials 29:4091–4099
34. Brunner TJ, Grass RN, Bohner M, Stark WJ (2007) Effect of particle size, crystal phase and crystallinity on the reactivity of tricalcium phosphate cements for bone reconstruction. J Mater Chem 17:4072–4078
35. Meyer JL, Eanes ED (1978) A thermodynamic analysis of the amorphous to crystalline calcium phosphate transformation. Calcif Tissue Res 25:59–68
36. Schneider OD, Weber F, Brunner TJ, Loher S, Ehrbar M, Schmidlin PR et al (2009) In vivo and in vitro evaluation of flexible, cottonwool-like nanocomposites as bone substitute material for complex defects. Acta Biomater 5:1775–1784
37. Bahr M, Bonhoeffer F (1994) Perspectives on axonal regeneration in the mammalian CNS. Trends Neurosci 17:473–479

38. Filbin MT (2003) Myelin-associated inhibitors of axonal regeneration in the adult mammalian CNS. Nat Rev Neurosci 4:703–713
39. Zhang S, Gelain F, Zhao X (2005) Designer self-assembling peptide nanofiber scaffolds for 3D tissue cell cultures. Semin Cancer Biol 15:413–420
40. Silva GA, Czeisler C, Niece KL, Beniash E, Harrington DA, Kessler JA et al (2004) Selective differentiation of neural progenitor cells by high-epitope density nanofibers. Science 303:1352–1355
41. Davis ME, Hsieh PCH, Takahashi T, Song Q, Zhang S, Kamm RD et al (2006) Local myocardial insulin-like growth factor 1 (IGF-1) delivery with biotinylated peptide nanofibers improves cell therapy for myocardial infarction. Proc Natl Acad Sci 103:8155–8160
42. Ellis-Behnke RG, Liang Y-X, You S-W, Tay DKC, Zhang S, So K-F et al (2006) Nano neuro knitting: peptide nanofiber scaffold for brain repair and axon regeneration with functional return of vision. Proc Natl Acad Sci U S A 103:5054–5059
43. Guo J, Su H, Zeng Y, Liang YX, Wong WM, Ellis-Behnke RG (2007) Reknitting the injured spinal cord by self-assembling peptide nanofiber scaffold. Nanomed: Nanotechnol, Biol Med 3:311–321
44. Guo J, Leung KK, Su H, Yuan Q, Wang Q, Chu TH (2009) Self-assembling peptide nanofiber scaffold promotes the reconstruction of acutely injured brain. Nanomed: Nanotechnol, Biol Med 5:345–351
45. Yu ACH, Lee YL, Eng LF (1993) Astrogliosis in culture: I. The model and the effect of antisense oligonucleotides on glial fibrillary acidic protein synthesis. J Neurosci Res 34:295–303
46. Tysseling-Mattiace VM, Sahni V, Niece KL, Birch D, Czeisler C, Fehlings MG et al (2008) Self-assembling nanofibers inhibit glial scar formation and promote axon elongation after spinal cord injury. J Neurosci 28:3814–3823
47. Tsai PP, Schreuder-Gibson H, Gibson P (2002) Different electrostatic methods for making electret filters. J Electrost 54:333–341
48. Huang Z-M, Zhang YZ, Kotaki M, Ramakrishna S (2003) A review on polymer nanofibers by electrospinning and their applications in nanocomposites. Compos Sci Technol 63:2223–2253
49. Li W-J, Laurencin CT, Caterson EJ, Tuan RS, Ko FK (2002) Electrospun nanofibrous structure: a novel scaffold for tissue engineering. J Biomed Mater Res 60:613–621
50. Wang W, Itoh S, Matsuda A, Aizawa T, Demura M, Ichinose S et al (2008) Enhanced nerve regeneration through a bilayered chitosan tube: the effect of introduction of glycine spacer into the CYIGSR sequence. J Biomed Mater Res Part A 85A:919–928
51. Kubinová Š, Syková E (2010) Nanotechnology for treatment of stroke and spinal cord injury. Nanomedicine 5:99–108
52. Panseri S, Cunha C, Lowery J, Del Carro U, Taraballi F, Amadio S et al (2008) Electrospun micro- and nanofiber tubes for functional nervous regeneration in sciatic nerve transections. BMC Biotechnol 8:39
53. Rochkind S, Shahar A, Fliss D, El-Ani D, Astachov L, Hayon T et al (2006) Development of a tissue-engineered composite implant for treating traumatic paraplegia in rats. Springer, Berlin/Heidelberg, pp 234–245
54. Stout DA, Webster TJ (2012) Carbon nanotubes for stem cell control. Mater Today 15: 312–318
55. Orgill D, Blanco C (2009) Biomaterials for treating skin loss. CRC Press, Woodhead Publishing Limited, Sawston, Cambridge, UK, Boca Raton
56. Midwood KS, Williams LV, Schwarzbauer JE (2004) Tissue repair and the dynamics of the extracellular matrix. Int J Biochem Cell Biol 36:1031–1037
57. Zheng Z, Yin W, Zara JN, Li W, Kwak J, Mamidi R et al (2010) The use of BMP-2 coupled – nanosilver-PLGA composite grafts to induce bone repair in grossly infected segmental defects. Biomaterials 31:9293–9300
58. Federle MP (1999) Traumatic Injury to the bowel and mesentery. In: Margulis AR (ed) Modern imaging of the alimentary tube. Springer, Berlin/Heidelberg, p 271
59. Blaser MJ, Smith PD, Ravdin JI et al (eds) (2002) Infections of the gastrointestinal tract. Lippincott Williams & Wilkins, Philadelphia

60. Fujie T, Saito A, Kinoshita M, Miyazaki H, Ohtsubo S, Saitoh D et al (2010) Dual therapeutic action of antibiotic-loaded nanosheets for the treatment of gastrointestinal tissue defects. Biomaterials 31:6269–6278

61. Fujie T, Kinoshita M, Shono S, Saito A, Okamura Y, Saitoh D et al (2010) Sealing effect of a polysaccharide nanosheet for murine cecal puncture. Mosby, Surgery. 148(1):48–58

62. Hu W, Huang ZM, Liu XY (2010) Development of braided drug-loaded nanofiber sutures. Nanotechnology 21:315104

63. Hong Y, Fujimoto K, Hashizume R, Guan JJ, Stankus JJ, Tobita K et al (2008) Generating elastic, biodegradable polyurethane/poly(lactide-co-glycolide) fibrous sheets with controlled antibiotic release via two-stream electrospinning. Biomacromolecules 9:1200–1207

64. Auffan M, Rose J, Bottero JY, Lowry GV, Jolivet JP, Wiesner MR (2009) Towards a definition of inorganic nanoparticles from an environmental, health and safety perspective. Nature Nanotechnology 4, 634–641

65. Greenhalgh K, Turos E (2009) In vivo studies of polyacrylate nanoparticle emulsions for topical and systemic applications. Nanomed-Nanotechnol Biol Medi 5:46–54

66. Hebeish A, El-Rafie MH, El-Sheikh MA, Seleem AA, El-Naggar ME (2014) Antimicrobial wound dressing and anti-inflammatory efficacy of silver nanoparticles. Int J Biol Macromol 65:509–515

67. Tian J, Wong KKY, Ho C-M, Lok C-N, Yu W-Y, Che C-M et al (2007) Topical delivery of silver nanoparticles promotes wound healing. ChemMedChem 2:129–136

68. Edwards-Jones V (2009) The benefits of silver in hygiene, personal care and healthcare. Lett Appl Microbiol 49:147–152

69. Ashavani K, Praveen Kumar V, Pulickel MA, George J (2008) Silver-nanoparticle-embedded antimicrobial paints based on vegetable oil. Nat Mater 7:236–241

70. Melo PS, Marcato PD, Huber SC, Ferreira IR, de Paula LB, Almeida ABA et al (2011) Nanoparticles in treatment of thermal injured rats: is it safe? J Phys Conf Ser 304:012027

71. Hashi CK, Zhu Y, Yang G-Y, Young WL, Hsiao BS, Wang K et al (2007) Antithrombogenic property of bone marrow mesenchymal stem cells in nanofibrous vascular grafts. Proc Natl Acad Sci 104:11915–11920

72. Segers VFM, Tokunou T, Higgins LJ, MacGillivray C, Gannon J, Lee RT (2007) Local delivery of protease-resistant stromal cell derived factor-1 for stem cell recruitment after myocardial infarction. Circulation 116:1683–1692

73. Holan V, Zajicova A, Lencova A, Pokorna K, Svobodova E, Krulova M (2009) Treatment of ocular surface injuries by the transfer of limbal and mesenchymal stem cells growing on nanofibrous scaffolds. Acta Ophthalmol 87, s244, page 0, 2009

74. Shimazaki J, Higa K, Kato N, Satake Y (2009) Barrier function of cultivated limbal and oral mucosal epithelial cell sheets. Invest Ophthalmol Vis Sci 50:5672–5680

75. Cardo D, Horan T, Andrus M, Dembinski M, Edwards J, Peavy G et al (2004) National Nosocomial Infections Surveillance (NNIS) system report, data summary from January 1992 through June 2004, issued October 2004. Am J Infect Control 32:470–485

76. O'Grady NP, Alexander M, Dellinger EP, Gerberding JL, Heard SO, Maki DG et al (2002) Guidelines for the prevention of intravascular catheter-related infections. Centers for disease control and prevention. MMWR Recomm Rep 51:1–29

77. Dimick JB, Pelz RK, Consunji R, Swoboda SM, Hendrix CW, Lipsett PA (2001) Increased resource use associated with catheter-related bloodstream infection in the surgical intensive care unit. Arch Surg 136:229–234

78. Rello J, Ochagavia A, Sabanes E, Roque M, Mariscal D, Reynaga E et al (2000) Evaluation of outcome of intravenous catheter-related infections in critically ill patients. Am J Respir Crit Care Med 162:1027–1030

79. Mermel LA (2000) Prevention of intravascular catheter-related infections (vol 132, pg 391, 2000). Ann Intern Med 133:395

80. Bozic KJ, Kurtz SM, Lau E, Ong K, Vail TP, Berry DJ (2009) The epidemiology of revision total hip arthroplasty in the United States. J Bone Joint Surg Am 91:128–133

81. Bozic KJ, Kurtz SM, Lau E, Ong K, Chiu V, Vail TP et al (2010) The epidemiology of revision total knee arthroplasty in the United States. Clin Orthop Relat Res 468:45–51
82. Stanton T, Haas J, Phillips M, Immerman I (2011/2010) Study points to savings with infection-screening program before TJR. AAOS Now, Rosemont
83. Donlan RM (2001) Biofilms and device-associated infections. Emerg Infect Dis 7:277–281
84. Martin P, Leibovich SJ (2005) Inflammatory cells during wound repair: the good, the bad and the ugly. Trends Cell Biol 15:599–607
85. Jones RN (1994) The antimicrobial activity of cefotaxime – comparative multinational hospital isolate surveys covering 15 years. Infection 22:S152–S160
86. Altman GH, Diaz F, Jakuba C, Calabro T, Horan RL, Chen J et al (2003) Silk-based biomaterials. Biomaterials 24:401–416
87. Hu W, Huang ZM (2010) Biocompatibility of braided poly(L-lactic acid) nanofiber wires applied as tissue sutures. Polym Int 59:92–99
88. Koller R, Miholic J, Jakl RJ (1997) Repair of incisional hernias with expanded polytetrafluoroethylene. Eur J Surg 163:261–266
89. Liu X, Lin T, Fang JA, Yao G, Zhao HQ, Dodson M et al (2010) In vivo wound healing and antibacterial performances of electrospun nanofibre membranes. J Biomed Mater Res Part A 94A:499–508
90. Hsu SH, Tseng HJ, Lin YC (2010) The biocompatibility and antibacterial properties of waterborne polyurethane-silver nanocomposites. Biomaterials 31:6796–6808
91. Huey-Shan H, Shan-hui H (2007) Biological performances of poly(ether)urethane–silver nanocomposites. Nanotechnology 18:475101
92. Li BY, Jiang BB, Boyce BM, Lindsey BA (2009) Multilayer polypeptide nanoscale coatings incorporating IL-12 for the prevention of biomedical device-associated infections. Biomaterials 30:2552–2558
93. Fujie T, Matsutani N, Kinoshita M, Okamura Y, Saito A, Takeoka S (2009) Adhesive, flexible, and robust polysaccharide nanosheets integrated for tissue-defect repair. Adv Funct Mater 19:2560–2568
94. Sivakumar M, Kumar Dhanadurai KS, Rajeswari S, Thulasiraman V (1995) Failures in stainless steel orthopaedic implant devices: a survey. J Mater Sci Lett 14:351–354
95. Allan SM, Rothwell NJ (2003) Inflammation in central nervous system injury. Philos Trans R Soc Lond Ser B-Biol Sci 358:1669–1677
96. McGraw J, Hiebert GW, Steeves JD (2001) Modulating astrogliosis after neurotrauma. J Neurosci Res 63:109–115
97. Cunha C, Panseri S, Antonini S (2011) Emerging nanotechnology approaches in tissue engineering for peripheral nerve regeneration. Nanomed: Nanotechnol Biol Med 7:50–59
98. Meng H, Chen L, Ye Z, Wang S, Zhao X (2009) The effect of a self-assembling peptide nanofiber scaffold (peptide) when used as a wound dressing for the treatment of deep second degree burns in rats. J Biomed Mater Res B Appl Biomater 89B:379–391
99. Zhou DS, Zhao KB, Li Y, Cui FZ, Lee IS (2006) Repair of segmental defects with nano-hydroxyapatite/collagen/PLA composite combined with mesenchymal stem cells. J Bioact Compat Polym 21:373–384
100. Ding T, Luo Z-J, Zheng Y, Hu X-Y, Ye Z-X (2010) Rapid repair and regeneration of damaged rabbit sciatic nerves by tissue-engineered scaffold made from nano-silver and collagen type I. Injury 41:522–527

Chapter 4
Nanomaterial Design and Computational Modeling

Zhengzheng Chen, Rong Chen, and Bin Shan

4.1 Introduction

Nanomaterials and nanotechnology may lead to breakthroughs in various fields such as VLSI circuits [1], energy storage solutions [2, 3], environmental protection [4, 5], and biomedical applications [6, 7]. Despite the incredible advances in characterization tools and techniques, there seems to be greater than ever needs to be able to carry out computational simulations with atomistic resolution for nanomaterials. This is in part due to the fact that nanoscale properties are extremely difficult to measure or manipulate, but more importantly, such properties are probably very sensitive to subtle environmental changes and perturbations, making repeated measurements more challenging. This is where computational modeling has much to offer in the booming nanomaterials design and nanomedicine, in that it supplies "virtue experimental methods" to investigate mechanisms of phenomena and even to design artificial structures in order to get desirable properties [8–10]. One good example is the design of nitrogen doped nanotube as chemical sensor which would have potential applications in biomedical fields. It was first proposed based on the gas response sensitivity analysis from first-principles calculations [11] and then a year later, confirmed by experiments, where fabricated CNx nanotubes have rapid sensing capabilities to low concentrations of toxic gases such as ammonia, acetone and OH groups [12]. Another example is the explanation of the presence of the strongest grain-size in nanocrystal metals by molecular dynamics simulations [13]. Computational simulations are able to present detailed insight of phenomena which are very difficult or even impossible to be obtained by experiments. It is clear that

Z. Chen • R. Chen • B. Shan (✉)
Department of Materials Science & Engineering, Huazhong University of Science and Technology, Wuhan 430074, China
e-mail: bshan@mail.hust.edu.cn

© Springer Science+Business Media New York 2014
Y. Ge et al. (eds.), *Nanomedicine*, Nanostructure Science and Technology,
DOI 10.1007/978-1-4614-2140-5_4

computational modeling provides invaluable information in both cases and it is computational techniques that give deep insight into the structure-property and structure-functionality relationship in nanomaterials.

Another aspect of the rising interest in nanomaterials design and computational modeling is that with the ever increasing power of CPUs and better designed parallel algorithms, the capability of computer simulation has greatly expanded. Nowadays, modern supercomputers has tera-flop computational power and full quantum mechanical simulations can tackle molecular systems consisting millions of atoms, while state-of-the-art molecular dynamics code can handle billions of atoms. This puts nanomaterials modelling squarely into experimentally accessible regions and the ability of mimicking real experimental systems has made computer simulations a complementary tool for understanding phenomena on the nanoscale.

Over the past few decades, we see a booming interest in computational modeling and the emergence of a number of new theoretical and numerical techniques. In order to tackle nanomaterials systems on different time and size scales, computational modeling has also developed several branches that aim to suitably describe various properties of assorted systems. The number one rule of thumb for computational simulation is that one has to choose the most suitable simulation method for a specific problem. Generally speaking, based on accuracy and size constraints, simulation methods can be categorized as First-Principles (FP) method (accurate, 10–1,000 atoms), tight binding (TB) (approximate electronic structure information, 10^2–10^5 atoms), and molecular dynamics (MD) (empirical potential, $>10^6$ atoms) (Fig. 4.1), which cover a wide area from sophisticated electronic structures to massive bulk properties. In the following sections, the fundamentals of these three simulation techniques are outlined and their applications in nanomaterials modeling are further demonstrated in details.

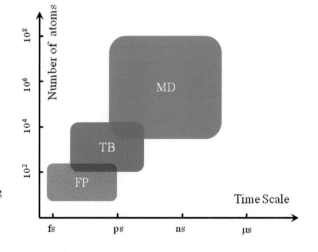

Fig. 4.1 Diagram illustrating suitable system sizes and time scales of FP, TB, and MD methods

4.2 First-Principles Methods

First-Principles methods, as shown in its own name, obtain electronic structure of materials basing on the very foundation of quantum mechanics, Schrödinger equation. Except several basic constants such as Plank constant, atomic mass, Bohr radius and certain approximations required to simplify numerical complexity, such as Born-Oppenheimer approximation, local density approximation, First-Principles methods calculate essential quantities directly without any presetting or empirical parameters. Therefore, this kind of methods has high accuracies, and can be used on most materials. On the other hand, due to the computationally demanding self-consistent solution procedures, FP methods are very time-consuming, and can only treat relatively small systems.

4.2.1 Born-Oppenheimer Approximation

The one single most important approximation employed in most FP calculations is the Born-Oppenheimer approximation, which effectively decouples the electronic and nuclear degree of freedom, and thus greatly simplifying the numerical solutions. By employing Hartree atomic unit, we can express the Hamiltonian of an N-ion-n-electron system as

$$\hat{H} = -\sum_{I=1}^{N} \frac{\nabla_I^2}{2M_I} - \sum_{i=1}^{n} \frac{\nabla_i^2}{2} + \frac{1}{2} \sum_{I,J,I\neq J}^{N} \frac{Z_I Z_J}{R_{IJ}} + \frac{1}{2} \sum_{i,j,i\neq j}^{n} \frac{1}{r_{ij}} - \sum_{I=1}^{N} \sum_{i=1}^{n} \frac{Z_I}{r_{il}} \tag{4.1}$$

Capital and lower case letters indicate ions and electrons, respectively. Since ions are ~10^4 heavier than electrons ($M_I \sim 10^4$), it is safe to say that electrons move much faster than ions do.

In other words, at every moment that ions move to a new configuration, electrons will instantaneously relax to their new ground state. Therefore, the movements of electrons and ions can be separated. In mathematical terms, this separation is realized by expressing the total wavefunction as follows:

$$\Phi\left(\{\vec{r}_i\};\{\vec{R}_I\}\right) = \phi\left(\{\vec{r}_i\}\right)\Big|_{\{\vec{R}_I\}} \chi\left(\{\vec{R}_I\}\right) \tag{4.2}$$

$\phi\left(\{\vec{r}_i\}\right)\Big|_{\{\vec{R}_I\}}$ is electronic wavefunction with ionic configuration as $\{\vec{R}_I\}$, and $\chi\left(\{\vec{R}_I\}\right)$ is ionic wavefuction, respectively. Furthermore, $\phi\left(\{\vec{r}_i\}\right)\Big|_{\{\vec{R}_I\}}$ satisfies Schrödinger equation:

$$\left\{ -\sum_{i=1}^{n} \frac{\nabla_i^2}{2} + \frac{1}{2} \sum_{i,j,i\neq j}^{n} \frac{1}{r_{ij}} - \sum_{I=1}^{N} \sum_{i=1}^{n} \frac{Z_I}{r_{il}} \right\} \phi\left(\{\vec{r}_i\}\right)\Big|_{\{\vec{R}_I\}} = \varepsilon_{elec}\left(\{\vec{R}_I\}\right)\phi\left(\{\vec{r}_i\}\right)\Big|_{\{\vec{R}_I\}} \tag{4.3}$$

Equations 4.2 and 4.3 are called Born-Oppenheimer approximation. It realizes the separation of movements of electrons and ions, and serves as the very foundation of most modern computational quantum chemistry/physics methods. Combining the above three equations, we can obtain the equation of $\chi\left(\left\{\vec{R}_I\right\}\right)$:

$$\left\{-\sum_{M=1}^{N}\frac{\nabla_I^2}{2M_I}+\frac{1}{2}\sum_{I,J,I\neq J}^{N}\frac{Z_I Z_J}{R_{IJ}}+\varepsilon_{elec}\left(\left\{\vec{R}_I\right\}\right)\right\}\chi\left(\left\{\vec{R}_I\right\}\right)=\varepsilon\chi\left(\left\{\vec{R}_I\right\}\right) \qquad (4.4)$$

Equation 4.4 shows that ions move in a potential field ε_{elec}. ε_{elec} is therefore also called "Born-Oppenheimer potential surface". From then on, we focus our discussion on solving the electronic wavefunction $\phi\left(\left\{\vec{r}_i\right\}\right)\big|_{\left\{\vec{R}_I\right\}}$. More information could be found in Grosso and Parravicini's book [15].

4.2.2 Density Functional Theory

Though Born-Oppenheimer approximation simplifies Schrödinger equation in solid systems, Eq. 4.3 is still very difficult to solve. This is mainly because the coupling term $1/r_{ij}$ which makes Eq. 4.3 a non-linear n-body coupled equation. The commonly used technique to overcome the complication of Eq. 4.3 is to transfer this n-body problem to a single body problem. The rigorous demonstration is the density functional theory (DFT) [16, 17]. DFT does not consider concrete electronic orbital configurations, but focuses on the relationship between the total energy and the charge distribution of the system. Further, employing the variation principle with constrained conditions, DFT reformulates Eq. 4.3 into a single-body equation describing the state of a single electron moving in an effective potential field, while all many body interactions are lumped into a so called "exchange-correlation" functional. As Hohenberg and Kohn pointed out, the ground energy of a system can be expressed as a universal functional as its ground charge distribution [16]:

$$E^{HK}\left[\rho\left(\vec{r}\right);V_{ext}\left(\vec{r}\right)\right]=T\left[\rho\left(\vec{r}\right)\right]+E_{ee}\left[\rho\left(\vec{r}\right)\right]+\int V_{ext}\left(\vec{r}\right)\rho\left(\vec{r}\right)d\vec{r}+E_{II} \qquad (4.5)$$

Terms on the right side are kinetic energy, electron-electron interaction, electron-ion interaction, and ion-ion interaction, respectively. Let us ignore the last term by now for it is a constant shift for a given atomic configuration. One can in principle get the minimum of E^{HK} by taking the variation of Eq. 4.5 with respect to ρ for a given external potential V_{ext}. Unfortunately, this is not a practical way at least in the near future since we have no idea about the kinetic functional for an interacting electron-gas system. To overcome this difficulty, Kohn and Sham formulated the KS equation by mapping a non-interacting electron gas system whose charge distribution $\rho_0\left(\vec{r}\right)$ is identical to the ground charge distribution $\rho\left(\vec{r}\right)$ to the real system [17]. The key point of this ansatz is that $\rho_0\left(\vec{r}\right)$ is able to be expresses as the summation of single electron wavefunctions $\varphi_j\left(\vec{r}\right)$:

$$\rho_0(\vec{r}) = \sum_i \varphi_i^*(\vec{r})\varphi_i(\vec{r}) \tag{4.6}$$

and so is $\rho(\vec{r})$. The kinetic energy functional of a non-interacting electron gas $T_0[\rho(\vec{r})]$ can be analytically calculated. We can then take this advantage. Furthermore, Kohn and Sham calculated Hartree interaction $E_H[\rho(\vec{r})]$ instead of $E_{ee}[\rho(\vec{r})]$:

$$E_H[\rho(\vec{r})] = \int \frac{\rho(\vec{r})\rho(\vec{r}')}{|\vec{r}-\vec{r}'|} d\vec{r} d\vec{r}' \tag{4.7}$$

Apparently, $T_0[\rho(\vec{r})] + E_H[\rho(\vec{r})]$ differs from $T[\rho(\vec{r})] + E_{ee}[\rho(\vec{r})]$ in Eq. 4.5. To compensate this difference, one needs an extra term:

$$E_{xc}[\rho] = T[\rho] + E_{ee}[\rho] - T_0[\rho] - E_H[\rho] \tag{4.8}$$

The importance of $E_{xc}[\rho]$ is that it contains all the effects from many-body interactions. Combining Eqs. 4.5, 4.6, 4.7, and 4.8, and taking the variation of the total energy with respective to $\varphi_j(\vec{r})$ under the constrain condition:

$$\sum_j \langle \varphi_j | \varphi_j \rangle = N \tag{4.9}$$

we can obtain the equation which ground state wavefunction $\varphi_j(\vec{r})$ satisfies:

$$\left[-\frac{\nabla^2}{2} + V_H(\vec{r}) + V_{ext}(\vec{r}) + V_{xc}(\vec{r}) \right] \varphi_j(\vec{r}) = \varepsilon_j \varphi_j(\vec{r}) \tag{4.10}$$

Explicitly, we define exchange-correlation potential as:

$$V_{xc}(\vec{r}) = \frac{\delta E_{xc}[\rho(\vec{r})]}{\delta \rho(\vec{r})} \tag{4.11}$$

Equation 4.10 is Kohn-Sham (KS) equation. By using KS equation, the ground energy could be rewritten as

$$E_0 = \sum_j \varepsilon_j - \frac{1}{2} \int \frac{\rho(\vec{r})\rho(\vec{r}')}{|\vec{r}-\vec{r}'|} d\vec{r} d\vec{r}' + E_{xc}[\rho(\vec{r})] - \int V_{xc}(\vec{r})\rho(\vec{r}) d\vec{r} \tag{4.12}$$

The first term in the right side is called band structure energy, and other three terms are called double counting terms.

One should understand that an exact analytical formula for the exchange-correlation energy $E_{xc}[\rho]$ is generally unavailable for most cases. Its correlation component, however, can be numerically obtained by quantum Monte-Carlo (QMC) method, which has been done by Ceperley and Alder [18]. Several research groups fit their data by different analytic functions and incorporate these functions into KS equation [18–22].

4.2.3 Self-Consistent Field Processes in DFT

Reviewing Eq. 4.10, a paradox may be found: To build up V_H and V_{ext} of Hamiltonian in Eq. 4.10, one has to know $\rho(\vec{r})$ in advance. At the same time, $\rho(\vec{r})$ needs to be solved. In other words, $\varphi_j(\vec{r})$ appears at both side of Eq. 4.10. Therefore, KS equation needs to be solved self consistently. First, an initial guess of $\varphi_j(\vec{r})$ or $\rho(\vec{r})$ is able to be presented. Second, the Hamiltonian is built up and Eq. 4.10 is solved. Third, $\rho(\vec{r})$ is updated according to the output and input of the current step, re-construct Hamiltonian. The above steps are repeated until the convergence criterion is satisfied. Usually, the criterion is chosen as the change of input and output values of total energy at current step. The whole process is called self-consistent field (SCF) calculations. After a set of high-quality eigenfunctions is obtained, the electronic structures could be constructed and analyzed, i.e. charge density, energy band structures, and density of states, etc. Details of SCF are far beyond the scope of this chapter, more details could be found in the books of Martin [23] and Knhanoff [24], respectively.

4.2.4 Examples

Since DFT methods, or more general, First-Principles methods, do not depend on availability of parameters for a given system. This character makes them very suitable to study materials which either is novel or has strong quantum effects [25, 26]. More importantly, as shown in Eqs. 4.6 and 4.10, the ground charge distribution is sensitive to concrete external potentials, arbitrary structural defects and/or alloying elements in principle are able to introduce unique electronic structures and even properties. Therefore, First-Principles methods can be employed to predict or even design new class of materials with desirable properties, which is a very important and vibrant aspect of modern computational materials science.

As an example, the study on the diffusion of an adatom on the Sn-alloying Cu(111) surface is hereby presented [14]. Figure 4.2 shows the potential energy surface (PES) of a Cu adatom on such an alloying surface. Clearly, Sn atoms disturb the profile of PES in two ways: (1) they increase the value of PES at their sites, and (2) they introduce forbidden region around each of them by transferring local valleys to slopes, which are shown as green areas in Fig. 4.2. These two features make

Fig. 4.2 The landscape of total energy of the system with single Cu adatom on the Cu(111) surface alloyed by 2 Sn atoms. The brighter (*darker*) area indicates weak (*strong*) adsorption sites of the Cu adatom. The lowest energy is set to zero [14]

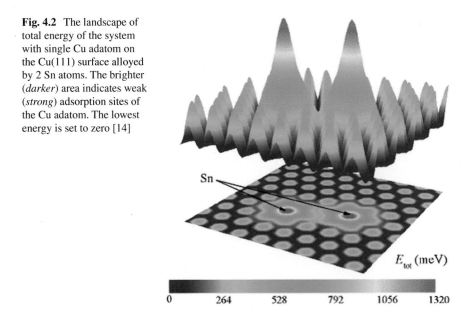

E_{tot} (meV)

| 0 | 264 | 528 | 792 | 1056 | 1320 |

Sn atoms diffusion blockers to a Cu adatom because it is very energetically unfavorable for a Cu adatom approaching sites occupying by Sn atoms. This phenomenon can be attributed to the fact that Sn atoms are larger than Cu atoms and therefore protrude from the surface. Besides the geometrical factor, different electronic configurations between Sn atoms and Cu atoms also contribute to the blocking effect of Sn. Figure 4.3 shows two minimum energy paths (MEP). When a Cu adatom approaches Sn sites, it climbs uphill. Not only local minimums, but also migration barriers rise up. This feature is more apparent along the path which goes through the two Sn atoms (Fig. 4.3b). To explain MEPs, Fig. 4.3c presents local density of states (LDOS) of the surface layer at Fermi energy $D(E_F)$. $D(E_F)$ is higher between the two Sn atoms because of the contribution from p-orbitals of Sn, and thus strengthens binding interaction between the surface and the Cu adatom according to Newns-Anderson model. The results of this First-Principles simulation are in good agreement with experimental observation in which alloying with Sn increases the serving lifetime of Cu interconnects by 10 times [27].

4.3 Tight Binding

Tight binding method (TB) is a wide-used semi-empirical computational method. Based on a set of well-chosen basis functions, TB builds up Hamiltonian matrix H of a given system and gets eigenvalues and eigen-wavefunctions by diagonalizing H. Further information of electronic structure, i.e. charge density, band structure, and optimal adsorption spectrum etc. can then be obtained. Different from

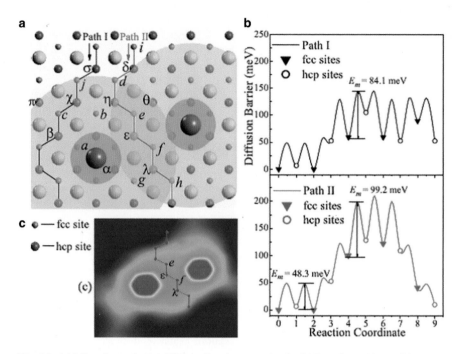

Fig. 4.3 (a) Migration paths I and II for a Cu adatom on the Cu (111) surface with two 4th nearest neighbor Sn surface atoms. *Yellow* (*gray*) *circles* denote the Cu (Sn) atoms, while the *green* (*red*) *circles* denote hcp (fcc) sites, denoted by Greek and Latin letters, respectively. (**b**) Migration energy landscape versus reaction coordinates for pathways I and II, respectively. (**c**) LDOS contours at the Fermi level, with bright (*dark*) color indicating high (low) values of LDOS [38]

first-principles methods, elements in H in TB are not directly calculated through SCF. They are expressed as functions of atomic positions with a set of predetermined parameters. With high-quality parameters, TB can accurately simulate systems of $10^3 \sim 10^4$ atoms. This is important since systems with this size could contain complicated atomic structures or functional group. Therefore, TB method is attractive for Nano-material simulations because objects in this area usually have artificial structure and peculiar electronic structures which need to be identified.

4.3.1 Linear Combination of Atomic Orbitals

Theoretical foundation of TB can be viewed by linear combination of atomic orbitals (LCAO) method [28]. Suppose there is a system containing N atoms and there are n_i orbitals belonging to the i-th atom. One eigen-wavefunction can be expressed as

$$\Psi\left(\vec{r}\right) = \sum_{i,\alpha} c_{i\alpha}\phi_{\alpha}\left(\vec{r}-\vec{R}_i\right) \tag{4.13}$$

where i and α are indexes of atoms and orbitals, respectively. Accordingly, the energy of the system E is

$$E = \frac{\left\langle\Psi\left|\hat{H}\right|\Psi\right\rangle}{\left\langle\Psi|\Psi\right\rangle} = \frac{\sum_{i\alpha,j\beta} c_{i\alpha}^{*}c_{j\beta}\left\langle\phi_{i\alpha}\left|\hat{H}\right|\phi_{j\beta}\right\rangle}{\sum_{i\alpha,j\beta} c_{i\alpha}^{*}c_{j\beta}\left\langle\phi_{i\alpha}\left|\phi_{j\beta}\right\rangle} \tag{4.14}$$

The variation of E with respect to $c_{i\alpha}^{*}$ can be easily calculated:

$$\frac{\delta E}{\delta c_{i\alpha}^{*}} = \frac{1}{\sum_{i\alpha,j\beta} c_{i\alpha}^{*}c_{j\beta}\left\langle\phi_{i\alpha}|\phi_{j\beta}\right\rangle}\left[\sum_{j\beta} c_{j\beta}\left\langle\phi_{i\alpha}\left|\hat{H}\right|\phi_{j\beta}\right\rangle - Ec_{j\beta}\left\langle\phi_{i\alpha}|\phi_{j\beta}\right\rangle\right] \tag{4.15}$$

To obtain the lowest value of E, Eq. 4.15 should equal to 0 for any i and α. Therefore, we can straightforward obtain the secular equation of $c_{j\beta}$:

$$\left|H_{i\alpha,j\beta} - ES_{i\alpha,j\beta}\right| = 0 \tag{4.16}$$

Equation 4.16 is called generalized eigenvalue equation. $H_{i\alpha,j\beta}$ and $S_{i\alpha,j\beta}$ are elements of Hamiltonian matrix and overlapping matrix expanded by $\{\phi_{i\alpha}\}$, respectively. If $\{\phi_{i\alpha}\}$ is a set of orthogonal functions, Eq. 4.16 is reduced to

$$\left|H_{i\alpha,j\beta} - E\delta_{i\alpha,j\beta}\right| = 0 \tag{4.17}$$

which has been familiar with in Sect. 4.2. Eqs. 4.16 and 4.17 are called LCAO method. If $\{\phi_{i\alpha}\}$ are not chosen as actual atomic orbitals, this method is usually named as "tight binding".

4.3.2 Slater-Koster Two-Center Approximation

According to above discussions, the key step in TB method is to construct Hamiltonian matrix H. Appropriate approximations are able to essentially simplify the construction of elements of H. In 1954, Slater and Koster suggested two-center approximation in their classic paper [29], which expresses all Hamiltonian elements with a limited number of two-center integrations. The total number of all these basic integrations is around 30. Therefore, Slater-Koster two-center approximation makes TB method to be practical and is the foundation of modern TB simulating software packages.

Two points of Slater-Koster two-center approximation (SK approximation for short) should be emphasized. First, let us specifically write down the expression of one element of H for a periodic system i.e. crystal. In this case, the basic function is φ_α, the Bloch summation of atomic orbitals ϕ_α:

$$\varphi_{i\alpha}\left(\vec{r}\right) = \frac{1}{\sqrt{N_{cell}}} \sum_{n=1}^{N_{cell}} \exp\left(i\vec{k}\cdot\vec{R}_i^n\right)\phi_\alpha\left(\vec{r}-\vec{R}_i^n\right) \tag{4.18}$$

The superscript n is the index of cell. $\vec{k}$ is the vector in reciprocal space. After some basic algebra calculations, we can obtain $H_{i\alpha,j\beta}\left(\vec{k}\right)$:

$$H_{i\alpha,j\beta}\left(\vec{k}\right) = \frac{1}{N_{cell}} \sum_{n,m=1}^{N_{cell}} \exp\left[i\vec{k}\cdot\left(\vec{R}_i^n - \vec{R}_i^m\right)\right] \times \int d\vec{r}\phi_\alpha^*\left(\vec{r}-\vec{R}_i^n\right)\hat{H}\phi_\alpha\left(\vec{r}-\vec{R}_i^m\right) \tag{4.19}$$

Because $\hat{H}$ is a function of positions of electrons and atoms, Eq. 4.19 can be categorized as three types: (1) on-site integration, in which three integrands, ϕ_α, $\hat{H}$ and ϕ_β are at the same center; (2) two center integration, in which two of the above integrands are at one center, and the other one is at another center; and (3) three center integration, in which each integrand is at its own center. Because of the local nature of ϕ_α and ϕ_β, in most case on-site integration has the largest value while the three center integration is the smallest. Therefore, the contribution to $H_{i\alpha,j\beta}\left(\vec{k}\right)$ is truncated up to two center integrations. Three center integrations and higher order ones are ignored. This truncation essentially decreases the number of terms in $H_{i\alpha,j\beta}\left(\vec{k}\right)$, and is the first point in SK approximation.

Two center integrations are usually understood as "bonding" between two orbitals centering at two atoms. They are explicit functions of the relative position $\vec{R}_{ij}$ between two atoms. Though they could be very different from each other since $\vec{R}_{ij}$ can be any vector, they always can be expressed as a combination of several basic two center terms. This is the second key point in SK approximation. These basic terms could be referred to "bonding terms". These terms are showed as V_{ss}, V_{sp}, V_{pp} and $V_{pp\pi}$, etc. The first two subscripts are the angular momentum numbers of two orbitals, and the third subscript indicates the bonding type, which depends on the relative orientations and symmetries of the two orbitals.

In Figs. 4.4 and 4.5, we show simple examples about how to express two center integrations in terms of bonding terms. One atom is at original point and another atom is at $\vec{R}$, the orientation cosine with respect to xyz axis are l, m and n, respectively. The s-s term is independent from orientation of $\vec{R}$: $\langle s|\hat{H}|s\rangle = V_{ss\sigma}$ (Fig. 4.4a). The p_y-p_y interaction can be decomposed as $\langle p_y|\hat{H}|p_y\rangle = m^2 V_{pp\sigma} + \left(1-m^2\right)V_{pp\pi}$ (Fig. 4.4b), and the s-p_y term is $\langle s|\hat{H}|p_y\rangle = mV_{sp\sigma}$, (Fig. 4.5).

Therefore, four basic bonding terms can be used to express all sp-type interactions. That is why SK approximation has made tremendous success. For higher order orbitals, i.e. d, f, and g, etc., one has to use angular momentum theory to calculate two center integrations, which is discussed in detail in a couple of references [30, 31].

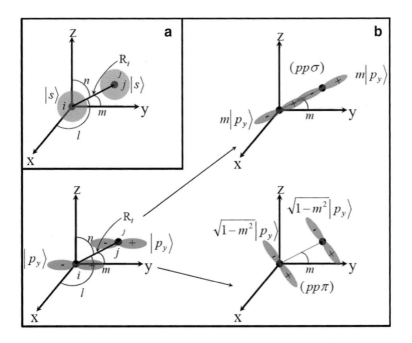

Fig. 4.4 Illustration of two-center approximation of *sp*-type interactions. (**a**) is s-s term and (**b**) is p_y-p_y term

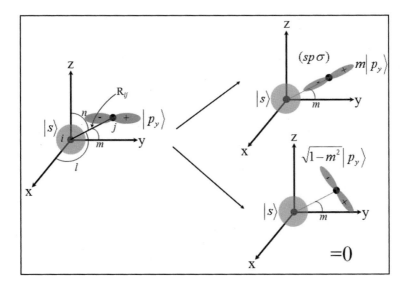

Fig. 4.5 Illustration of two-center approximation of s-p_y term

4.3.3   Total Energy in TB

By diagonalizing Hamiltonian matrix, a set of eigenvalues can be obtained. The corresponding energy of the system is then can be expressed as

$$E_{tot} = 2\sum_{\lambda}\varepsilon_{\lambda}f\left(\varepsilon_{\lambda}\right) \tag{4.20}$$

λ is the band index, the 2 comes from the spin-degeneracy of each band, and $f(\varepsilon_{\lambda})$ is Fermi-Dirac distribution at 0 K. However, the TB method might be challenged on its accuracy since the above equation takes into account only on-site terms and "bonding" contributions. E_{tot} should contain Coulomb repulsion of electrons and ions. Therefore, Eq. 4.20 severely overestimates cohesive energy of the system. Practically, Coulomb repulsion is dealt by adding an extra term E_{rep}, we have

$$E_{tot} = 2\sum_{\lambda}\varepsilon_{\lambda}f\left(\varepsilon_{\lambda}\right)+E_{rep} = 2\sum_{\lambda}\varepsilon_{\lambda}f\left(\varepsilon_{\lambda}\right)+\frac{1}{2}\sum_{i,j}A\exp\left(-R_{ij}/R_{0}\right) \tag{4.21}$$

A and R_0 are parameters which are determined by experiments or DFT calculations.

4.3.4   Examples

Compared to full self-consistent first-principles calculations, TB method does have limitation of generality and transferability of parameters. However, its transparent physical picture and its approximate but reasonable way of describing electronic structures make it a very useful tool in analyzing the structure-electronic relationship in nanomaterials. An example of how TB method is used to design a quantum dot based on a single nanotube is illustrated here (Fig. 4.6) [32].

As is well known, a nanotube can be thought of a graphene sheet rolled into a cylindrical tube. It has a tunable bandgap that is highly dependent on its topological structures, and this very unique electronic property may find many applications in nanoelectronics. For example, by changing the chirality of a single nanotube using topological defects, a variety of metal-semiconductor, metal-metal, and semiconductor-semiconductor junctions can be generated. Quantum dot (QD) can be fabricated on a single wall nanotube (SWNT) by the mechanical deformation. As shown in Fig. 4.6a, kinks on a semi-conductive SWNT create dips on energy band gap. Together with the information of eigenstate wavefunctions, one can conclude that these kinks equivalently behavior like acceptor QDs. This could be an effective and simple way of creating room-temperature quantum dot devices. On the other hand, for a metallic SWNT, the response of electronic structure to the deformation is not very sensitive (Fig. 4.6b).

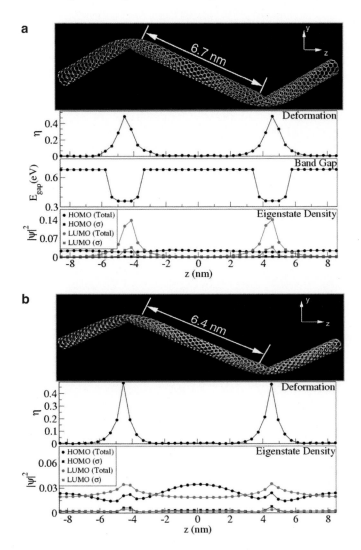

Fig. 4.6 The geometry of tubes used to create nanotube-based quantum wells. The two bending regions are separated by an undeformed segment (**a**) 6.7 nm long in the (10,0) tube and (**b**) 6.4 nm long in the (9,0) SWNT [32]

4.4 Molecular Dynamics

Though molecular dynamics (MD) method perform atomic simulations, it has no quantum mechanics background, which is different from DFT and TB methods. MD treats atoms as classic particles. Movements of atoms are determined by Newton equations. The atomic interaction is presented by empirical potentials, which shows as either analytical functions with parameters or data on grids.

Therefore, MD method does not perform SCF calculations or diagonalization of large matrix, and can be employed on study dynamical evolution of large systems ($10^6 \sim 10^8$ atoms, as shown in Fig. 4.1) in complicated loading conditions, i.e., stress, temperature, and ion bombardment, etc. [33–35]

4.4.1   Empirical Potentials

4.4.1.1   Lennard-Jones Potential

Lennard-Jones potential is a famous type of pair-potential. It describe atomic interactions as

$$V(r) = 4\varepsilon \left[\left(\frac{\sigma}{r} \right)^{12} - \left(\frac{\sigma}{r} \right)^{6} \right] \tag{4.22}$$

ε and σ are parameters which are determined by fitting important properties, i.e. equilibrium distance, binding energy, etc. In modern MD simulations, Lennard-Jones potential is usually used to describe interactions between gas atoms.

4.4.1.2   Embedded Atomic Method

Embedded atomic method (EAM) presented better simulating results on bulk materials than pair-potential [36]. Besides pair potentials, EAM introduces an addition term. Thus the total energy of the system is expressed as

$$E_{\text{tot}} = \frac{1}{2} \sum_{i,j} V\left(R_{ij} \right) + \sum_{i} F\left[\sum_{j} \rho\left(R_{ij} \right) \right] \tag{4.23}$$

$F[\rho]$ is called "embedded energy", which means the energetic gain when an atom is put into electron gas contributed by other atoms in vicinity. Clearly, EAM has concepts similar to "atomic bonds" and density functional. $V(R)$, $\rho(R)$ and $F[\rho]$ are usually functions with $10 \sim 20$ parameters. One needs fit these parameters to reproduce lattice constant, cohesive energy, vacancy formation energy, surface energy, several elastic constants, and migration energy of a point defect, etc. Recently, Ercolessi and Adams developed the *force-matching* method [37]. This method employs first-principles methods to obtain force acting on each atom in a serial of reference configurations. It then defines a target function as follows

$$F(\mathbf{p}) = F_f(\mathbf{p}) + F_c(\mathbf{p}) = \left(3 \sum_{k=1}^{M} N_k \right)^{-1} \sum_{k=1}^{M} \sum_{i=1}^{N_k} \left| \mathbf{f}_{ki}(\mathbf{p}) - \mathbf{f}_{ki}^{\text{ref}} \right|^2 + \sum_{r=1}^{N_c} W_r \left| A_r(\mathbf{p}) - A_r^{\text{ref}} \right|^2 \tag{4.24}$$

M is the number of reference configurations, N_k is the number of atoms in the k-th configuration. $\mathbf{f}$ is the atomic force, A_r is an abovementioned property of the material, W_r is the assigned weight, and $\mathbf{p}$ is the set of parameters. By minimizing Eq. 4.24, one can obtain desirable EAM potentials [39]. Different groups have applied the *force-matching* method to get high-qualified potentials for several metals and even binary systems [34, 40], which is an approval to the fidelity of this method.

4.4.2   Integrator of Motion Equations

With a given reliable empirical potential, MD calculates the force on each atom as the negative derivative of potential energy with respect to the position of the atom, then update the velocity and the position of each atom according to Newton equation. Therefore, a mathematical interpretation of MD is to solve a second order ordinary partial equation:

$$\frac{d^2\vec{r}}{dt^2} = f(t,\mathbf{r},\mathbf{v}) \quad (\mathbf{r}_0,\mathbf{v}_0) \tag{4.25}$$

Appropriate differential algorithms are essential for MD simulations.

4.4.2.1   Verlet algorithm and Prediction-Correction Algorithm

The Verlet algorithm and prediction-correction algorithm are discussed here since they are widely used and are basis of other advanced algorithms as well. Given position $\mathbf{r}(t)$, velocity $\mathbf{v}(t)$, and force $\mathbf{f}(t)$ at time t, we can get $\mathbf{r}(t+\Delta t)$ at $t+\Delta t$ and $\mathbf{r}(t-\Delta t)$ at $t-\Delta t$ through Taylor expansion:

$$\vec{r}(t+\Delta t) = \vec{r}(t) + \vec{v}(t)\cdot\Delta t + \frac{\vec{f}(t)}{2m}\cdot\Delta t^2 + \frac{\Delta t^3}{6}\dddot{\vec{r}} + O(\Delta t^4) \tag{4.26}$$

$$\vec{r}(t-\Delta t) = \vec{r}(t) - \vec{v}(t)\cdot\Delta t + \frac{\vec{f}(t)}{2m}\cdot\Delta t^2 - \frac{\Delta t^3}{6}\dddot{\vec{r}} + O(\Delta t^4) \tag{4.27}$$

By summing Eqs. 4.26 and 4.27, we have

$$\vec{r}(t+\Delta t) = 2\vec{r}(t) - \vec{r}(t-\Delta t) + \frac{\vec{f}(t)}{m}\cdot\Delta t^2 + O(\Delta t^4) \tag{4.28}$$

$$\vec{v}(t) = \frac{\vec{r}(t+\Delta t) - \vec{r}(t-\Delta t)}{2} + O(\Delta t^2) \tag{4.29}$$

Equations 4.28 and 4.29 are called Verlet algorithm. Since positions and velocities at the time t can be obtained simultaneously, Verlet algorithm can be used to obtain the total energy of the system. Another key feature of Verlet algorithm is the time reversibility, which means if we suddenly flip the velocity of each atom at time $t = n\Delta t$, the system will go back to the initial positions along the same trajectory after n steps. Therefore, the total energy of the system is conserved in Verlet algorithm. Detailed analysis demonstrates that this feature comes from Liouville equation of conserve force systems [41]. Clearly, Verlet algorithm is ideal for micro-canonical ensembles. Main limitation of Verlet algorithm is that fluctuation of energy is large in a short period since velocities has only accuracy to the order of Δt^2, as shown in Eqs. 4.28 and 4.29.

As shown in Eq. 4.25, MD simulation is to solve a second-order ODE. The prediction-correction (PC) algorithm can be performed to get the trajectory of the system. First, the position and the velocity at time $t + \Delta t$ as the linear combination of forces of previous k steps are expressed, and the linear coefficients could be set to equal to the corresponding coefficients of Taylor expansion up to the term of Δt^k, hence the position and velocity are updated as below:

$$\vec{r}(t+\Delta t) = \vec{r}(t) + \vec{v}(t)\cdot\Delta t + \Delta t^2 \sum_{i=1}^{k-1}\alpha_i \vec{f}\left[t+(1-i)\Delta t\right]$$

$$\vec{v}(t+\Delta t) = \frac{\vec{r}(t)-\vec{r}(t-\Delta t)}{\Delta t} + \Delta t \sum_{i=1}^{k-1}\alpha_i' \vec{f}\left[t+(1-i)\Delta t\right] \tag{4.30}$$

This is called prediction step. Second, one correction step needs to be performed. Calculate $\vec{f}(t+\Delta t)$ and take it as a new point. Then re-estimate $\vec{r}(t+\Delta t)$ and $\vec{v}(t+\Delta t)$:

$$\vec{r}(t+\Delta t) = \vec{r}(t) + \vec{v}(t)\cdot\Delta t + \Delta t^2 \sum_{i=1}^{k-1}\beta_i \vec{f}\left[t+(2-i)\Delta t\right]$$

$$\vec{v}(t+\Delta t) = \frac{\vec{r}(t)-\vec{r}(t-\Delta t)}{\Delta t} + \Delta t \sum_{i=1}^{k-1}\beta_i' \vec{f}\left[t+(2-i)\Delta t\right] \tag{4.31}$$

Table 4.1 presents PC coefficients with $k=4$. PC algorithm is suitable to complex simulations due to its flexibility. However, total energy of the system is not a conserving quantity in PC algorithm because it is not time-reversible. This is not a severe problem in canonical ensemble simulations since the total energy needs to be manipulated constantly.

4.4.3 Examples

An example of the MD simulation on interaction of a <111>/2 screw dislocation and Cu-precipitate in BCC Fe matrix is presented here [38]. The whole system contains 576,000 atoms. As shown in Fig. 4.7, the diameter of a spherical Cu-precipitate

Table 4.1 PC coefficients with $k=4$

$k=4$		1	2	3
Prediction	α_i	19/24	−10/24	3/24
	α_i'	27/24	−22/24	7/24
Correction	β_i	3/24	10/24	−1/24
	β_i'	7/24	6/24	−1/24

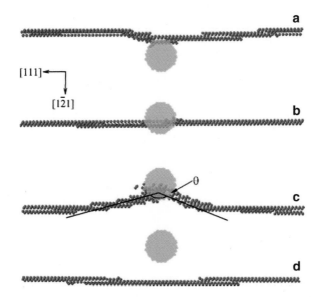

Fig. 4.7 MD snapshots of the dislocation core interacting with the 2.3 nm Cu precipitate as the dislocation glides on the $\left(10\bar{1}\right)$ plane along the $\left[1\bar{2}1\right]$ direction at (**a**) 243 ps, (**b**) 272 ps, (**c**) 312 ps, and (**d**) 320 ps, respectively. *Green* and *red circles* represent Cu and Fe atoms, respectively. [38]

with BCC structure is 2.3 nm. Under an external stress $=700$ MPa, the dislocation can penetrate into the Cu precipitate, as shown in Fig. 4.7a, b. However, the dislocation becomes pinned as it approaches the opposite precipitate-matrix interface, and hence is unable to glide outside the precipitate. Upon increasing the external stress, the dislocation line outside the precipitate continues to glide forward while the short dislocation line segments within the precipitate remains pinned, resulting to a bowing out of the dislocation. The bow-out angle, θ gradually decreases upon increasing the external stress from 180° at $=700$ MPa until it reaches the critical value of $\theta_c = 144°$ under 1,000 MPa shear stress, where the dislocation suddenly detaches from the Cu precipitate and the dislocation line renders straight.

Figure 4.8 presents the dislocation core structures during the pinning process as shown in Fig. 4.7. Note that dislocation core in Cu-precipitate spreads along three directions (polarized core), while spreads along six directions in BCC Fe matrix (non-polarized core). When the dislocation reaches to the boundary, it stops moving and transfers its structure from polarized core to non-polarized core (Fig. 4.8b, c).

Fig. 4.8 The dislocation core structure during the detachment process in Fig. 4.1 at (**a**) 280 ps, (**b**) 297 ps, (**c**) 303 ps, and (**d**) 315 ps. The *red* and *green spheres* represent Fe and Cu atoms, respectively. The dislocation glides along the $\left[1\bar{2}1\right]$ direction under an external stress of 1,000 MPa [38]

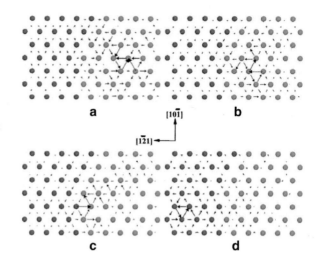

The transferring process corresponds the pinning. And the energy cost during the process is supplied by bowing-out of dislocation core. These results reveal that the dislocation/precipitate detachment process is accompanied with a polarized → nonpolarized core transition, which may be responsible for the pinning effect, the above discussion presents a plausible precipitate-size induced strengthening mechanism.

4.5 Conclusions and Future Outlook

Computational simulations with atomistic resolution for nanomaterials has received an increasing interest due to the fact that nanoscale properties are extremely difficult to measure or manipulate, but more importantly, such properties are probably very sensitive to subtle environmental changes and perturbations, making repeated measurements more challenging. In this chapter, the theoretical background of three types of widely-used material simulation methods: first-principles method, tight binding method, and molecular dynamics are introduced, followed by a detailed discussion including further theories with necessary mathematical treatments and examples for each method. A raw picture of applications of atomic simulation methods on material sciences could thus be generated.

Although there are some essential limitations for each kind of simulation methods, virtual material modeling/design has been made massive successes in the past four decades. Correction, extension and even renovation of current simulation methods have been or will be introduced in order to match the rapid development of material sciences. There are reasons to believe that virtual material modeling/design would become more and more important in both sciences and technologies in the near future.

References

1. Heath J, Keukes P, Snider G et al (1998) A defect-tolerant computer architecture: opportunities for nanotechnology. Science 280:1716–1721
2. Hu L, Choi J, Yang Y et al (2009) Highly conductive paper for energy-storage devices. Proc Natl Acad Sci 106:21490–21494
3. Jeon I, Choi HJ, Choi M et al (2013) Facile, scalable synthesis of edge-halogenated graphene nanoplatelets as efficient metal-free eletrocatalysts for oxygen reduction reaction. Sci Rep 3:1810
4. Derfus A, Chan W, Bhatia S (2004) Probing the cytotoxicity of semiconductor quantum dots. Nano Lett 4:11–18
5. Wang W, McCool G, Kapur N et al (2012) Mixed-phase oxide catalyst based on Mn-mullite (Sm, Gd)Mn_2O_5 for NO oxidation in Diesel exhaust. Science 337:832
6. Yang W, Thordarson P, Gooding J et al (2007) Carbon nanotubes for biological and biomedical applications. Nanotechnology 18:412001
7. Bhattacharya P, Du D, Lin Y (2014) Bioinspired nanoscale materials for biomedical and energy applications. J R Soc Interface 11(95):20131067
8. Decuzzi P, Schrefler B, Liu WK (2014) Nanomedicine. Comput Mechan 53(3):401–402
9. Dell'Orco D, Lundqvist M, Linse S et al (2014) Mathematical modeling of the protein corona: implications for nanoparticulate delivery systems. Nanomedicine 9(6):851–858
10. Redler RL, Shirvanyants D, Dagliyan O et al (2014) Computational approaches to understanding protein aggregation in neurodegeneration. J Mol Cell Biol 6(2):104–115
11. Peng S, Cho K (2003) Ab initio study of doped carbon nanotube sensors. Nano Lett 3:513–517
12. Villalpando-Paez F, Romero A, Munoz-Sandoval E et al (2004) Fabrication of vapor and gas sensors using films of aligned CNx nanotubes. Chem Phys Lett 386:137–143
13. Schiøtz J, Di Tolla F, Jacobsen F (1998) Softening of nanocrystalline metals at very small grain sizes. Nature 391:561–563
14. Chen Z, Kioussis N, Tu KN et al (2010) Inhibiting adatom diffusion through surface alloying. Phys Rev Lett 105:015703
15. Grosso G, Parravicini G (2000) Solid state physics. Academic, San Diego
16. Hohenber P, Kohn W (1964) Inhomogeneous electron gas. Phys Rev 136:B864
17. Kohn W, Sham L (1965) Self-consistent equations including exchange and correlation effects. Phys Rev 140:A1133
18. Ceperley D, Alder B (1980) Ground state of the electron gas by a stochastic method. Phys Rev Lett 45:566–569
19. Perdew J, Zunger A (1981) Self-interaction correlation to density-functional approximations for many-electron systems. Phys Rev B 23:5048–5079
20. Perdew J, Wang Y (1992) Accurate and simple analytic representation of the electron-gas correlation energy. Phys Rev B 45:13244–13249
21. Perdew J, Burke K, Ernzerhof M (1996) Generalized gradient approximation made simple. Phys Rev Lett 77:3865–3868
22. Lee C, Yang W, Parr R (1988) Development of the Colle-Salvetti correlation-energy formula into a functional of the electron density. Phys Rev B 37:785–789
23. Martin R (2004) Electronic structure basic theory and practical methods. Cambridge University Press, Cambridge, MA
24. Kohanoff J (2006) Electronic structure calculations for solids and molecules. Cambridge University Press, New York
25. Bevan K, Zhu W, Stocks G et al (2012) Local fields in conductor surface electromigration: a first-principles study in the low-bias ballistic limit. Phys Rev B 85:235421
26. Winterfeld L, Agapito L, Li J et al (2013) Strain-induced topological insulator phase transition in HgSe. Phys Rev B 87:075143

27. Yan M, Suh J, Ren F et al (2005) Effect of Cu₃Sn coatings a electromigration lifetime improvement of Cu dual-damascene interconnects. Appl Phys Lett 87:211103
28. Harrison W (1999) Elemental electronic structure. World Publishing, Singapore
29. Slater J, Koster G (1954) Simplified LCAO method for the periodic potential problem. Phys Rev 94:1498–1524
30. Sharma R (1979) General expressions for reducing the Slater-Koster linear combination of atomic orbitals integrals to the two-center approximation. Phys Rev B 19:2813–2823
31. Podolskiy A, Vogl P (2004) Compact expressions for the angular dependence of tight-binding Hamiltonian matrix elements. Phys Rev B 69:233101
32. Shan B, Lakatos G, Peng S et al (2005) First-principles study of band-gap change in deformed nanotubes. Appl Phys Lett 87:173109
33. Brown J, Ghoniem N (2010) Reversible-irreversible plasticity transition in twinned copper nanopillars. Acta Mater 58:886–894
34. Starikov S, Insepov Z, Rest J et al (2011) Radiation-induced damage and evolution of defects in Mo. Phys Rev B 84:104109
35. Trautt Z, Adland A, Karma A et al (2012) Coupled motion of asymmetrical tilt grain boundaries: molecular dynamics and phase field crystal simulations. Acta Mater 60:6528–6546
36. Daw M, Baskes M (1984) Embedded-atom method: derivation and application to impurities, surfaces, and other defects in metals. Phys Rev B 29:6443–6453
37. Ercolessi F, Adams J (1994) Interatomic potentials from first-principles calculations: the force-matching method. Europhys Lett 26:583–588
38. Chen Z, Kioussis N, Ghoniem N (2009) Influence of nanoscale Cu precipitates in α-Fe on dislocation core structure and strengthening. Phys Rev B 80:184104
39. Brommer P, Gähler F (2007) Potfit: effective potentials from ab initio data. Modell Simul Mater Sci Eng 15:295–304
40. Mendelev M, Han S, Srolovitz D et al (2003) Development of new interatomic potentials appropriate for crystalline and liquid iron. Phil Mag 83:3977–3994
41. Frenkel D, Smit B (2002) Understanding molecular simulation from algorithms to applications. Academic, San Diego

Chapter 5
Medical Nanomaterials

Steven D. Perrault

5.1 Introduction

Proponents of nanotechnology claim that it will make broad contributions to medical technology over the coming years. But an outsider could ask; why would nanotechnology be so central to a new generation of devices and medicines? What is it about nanometer-scale materials that could provide an improvement on the current state-of-the-art? How can they fulfill current needs within medical practice, and improve how we are able to detect and treat complex diseases such as cancer?

To answer these questions, it's necessary to first understand why new medical technologies are required, and whether it's worth investing money and research into replacing the current technologies. We'll begin this chapter by considering why we need new medical technologies, and what the potential market for nanomedicine might be. In the same context, we can look at what the current standards are for medical technologies. From there, it should start to become clear where nanotechnology can find a home and where it may not be appropriate. We'll then look at how nanomaterials can be systematically organized and described, and the classes of materials that are common in nanomedicine. We can then look deeper at some of these, discuss how they are used in research and how they are being developed towards clinical applicability. This chapter will review some of the most common nanomaterials being developed for medical technologies. More importantly, it will try to provide a framework so that the reader understands why certain directions and materials are being pursued.

S.D. Perrault (✉)
Wyss Institute, Center for Life Sciences, Harvard University,
3 Blackfan Circle, 524-1B, Boston, MA, USA
e-mail: steven.perrault@wyss.harvard.edu

© Springer Science+Business Media New York 2014
Y. Ge et al. (eds.), *Nanomedicine*, Nanostructure Science and Technology,
DOI 10.1007/978-1-4614-2140-5_5

Fig. 5.1 Number of "nanotechnology" publications per year between 1990 and 2010

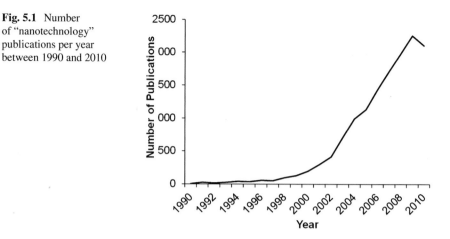

It's worth saying that at the time of writing this, nanotechnology research is likely at the start of a very long road. A search for "nanotechnology" studies in a journal database shows only a handful of papers published prior to 1990, breaking above 10 per year in 1991, and 100 per year only in 1999 (Fig. 5.1). The field then grew rapidly, reaching 1,000 papers published per year in 2005 and twice that just 4 years later. As of 2012, there is an enormous library of well-characterized nanomaterials available, and a small number that can be purchased commercially. In general, not all aspects of the nanomaterials have been characterised. As our ability to design and assemble more complex nano-scale devices improves, the number of studies, applications and products could grow far beyond what anyone today is imagining. I present some early examples of multi-component medical nano-devices and molecular engineering approaches to their assembly at the end of the chapter.

5.2   The Need for New Medical Technologies

As of 2008, the leading causes of global deaths included a large number of chronic and infectious diseases (e.g. cardiovascular disease, diabetes, cancer, HIV/AIDS, malaria). Because these diseases place such a large burden on individuals and on health care systems, it makes sense to invest in research that aims to reduce the burden. Some of these diseases are entirely preventable diseases, suggesting that more investment is needed in education as well as technology. The HIV epidemic has shown improvement in the past few years thanks in part to preventative programs, as well as to reduced transmission rates from anti-retroviral therapies [1], and there's hope that researchers will discover a vaccine that successfully blocks infection. Other diseases can already be accurately diagnosed and well-managed using current technology, even if they aren't completely treatable (e.g., diabetes). And then there are diseases such as cancer, where we have had very little success at reducing burden.

Ominously, the World Health Organization is predicting a dramatic rise in the global number of cancer cases over the next two decades. This is owing to increased tobacco use in emerging economies, older and larger populations over much of the globe, and decreases in other types of mortality [2].

Cancer is in many ways the most challenging of these, owing to its biological complexity [3, 4]. As our appreciation of its complexity has improved, it has become apparent that many of the molecular diagnostic tools that we need will have to be capable of measuring large panels of molecules simultaneously, rather than detecting a single gene or protein. We need to detect genetic mutations, epigenetic modifications to chromosomes, levels of gene transcription into messenger RNA molecules, levels of translation into proteins, and post-translation modifications. To fulfill these needs we require new technologies that can measure large numbers of genes or proteins in a manner that is relevant to clinical practice, and it is hoped that the properties of nanomaterials can contribute to this goal.

Similarly, we need a new generation of therapeutics designed to impact specific molecular pathways that are known to be central to a disease. Vaccines are one example of biologics that have been around for decades, and which have had an enormously positive impact without any need for nanotechnology. More recently, a number of antibody- and protein-based biologics have been developed to treat forms of cancer [5] and arthritis [6]. As well, gene therapy [7] has shown success against diseases such as severe combined immunodeficiency [8]. Nanotechnology may not be useful for some of these, but for others it may be absolutely necessary. Success with gene therapy has been limited to cells that are relatively easy to access, such as those of the lung, or that can be harvested, transformed and re-implanted. Otherwise, gene therapy will require a vehicle to deliver the genetic "drug" to sites deeper within the body. Therapy using RNA-inhibition faces similar challenges, but a recent and exciting study demonstrated the first success of using it against cancer via an optimized nanoparticle vehicle [9, 10].

In summary, we need new medical technologies that will allow us to put our knowledge to use in diagnosing and treating disease on a molecular and cellular scale. Most current clinical technologies have not kept pace with research, but there are notable exceptions such as the microarray and some biologic therapeutics. Developing the technologies needed to enable an era of personalized medicine will allow clinicians to better predict who many suffer from an illness and allow for prevention, or to match a therapy with a patient and thereby achieve the best possible outcome.

5.3 The Advantage of Nanotechnology

Now that we've started to consider what is needed to exploit our molecular-biomedical understanding of diseases, we can ask how nanotechnology might contribute. Nanotechnology involves the engineering of materials having at least one dimension on a scale of 1–100 nm. These materials provide a number of specific advantages that can fulfill some duplicate of the requirements of a new generation of medical technologies [11].

To begin with, the scale of the materials used in nanotechnology overlaps with the scale of biomolecules and sub-cellular structures. For example, an immunoglobulin G antibody molecular has a molecular weight of 152,000 Da and a functional diameter of approximately 11 nm [12]. The ribosomes responsible for synthesizing the polypeptide chains of proteins are approximately 25 nm in diameter. This overlap in scale is a first important advantage of nanotechnology, particularly for in vivo applications. Because of this, we can make nanomaterials that are suitable for use in normal physiological environments. For example, a nanoparticle designed for intravenous injection will have a comparable size to the native biomolecules that are normally present in blood. This means that they are unlikely to become stuck and obstruct small vesicles, which was a problem in earlier research that aimed to develop micron-sized particles as drug delivery vehicles [13]. The size of nanomaterials also allows us to engineer a highly-specific interaction between it and target molecules or structures, which is useful for both in vitro and in vivo diagnostic and therapeutic applications [14]. In some cases, the interaction may need to be optimized to increase its strength or the number of molecules involved, which we can achieve by optimizing the molecules on the material's surface. This is possible in part because the size of the material is similar and can be tuned such that it displays the molecules in an appropriate orientation for the target. The size of nanomaterials therefore allows us to produce devices with increased sophistication and a more refined interaction with target biological molecules and structures that are central to disease.

A second major advantage is that nanomaterials provide a platform that can be engineered through a seemingly infinite number of modifications. As mentioned above, this can include addition of biomolecules on the surface of a material to define a binding interaction with some target molecule or structure. In other cases, modifications may take the form of a polymer layer. The addition of poly(ethylene glycol) to the surface of a material is the most common approach for preventing non-specific adsorption of proteins to its surface or for increasing solubility of a poorly soluble drug, and is therefore used on many implantable or injectable devices, as well as on biological therapeutics [15]. Other modifications include addition of a layer of metal or polymer that provides a functional property, such as those described in the next paragraph. The large numbers of modifications that can be made to any nanomaterial mean that its properties can be highly optimized towards a given application, greatly expanding its potential usefulness.

A last major advantage of nanomaterials is that many of them gain useful functional properties due to their size falling within the quantum realm (below 100 nm). Some of these properties are explained by quantum mechanics, such as the interesting optical properties of fluorescent semiconductor nanoparticles (quantum dots) [16]. Recently, Peng et al. [17] has reported efficient photoluminescent behaviour observed on graphene quantum dots (Fig. 5.2). The luminescence can be varied by controlling relevant process parameters [17]. Other materials gain catalytic properties owing to their high surface-to-volume ratio, such as silver nanoparticles which display anti-microbial activity (Fig. 5.3) [18, 19]. These electronic, optical and catalytic functions are perhaps the most exciting properties that nanomaterials have to offer.

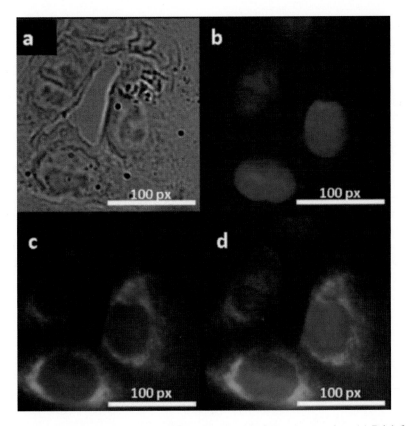

Fig. 5.2 Human breast cancer cells (T47D) exposed to graphene quantum dots. (**a**) Brightfield image of cancer cells. (**b**) Nuclei stained in *blue*, (**c**) *green* fluorescence shows accumulation of quantum dots in the cytoplasm, (**d**) overlay. [17]

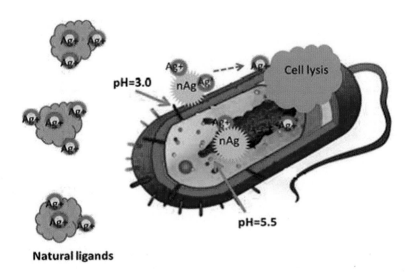

Fig. 5.3 Interaction between silver nanoparticle and the bacterial cell [18]

They provide the potential for a range of diagnostic and therapeutic devices that would otherwise be unimaginable, such as photothermal gold nanorods [20, 21] or nanoshells that can be used for localized ablation of tissue.

Based on these advantages, we can see where nanotechnology can provide gains in medical technology. We can achieve greater control over biomolecular interactions, which will be central to advances in personalized medicine. The material itself provides a platform that can be highly engineered towards a specific goal, such as modifying its in vivo behavior. Finally, we can exploit the electronic, optical and catalytic properties of nanomaterials provide functions to devices that would otherwise be unavailable. We can also see that it may not be useful in every application. If a therapeutic already has favorable kinetics, creating a nanoparticle formulation will unnecessarily increase its complexity and cost. If a diagnostic device is already highly sensitive and meets our needs using current technology, there would be no rationale for developing a nanotechnology-based alternative. Hopefully this has begun to build a framework for what we should expect from nanomedicine.

5.4   The Market for Medical Nanotechnologies

Nanotechnology is a fascinating area of research, and the materials themselves are interesting enough to justify some investment of research time and effort. Beyond this, we've seen that there are clear needs for a new generation of medical devices, and that nanotechnology has specific advantages that will be useful for some of these. There is obvious synergy between these advantages and the goals of personalized medicine, which is to predict, treat, and prevent disease on an individual basis [22, 23]. There is also the opportunity to reformulate many conventional drugs within a nanoparticle vehicle to improve their pharmacokinetics and specific. In almost all cases, we would be aiming to integrate our molecular knowledge of disease with nanomaterials to improve our ability to detect and treat disease.

There is of course already a very large market for medical diagnostics and therapeutics. One report estimated annual spending on in vitro diagnostics in the United States to be $17.6 billion as of 2009 [24]. Much more significant is the market for pharmaceuticals and particularly for oncology therapeutics, which was estimated at $104.1 billion in the United States in 2006 [2], nearly six times the amount spent on all medical diagnostics. Based on this, it's not difficult to believe that medical diagnostics and therapeutics will soon be a trillion-dollar industry, if it isn't already.

How much of this involves nanotechnology? At the moment there are very few marketed diagnostic devices that are reliant on nanotechnology. The most common is likely the lateral flow assay, such as a typical home pregnancy test. This uses antibodies bound to either gold nanoparticles (or alternatively a dye molecule) to determine the presence or absence of a compound. However, each test uses very little material and it is unlikely to amount to a very deep market. The optical properties of gold nanoparticles have also been used to develop a technology platform called Verigene [25], marketed by Nanosphere, which can be adapted for detecting a wide variety of genetic markers. Nanosphere reported an income of \$2-million in 2010, and potentially owns the largest share of the nanotechnology-based diagnostic market. Many other medically-related nanomaterials are also being sold, such as quantum dots (Invitrogen) and gold nanorods (Nanopartz), but for research purposes only.

By far the largest market for medical nanotechnology is currently in nanoparticle-formulated oncology therapeutics [26]. These have been used against cancer for well over a decade now, but account for only \$5.6 billion of spending. The formulations currently used in clinics are first generation, and are composed of liposomes, pegylated liposomes, or protein particles [27]. They are relatively low-technology and low-engineering designs in comparison to what is being developed in research labs and even what is currently undergoing clinical trials.

The market for medical technologies is enormous and nanomedicine makes up a very small portion of this. It seems that there are many opportunities for growth. Most of the highest potential nanomaterials and ideas are still at the research stage, and won't begin to make a serious dent in the diagnostic or therapeutic market for another decade. However, we can expect that they will begin to displace older technologies and medicines as we improve our ability to engineer them, exploit their properties and scale-up production.

5.5 Medical Nanomaterials

As mentioned above, there are many different types of nanomaterials, each of which can be modified in a seemingly infinite number of ways. It would therefore be overwhelming to attempt to create a comprehensive description of all the materials that could potentially be used as part of future nanomedical devices. Instead, we'll focus on some general classes and their specific advantages. To make it easier, we'll begin by looking at a recently published nomenclature system that provides a basis for understanding the wide range of available materials.

5.5.1 A Systematic Approach to Understanding Nanomaterials

Until recently there was no unified method available for naming or classifying nanomaterials. A nomenclature presented by Gentleman and Chan in 2009 [28] takes a hierarchical approach, in which a material is systematically classified based on its chemical class, geometry, core chemistry, ligand chemistry, and solubility. We'll make use of their approach to examine what properties define a nanomaterial.

A first major distinction between the many types of nanomaterials is by their different chemical classes. Nanomaterials can consist of either purely organic molecules, as is the case with liposomes, purely inorganic or metallic materials, or some hybrid of the two. This first distinction is important in the development of materials that may be used in vivo, as organic nanomaterials designed to biodegrade would generally be more biocompatible. At the same time, the interesting and useful optical and electronic properties are restricted to inorganic and metallic materials. For example the fluorescence properties of quantum dots [29] make for an excellent in vivo contrast agent [30], but there are concerns about their toxicity and long-term persistence within the body [31, 32]. Carbon-based fullerene materials such as multiwall nanotubes are an obvious exception to this differentiation based on organic and inorganic material function. Although they are composed purely of carbon, they offer some interesting electronic properties that are more similar to metallic nanomaterials [33].

The second consideration for classifying a nanomaterial is geometry. The size and shape of most nanomaterials is central to many of their important properties. For example, the optical properties of quantum dots are a product of their semiconductor core diameter [29]. The surface resonance plasmon of gold nanoparticles is also dependent on their diameter [34], as well as on their shape. Gold nanorods strongly absorb longer wavelength light than spheroid particles of the same volume, and translate much of that absorption into heat [35–38]. The catalytic properties of metal nanoparticles are dependent on surface area, and therefore on their size [39]. Geometry also determines how nanoparticles behave in vivo. It determines their access to various compartments within the body [13, 40–42], how quickly they will be recognized by the immune system [43], how they are excreted, and how they interact with cells [44, 45]. Geometry is therefore one of the most important parameters for how a nanomaterial is chosen and designed towards a particular medical application.

Next, we can consider both the core chemistry and the outer ligand chemistry. Some materials consist primarily of a single material or crystal structure (described in part by chemical class), but may also include modifications to that primary structure. For example, semiconductor materials are often doped with rare-earth metals in order to optimize their electronic properties [46, 47]. Ligand chemistry defines what is presented to the outside environment. Because of this, engineering the outer ligand can change the solubility of a material. This is particularly important for the many nanomaterials that are highly soluble in non-polar solvents (e.g. quantum dots), but poorly soluble in the aqueous environments in which they would be used in an interaction with biomolecules or cells [48]. The outer ligands also determine how a material interacts with biomolecules and cells, whether there is a specific and defined interaction (e.g., antibody against a viral protein), or whether the purpose of the ligand is to reduce non-specific interactions (e.g., to slow immune recognition). We can see that the core chemistry and outer ligand chemistry are also very important for determining what properties a material will have, and how it will behave within a given environment.

Now that we've seen the major design parameters that define a nanomaterial, we can look more specifically at a few materials more specifically, see how they've been developed over the years and how it is hoped they can contribute to nanomedicine.

5.5.2   An In-Depth Look at Various Nanomaterials

5.5.2.1   From Liposomes to Polymer Nanoparticles for Drug Delivery

Liposomes are generally not considered to fall within the realm of nanotechnology because their dimensions extend far beyond the nano-realm. Although they can be processed to have diameters as small as approximately 50 nm, liposomes of several hundred nanometers to several microns are commonly used for a variety of applications. As well, liposomes are not synthesized and engineered in the same sense as other nanomaterials, and do not offer the same types of interesting quantum properties as conventional nanomaterials. Nevertheless, liposomes are an important topic because they were one of the first particle systems used for drug delivery, a field that is very prominent in nanomedicine.

Liposomes were first scientifically synthesized five decades ago [49]. They are composed primarily of amphipathic phospholipid molecules that spontaneously self-organize into bilayer membranes when in an aqueous environment. The phospholipids can be from a biological source or synthetic, and their mixture within the liposome formulation can be modified to control properties such as membrane fluidity, curvature, charge and stability. Self-assembly of the phospholipids into a bilayer membrane causes encapsulation of a volume of the aqueous buffer as a cavity inside of the membrane, much like cell and organelle membranes that are present in biological systems. Agents that cannot readily diffuse across the membrane can therefore be encapsulated in the inner cavity of the liposome, which can then act as a vehicle for that agent. The liposome design then determines parameters that are important to pharmacokinetics, such as circulation half-life, rather than it being dependent on the agent itself. They are therefore an obvious choice for use as drug delivery vehicles, and have been used extensively to encapsulate and alter the pharmacokinetic behavior of many cancer therapeutics.

Through their size, liposomes are also effective at exploiting an inherent property of tumors. Vascularized tumours typically do not contain well-formed and mature blood vessels comparable to other tissues. Instead, tumour vessels are tortuous in their architecture [50], immature and porous [51, 52]. To improve accumulation of drugs in tumours, we can therefore synthesize particles which have a diameter large enough to remain within the blood vessels of healthy tissue, but small enough to leak from circulation into tumour tissue when passing through their leaky vessels. This property was first discovered by Matsumura and Maeda using dye-labeled albumin protein [53], and was afterwards characterized by Jain using liposomes [40].

Finally, liposomes were one of the first particle systems to have their surface modified with PEG in order to increase circulation half-life [54]. This was a major breakthrough in nanoparticle-based drug delivery because it decreased the fraction of a dose that ended up cleared by the immune system, increasing the fraction that could accumulate within tumours. It was these discoveries that led to the first "nanoparticle" formulations of anti-cancer agents, and to significant reductions in many of the side effects suffered by patients from earlier forms of therapeutics.

Although liposomes display these very useful properties, they are lacking in several areas. Their minimum diameters are approximately 50 nm and they have only moderate stability. Moreover, they do not allow for a great deal of engineering, which limits our ability to rationally design their properties to overcome specific barriers to drug delivery. Polymer-based particles are therefore an excellent alternative, as we are unlimited in the types of polymers that can be used, and how they can be engineered to provide specific, desirable properties. The most common polymer particles being pursued for drug delivery are composed of biodegradable polymers, such as poly(lactic-co-glycolic acid) (PLGA). These offer the same advantages as liposomes in terms of being able to act as vehicles for agents, take advantage of leaky tumour vasculature and avoid immune clearance by PEGylation. It is also possible to rationally design their permeability for the encapsulated agent, thereby allowing for control over release of the agent into the environment [55, 56]. If liposomes can be considered to have contributed to the first generation of nanoparticle drug formulations, polymer nanoparticles are likely to make up much of the second generation, and many polymer-based nanoparticles are now in clinical trials for drug delivery to tumours [27].

5.5.2.2 Gold Nanoparticles as Diagnostic Agents and Therapeutic Vehicles

Colloidal gold or gold nanoparticles have been produced for centuries. Their magnificent interaction with visible light has made them a favorite ink in stained glass and other types of art. The first scientific report of colloidal gold synthesis (and of any metallic nanoparticle) was by Michael Faraday, who published the work in 1857 [57]. Amazingly, his syntheses are still in suspension, and are held at the Royal Institution of Great Britain. More recently, gold nanoparticles have become a favourite nanomaterial for use in a broad range of applications, including as medical diagnostics and therapeutics. This is because they are relatively straight forward to synthesize over a broad range of sizes, it is very easy to modify their surface via adsorption of ligands (e.g., of proteins) or through a thiol-Au bond, and their surface plasmon can be exploited for diagnostic and therapeutic purposes.

5.5.2.2.1   Synthesis Methods

One of the most commonly used methods for synthesis is that published by Turkevich in 1951 [58] and further refined by Frens in 1973 [34]. This uses citrate as a reducing agent and stabilizing surface ligand. Reduction of gold ions in solution to metallic Au^0 results in formation of ordered crystals that grow into stabilized gold nanoparticles. Adsorption of an organic ligand (in this case the oxidized citrate molecules) to their surface stabilizes the particles, preventing them from aggregating and forming their thermodynamically-favored bulk material. The reducing conditions can be altered to control the rate of particle formation, and through this it is possible to control particle size. Using the Turkevich/Frens method, batches of gold nanoparticles having diameters between 15 and approximately 100 nm can be synthesized, although the quality of larger diameter batches is greatly reduced. An alternative method for synthesis of larger diameter gold nanoparticles makes use of seeded growth. In this approach, ionic gold is reduced in the presence of high quality (i.e. near-spherical and highly monodispersed in size) small diameter "seed" particles, which provide a template onto which the newly reduced Au^0 is added. By separating the reactions that are responsible for initial formation of nanoparticle crystals and for their growth, we could potentially produce batches that are more monodispersed in size and shape. Early attempts at seed-based growth used conditions that failed to suitably favour growth of the existing particle seeds over formation of new particles, resulting in two populations [59]. We were able to overcome this using hydroquinone as a reducing agent [60], which has a long history of use in photographic film processing, where it is used to selectively grow clusters of silver atoms into larger grains. Our method is able to grow very high quality batches of gold nanoparticles having diameters up to several hundred nanometers.

5.5.2.2.2   Diagnostic Applications

Gold nanoparticles display an interesting and useful surface plasmon resonance (a collective oscillation of valance-band electrons) which is different from that of bulk gold material. The interaction of gold nanoparticles with light is dependent on their diameter, or more specifically on the number of surface atoms relative to internal atoms within the crystal structure [61]. Larger particles display red-shifted plasmon maximum relative to smaller-sized particles, and recently developed hollow gold nanoshells display near-infrared surface resonance plasmon absorptions [62]. The plasmon is also highly sensitive to the particle's local external medium, such that changes to ligands on the particle surface or to the solvent results in a measurable shift in the absorption spectra. The dependence of the plasmon on the number of exposed surface atoms means that aggregation of individual nanoparticles into clusters causes a dramatic red-shifting of their surface plasmon resonance. This property provided the basis for development of elegant diagnostic devices in which aggregation of gold nanoparticles and the resulting colour shift is controlled by biomolecules present on the particle surface, and through their specific recognition of target molecules. This phenomenon was first demonstrated by Storhoff, Mirkin and Letsinger [63] for the detection of nucleic acid single-nucleotide polymorphisms, and has now been developed into the Verigene diagnostic device sold by Nanosphere. A second potentially useful property of the surface plasmon is its ability to enhance or quench the emission of a fluorophore [37, 64, 65]. Although still in the research stage, this can be exploited to mask the signal of a fluorescent contrast agent until released by a biological trigger in vivo, such as the presence of an enzyme [66].

5.5.2.2.3   Therapeutic Applications

Gold nanoparticles are a highly versatile material for use in nanomedical applications. Besides the various clever ways in which their properties have been used in diagnostic devices, they have also been used in various forms as therapeutic agents.

As mentioned above, gold nanoparticle display a unique surface plasmon resonance that is dependent on the number of surface-exposed atoms relative to those within the particle. For solid spherical gold nanoparticles, the maximum absorption of the plasmon is found in the range of 520–550 nm. However, if the particles are prepared in a manner such that they are hollow, the plasmon shifts into the near-infrared range [62]. Unlike solid spherical gold nanoparticles which heavily scatter light, they efficiently absorb and translate light into heat, giving rise to dramatic photothermal effects in the local environment. Gold nanorods, which are synthesized to produce an elongated aspect ratio, display a plasmon in the 650–800 nm range (aspect ratios of 2.5–3.5) and produce similar photothermal effects [35, 37, 38]. In this case, it is the oscillation of surface electrons along the length of the rod (longitudinal surface plasmon) that gives rise to the photothermal effects. This property of nanoshells and nanorods is being developed as a cancer therapeutic, in which the particles are targeted to tumours by systemic or local administration, and are then optically excited to thermally ablate tissue in a localized manner [67, 68]. Both gold nanorods and nanoshells are now commercially available, and gold nanoshells are being tested in clinical trials for thermal ablation by Nanospectra Biosciences, Inc. Finally, spherical gold nanoparticles have also been used for delivery of conventional molecular therapeutics and neoadjuvants to tumours [69–71]. In this case, the use of gold nanoparticles provides a highly tunable platform, allowing for a rational design approach to the vehicle design and a greater efficiency in tumour accumulation. Engineered gold nanoparticles carrying a potent anti-cancer agent (TNF-α) have been developed as Aurimune nanotherapeutic by CytImmune, and have recently completed Phase I clinical trials.

Gold nanoparticles are likely to remain a very prominent material within nanomedicine, owing to their versatility, biocompatibility and useful optical properties. They are one of the first nanomaterials to be integrated within a saleable diagnostic device, and are likely to contribute to next generation targeted cancer therapies.

5.5.2.3   Multi-component Nano-devices

Nanoparticle-based tumour targeting is one of the most prominent research areas of nanomedicine. It is a decades-old field, and there are already numerous nanoparticle-based formulations of cancer drugs in clinical use [27]. The targeting field has progressed significantly through a rational design, evidence-based approach. It evolved from using large particles that would obstruct small capillaries, to smaller PEGylated particles that could passively exploit the leaky vasculature of tumours [54], to actively targeting particles in which a biomolecule presented on the particle surface can specifically bind to antigens present within the target tissue [30, 72, 73], and finally to some of the functional nanomaterials that were described above. These advances have overcome some of the primary physiological limitations of drug delivery; the poor pharmacokinetics and tumour accumulation efficiency of many cancer therapeutics. Despite this success, there are additional in vivo barriers to targeting that reduce the effectiveness of therapeutics beyond the point where tumour can be completely eradicated within a patient. These barriers include the permeation of a vehicle or drug through the bulk of a tumour, specificity of targeting for deregulated cells over healthy cells, efficient delivery of the drug into the appropriate compartments within target cells, and the multi-drug resistance pathways that expel toxic drugs out of cells. From this we can see that we may have a long way to go to achieve truly effective drug delivery. Nevertheless, nanomedicine offers perhaps the best means of achieving this, because nanomaterials provide a platform that can be engineered and optimized using a rational approach to overcome these barriers.

In the last few years, researchers have begun to explore the possibility of using multi-component nano-devices, rather than single nanoparticles, to overcome some of the barriers to targeting. The first such example of an multi-component in vivo system was demonstrated by von Maltzahn and Bhatia in 2010. Their approach makes use of a tumour-homing nanoparticle, which can broadcast a homing signal from within the tumour via the native coagulation cascade. This homing signal then attracts a secondary nanoparticle component present in the circulation, increasing its accumulation within the tumour 40-times higher than conventional controls [74, 75].

A second multi-component system demonstrated by myself and Chan in the same year [76] uses in vivo assembly of a two component system to favourably alter tumour accumulation pharmacokinetics of a contrast agent [76]. The first component consists of a PEGylated gold nanoparticle that is systemically administered and passively accumulates in tumour tissue over a 24-h period. The particle size was engineered such that it was small enough to gain access to tumours through their leaky vasculature, but large enough to restrict permeation into the tumour's extracellular matrix [41]. This results in a large accumulation of nanoparticles just outside the tumour vasculature, highly accessible to agents in circulation. The particles were also engineered to present a biomolecule for assembly (biotin) on the periphery of their surface ligands. Contrast agents linked to a secondary assembly component (in this case streptavidin) can then leak from the vasculature and assemble onto the gold nanoparticles within the tumour. In control studies we showed that without assembly, the molecular contrast agent was small enough to rapidly diffuse through a tumour mass, decreasing its overall accumulation and limiting its diagnostic signal-over-noise. Therefore, by using in vivo assembly, we were able to achieve accumulation kinetics that might be comparable to an actively targeting system, but without requiring prior knowledge of antigens presented by the tumour tissue itself.

These studies are the first demonstrations of multi-component systems. In general, they take an approach in which the complexity of the nanoparticle targeting device is increased in order to improve targeting. This multi-component, higher complexity approach may become more prominent in generations of future nano-devices, whether for drug delivery of other nanomedicine applications. There are some non-trivial challenges to nanomaterial design and synthesis that limit the potential complexity and behaviours that we can achieve, but this author believes that multi-component systems could overcome some of the most important remaining barriers to targeting.

5.6   Summary and Future Outlook

As was mentioned at the start of this chapter, nanomedicine is likely near the start of a long journey. Very few of the most exciting ideas and applications have moved out of research labs and into clinical use, but we are already starting to see a few examples of nanomaterial-based diagnostics and therapeutics. There is clearly a need and a market for new medical devices, and nanomaterials offer some unique advantages that could go a long way towards improving disease detection and treatment. A major limitation to all nanomaterials that are prepared using conventional chemical synthesis is that they don't provide angstrom-level control over features that are central to nanomedicine, such as functionalization with biomolecules. As researchers overcome this and begin to design and assemble more complex devices with improved molecular-scale behaviour, we can expect to see major advances in the types of nanotechnology-based applications, and in our success at making measurable impacts.

References

1. World Health Organization (2009) New HIV infections reduced by 17 % over the past eight years. World Health Organisation. http://www.who.int/mediacentre/news/releases/2009/hiv_aids_20091124/en/index.html
2. National Cancer Institute (2010) Cancer trends progress report – 2009/2010 update. http://progressreport.cancer.gov
3. Moghimi S, Farhangrazi Z (2014) Just so stories: the random acts of anti-cancer nanomedicine performance. Nanomed Nanotechnol Biol Med. doi:10.1016/j.nano.2014.04.011
4. Toy R, Peiris P, Ghaghada K et al (2014) Shaping cancer nanomedicine: the effect of particle shape on the in vivo journey of nanoparticles. Nanomedicine 9(1):121–134
5. Adams G, Weiner L (2005) Monoclonal antibody therapy of cancer. Nat Biotechnol 23:1147–1157
6. Wong M, Ziring D, Korin Y et al (2008) TNFα blockade in human diseases: mechanisms and future directions. Clin Immunol 126:121–136
7. Fischer A, Hacein-Bey-Abina S, Cavazzana-Calvo M (2010) 20 years of gene therapy for SCID. Nat Immunol 11:457–460
8. Touzot F, Hacein-Bey-Abina S, Fischer A et al (2014) Gene therapy for inherited immunodeficiency. Expert Opin Biol Ther 14(6):789–798
9. Davis M, Jonathan Z, Chung Hang C et al (2010) Evidence of RNAi in humans from systemically administered siRNA via targeted nanoparticles. Nature 464:1067–1070
10. Fan Z, Fu P, Yu H et al (2014) Theranostic nanomedicine for cancer detection and treatment. J Food Drug Anal 22(1):3–17
11. Chaturvedi S, Dave PN (2014) Emerging applications of nanoscience. Mater Sci Forum 781:25–32

12. Goel A, Colcher D, Baranowska-Kortylewicz J et al (2000) Genetically engineered tetravalent single-chain Fv of the pancarcinoma monoclonal antibody CC49: improved biodistribution and potential for therapeutic application. Cancer Res 60:6964–6971
13. Ilium L, Davis S, Wilson C et al (1982) Blood clearance and organ deposition of intravenously administered colloidal particles: the effects of particle size, nature and shape. Int J Pharm 12:135–146
14. Krpetić Ž, Anguissola S, Garry D, Kelly PM, Dawson KA (2014) Nanomaterials: impact on cells and cell organelles. In: Nanomaterial. Springer, Dordrecht, pp 135–156. http://link.springer.com/book/10.1007%2F978-94-017-8739-0
15. Mathaes R, Winter G, Besheer A et al (2014) Influence of particle geometry and PEGylation on phagocytosis of particulate carriers. Int J Pharm 465(1):159–164
16. Lhuillier E, Keuleyan S, Guyot-Sionnest P (2012) Optical properties of HgTe colloidal quantum dots. Nanotechnology 23:175705
17. Peng J, Gao W, Gupta B et al (2012) Graphene quantum dots derived from carbon fibres. Nano Lett 12(2):844–849
18. Xiu Z, Zhang G, Puppala H et al (2012) Negligible particle-specific antibacterial activity of silver nanoparticles. Nano Lett 12:4271–4275
19. Lu Z, Rong K, Li J et al (2013) Size-dependent antibacterial activities of silver nanoparticles against oral anaerobic pathogenic bacteria. J Mater Sci Mater Med 24(6):1465–1471
20. Alkilany A, Thompson L, Boulos S et al (2012) Gold nanorods: their potential for photothermal therapeutics and drug delivery, tempered by the complexity of their biological interactions. Adv Drug Deliv Rev 64:190–199
21. Choi W, Sahn A, Kim Y et al (2012) Photothermal cancer therapy and imaging based on gold nanorods. Ann Biomed Eng 40(2):534–546
22. Moghimi S, Peer D, Langer R (2011) Reshaping the future of nanopharmaceuticals: ad Iudicium. ACS Nano 5(11):8454–8458
23. Harvey A, Brand A, Holgate S et al (2012) The future of technologies for personalised medicine. N Biotechnol 29(6):625–633
24. The Freedonia Group (2009) In vitro diagnostics: US industry study with forecasts for 2013 & 2018. The Freedonia Group, Cleveland
25. Giljohann D, Seferos D, Daniel W et al (2010) Gold nanoparticles for biology and medicine. Angew Chem Int Ed 49(19):3280–3294
26. Davis M, Chen Z, Shin D (2008) Nanoparticle therapeutics: an emerging treatment modality for cancer. Nat Rev Drug Discov 7:771–782
27. Wagner V, Dullaart A, Bock A et al (2006) The emerging nanomedicine landscape. Nat Biotechnol 24:1211–1217
28. Gentleman D, Chan W (2009) A systematic nomenclature for codifying engineered nanostructures. Small 5:426–431
29. Alivisatos A (1996) Semiconductor clusters, nanocrystals, and quantum dots. Science 271:933–937
30. Åkerman M, Chan W, Laakkonen P et al (2002) Nanocrystal targeting in vivo. Proc Natl Acad Sci U S A 99:12617–12621
31. Fischer H, Liu L, Pang K et al (2006) Pharmacokinetics of nanoscale quantum dots: in vivo distribution, sequestration, and clearance in the rat. Adv Funct Mater 16:1299–1305
32. Hauck T, Anderson R, Fischer H et al (2010) In vivo quantum-dot toxicity assessment. Small 6:138–144
33. Tans S, Devoret M, Dai H et al (1997) Individual single-wall carbon nanotubes as quantum wires. Nature 386:474–477
34. Frens G (1973) Controlled nucleation for regulation of particle-size in monodisperse gold suspensions. Nat Phys Sci 241:20–22
35. Cepak V, Martin C (1998) Preparation and stability of template-synthesized metal nanorod sols in organic solvents. J Phys Chem B 102:9985–9990
36. Link S, Mohamed M, El-Sayed M (1999) Simulation of the optical absorption spectra of gold nanorods as a function of their aspect ratio and the effect of the medium dielectric constant. J Phys Chem B 103:3073–3077

37. Link S, El-Sayed M (2000) Shape and size dependence of radiative, non-radiative and photo-thermal properties of gold nanocrystals. Int Rev Phys Chem 19:409–453
38. Jana N, Gearheart L, Murphy C (2001) Wet chemical synthesis of high aspect ratio cylindrical gold nanorods. The Journal of Physical Chemistry B 105:4065–4067
39. Subramanian V, Wolf E, Kamat P (2004) Catalysis with TiO_2/gold nanocomposites. Effect of metal particle size on the fermi level equilibration. J Am Chem Soc 126:4943–4950
40. Yuan F, Dellian M, Fukumura D et al (1995) Vascular permeability in a human tumor xeno-graft: molecular size dependence and cutoff size. Cancer Res 55:3752–3756
41. Perrault S, Walkey C, Jennings T et al (2009) Mediating tumor targeting efficiency of nanopar-ticles through design. Nano Lett 9:1909–1915
42. Cabral H, Matsumoto Y, Mizuno K et al (2011) Accumulation of sub-100 nm polymeric micelles in poorly permeable tumours depends on size. Nat Nanotechnol 6:815–823
43. Fang C, Shi B, Pei Y et al (2006) In vivo tumor targeting of tumor necrosis factor-α-loaded stealth nanoparticles: effect of MePEG molecular weight and particle size. Eur J Pharm Sci 27:27–36
44. Chithrani B, Ghazani A, Chan W (2006) Determining the size and shape dependence of gold nanoparticle uptake into mammalian cells. Nano Lett 6:662–668
45. Jiang W, Kim B, Rutka J et al (2008) Nanoparticle-mediated cellular response is size-dependent. Nat Nanotechnol 3:145–150
46. Boyer J, Vetrone F, Cuccia L et al (2006) Synthesis of colloidal upconverting $NaYF_4$ nanocrystals doped with Er^{3+}, Yb^{3+} and Tm^{3+}, Yb^{3+} via thermal decomposition of lanthanide trifluoroacetate precursors. J Am Chem Soc 128:7444–7445
47. Boyer J, Cuccia L, Capobianco J (2007) Synthesis of colloidal upconverting $NaYF_4$: Er^{3+}/Yb^{3+} and Tm^{3+}/Yb^{3+} monodisperse nanocrystals. Nano Lett 7:847–852
48. Chan W (1998) Quantum dot bioconjugates for ultrasensitive nonisotopic detection. Science 281:2016–2018
49. Horne R, Bangham A, Whittaker V (1963) Negatively stained lipoprotein membranes. Nature 200:1340
50. Baish J, Gazit Y, Berk D et al (1996) Role of tumor vascular architecture in nutrient and drug delivery: an invasion percolation-based network model. Microvasc Res 51:327–346
51. Hashizume H, Baluk P, Morikawa S et al (2000) Openings between defective endothelial cells explain tumor vessel leakiness. Am J Pathol 156:1363–1380
52. Morikawa S, Baluk P, Kaidoh T et al (2002) Abnormalities in pericytes on blood vessels and endothelial sprouts in tumors. Am J Pathol 160:985–1000
53. Matsumura Y, Maeda H (1986) New concept for macromolecular therapeutics in cancer che-motherapy: mechanism of tumoritropic accumulation of proteins and the antitumor agent smancs. Cancer Res 46:6387–6392
54. Klibanov A, Maruyama K, Torchilin V et al (1990) Amphipathic polyethyleneglycols effec-tively prolong the circulation time of liposomes. FEBS Lett 268:235–237
55. Cohen S, Yoshioka T, Lucarelli M et al (1991) Controlled delivery systems for proteins based on poly(lactic glycolic acid) microspheres. Pharm Res 8:713–720
56. Jain R (2000) The manufacturing techniques of various drug loaded biodegradable poly(lactide-co-glycolide) (PLGA) devices. Biomaterials 21:2475–2490
57. Faraday M (1857) Experimental relations of gold (and other metals) to light. Philos Trans R Soc Lond 147:145–181
58. Turkevich J, Stevenson P, Hillier J (1951) A study of the nucleation and growth processes in the synthesis of colloidal gold. Discuss Faraday Soc 11:55–75
59. Jana N, Gearheart L, Murphy C (2001) Evidence for seed-mediated nucleation in the chemical reduction of gold salts to gold nanoparticles. Chem Mater 13:2313–2322
60. Perrault S, Chan W (2009) Synthesis and surface modification of highly monodispersed, spherical gold nanoparticles of 50–200 nm. J Am Chem Soc 131:17042–17043
61. Mie G (1908) Articles on the optical characteristics of turbid tubes, especially colloidal metal solutions. Ann Phys 25:377–445

62. Oldenburg S, Jackson J, Westcott S et al (1999) Infrared extinction properties of gold nanoshells. Appl Phys Lett 75:2897–2899
63. Storhoff J, Elghanian R, Mucic R et al (1998) One-pot colorimetric differentiation of polynucleotides with single base imperfections using gold nanoparticle probes. J Am Chem Soc 120:1959–1964
64. Anger P, Bharadwaj P, Novotny L (2006) Enhancement and quenching of single-molecule fluorescence. Phys Rev Lett 96:113002
65. Kuehn S, Hakanson U, Rogobete L, Sandoghdar V (2006) Enhancement of single-molecule fluorescence using a gold nanoparticle as an optical nanoantenna. Phys Rev Lett 97(1):017402
66. Mu C, LaVan D, Langer R et al (2010) Self-assembled gold nanoparticle molecular probes for detecting proteolytic activity in vivo. ACS Nano 4:1511–1520
67. Dickerson E, Dreaden E, Huang X et al (2008) Gold nanorod assisted near-infrared plasmonic photothermal therapy (PPTT) of squamous cell carcinoma in mice. Cancer Lett 269:57–66
68. Hauck T, Ghazani A, Chan W (2008) Assessing the effect of surface chemistry on gold nanorod uptake, toxicity, and gene expression in mammalian cells. Small 4:153–159
69. Paciotti G, Myer L, Weinreich D et al (2004) Colloidal gold: a novel nanoparticle vector for tumor directed drug delivery. Drug Deliv 11:169–183
70. Visaria R, Griffin R, Williams B et al (2006) Enhancement of tumor thermal therapy using gold nanoparticle-assisted tumor necrosis factor-alpha delivery. Mol Cancer Ther 5:1014–1020
71. Goel R, Shah N, Visaria R et al (2009) Biodistribution of TNF-alpha-coated gold nanoparticles in an in vivo model system. Nanomedicine 4:401–410
72. Farokhzad O, Jon S, Khademhosseini A et al (2004) Nanoparticle-aptamer bioconjugates: a new approach for targeting prostate cancer cells. Cancer Res 64:7668–7672
73. Gao X, Cui Y, Levenson R et al (2004) In vivo cancer targeting and imaging with semiconductor quantum dots. Nat Biotechnol 22:969–976
74. Park J, von Maltzahn G, Xu M et al (2010) Cooperative nanomaterial system to sensitize, target, and treat tumors. Proc Natl Acad Sci U S A 107:981–986
75. von Maltzahn G, Park J, Lin K et al (2011) Nanoparticles that communicate in vivo to amplify tumour targeting. Nat Mater 10:545–552
76. Perrault S, Chan W (2010) In vivo assembly of nanoparticle components to improve targeted cancer imaging. Proc Natl Acad Sci 107:11194

Chapter 6
Multifunctional Nanoparticles for Theranostics and Imaging

Xue Xue and Xing-Jie Liang

Abbreviations

ROS	Reactive oxygen species
SWCNTs	Single-walled carbon nanotubes
DWCNTs	Double-walled carbon nanotubes
MWCNTs	Multi-walled carbon nanotubes
CNTs	Carbon nanotubes
Gd	Gadolinium
MCAO	Middle cerebral artery occlusion
CT	Computed tomography
MRI	Magnetic resonance imaging
GNPs or AuNPs	Gold nanoparticles
PET	Photon emission tomography
IONPs	Iron oxide nanoparticles
SIONPs	Superparamagnetic iron oxide nanoparticles
FR-α	Folate receptor-α
CTCs	Circulating tumor cells
Hb	Hemoglobin
NPs	Nanoparticles

X. Xue • X.-J. Liang (✉)
National Center for Nanoscience and Technology of China, CAS Key Laboratory
for Biomedical Effects of Nanomaterials and Nanosafety, No. 11, Beiyitiao Zhongguancun,
Beijing 100190, People's Republic of China
e-mail: liangxj@nanoctr.cn

© Springer Science+Business Media New York 2014
Y. Ge et al. (eds.), *Nanomedicine*, Nanostructure Science and Technology,
DOI 10.1007/978-1-4614-2140-5_6

6.1   Introduction

Nanotechnology is an interdisciplinary and multidisciplinary field, which depends on nanoscale materials, especially contributing to the development of nanomedicine [1]. Medical applications of nanomaterials include drug delivery system, proteins and peptides, nano-specific targeting, imaging and detection [2–5]. The continual identification and development of multifunctional nanoparticles is capable of target-specific delivery of therapeutic and/or imaging agents due to their appropriate features, including larger surface area, physical and chemical properties, structural diversity, etc. [6]. As previous studies have demonstrated, the progress in nanomedicine allows giving personalized treatments to patients with maximal therapeutic agents. In the past decade, nanomedicine has become an alternative and ideal possibility to undergo human healthy issue. Compared to the traditional theranostic strategies, nano-theranostics and imaging system has various advantageous aspects: (1) NPs can easily combine with more than one kind of additional imaging contrast enhancements or therapeutic moieties for simultaneous diagnosis and therapy, and high dosages of imaging agents or drugs can be loaded into NPs with simple physical or chemical conjugation; (2) Multifunctional NPs offer an opportunity to alter the absorbance, distribution, metabolism and excretion of drugs, reducing off-target toxicity and improving the therapeutic index. (3) The self-functionalization of nanoparticles, including pH sensitivity, hydrophobic/hydrophilic properties, electric charge, etc., can help to match the complex biochemical system. (4) With specific targeting moieties or physicochemical optimization of size and surface properties, NPs admit to precise and fast diagnosis through targeting disease sites for drug delivery and imaging. These features provide multifunctional NPs with great potential as innovative diagnostic and therapeutic systems. Furthermore, large mounts of detailed information, including environmental factors, genetic factors, individual differences in personality etc. should be carefully identified for the clinical field [7].

6.2   Design of Multifunctional Nanoparticle for Therapeutics and Diagnostics

Nanotheranostics was coined originally as a term to describe a treatment platform that fusing nanotechnology, therapeutics, and diagnostics based on the test results. The engineering of nanomedicine is required several advantageous therapeutic and diagnostic properties including low toxicity to healthy tissue, enhanced permeation and retention in the circulatory system, specific delivery of drugs to target sites, controlled releasing in pathognostic condition, etc. [8]. Traditional NPs like liposomes were initially developed for drug delivery systems. Recently, many newly nanoscale materials themselves have been broadly used and display distinguished therapeutic response and diagnostic properties in both the research setting as well as in a clinical setting, implying their enormous potential as medicinal candidates.

In 1985, fullerenes and their derivatives were discovered and extensively investigated as their unique physiochemical characters. However, In order to change their extremely high hydrophobicity, which hampers its direct biomedical evaluation and application, the modification of nano-C60 was developed and classified by several approaches: (i) the transfer fullerenes into physiological friendly media have been developed; (ii) chemical modification of the fullerene carbon cage; (iii) incorporation of fullerenes into water soluble micellar supramolecular structures; (iv) solvent exchange and long term stirring of pure C60 in water. These strategies considerably created and influenced various of functional fullerenes with the potential application in biomedicine, especially in the field of photodynamic therapy [9, 10], neuroprotection [11], apoptosis [12–14], reactive oxygen species (ROS) scavenging [15, 16], drug and gene delivery [17–19]. The biological activities of fullerenes are considerably influenced by their chemical modifications and light treatment. Gadolinium metallofullerenes $[Gd@C_{82}(OH)_{22}]_n$ were originally synthesized with the transition metal atom gadolinium (Gd) encapsulated in a C_{82} fullerene cage as a contrast agent for magnetic resonance imaging [20]. Recent research has shown that $[Gd@C_{82}(OH)_{22}]_n$ nanoparticles can also be utilized as a potential chemotherapeutic agent. When combining $[Gd@C_{82}(OH)_{22}]_n$ with conventional anticancer drug, $[Gd@C_{82}(OH)_{22}]_n$ may increase intracellular drug accumulation by nanoparticle-enhanced endocytosis appeared strong effect on circumventing tumor resistance in vivo and in vitro [21]. As an attractive theranostic agent, metallofullerene nanoparticle additionally showed a strong capacity to improve cellular immune response by activating T cells and macrophages, resulting in releasing Th1 cytokines and inducing the maturation of dendritic cells [22]. There were other kinds of vehicles for Gd contrast agent, like spherical, nonporous and monodisperse silica nanoparticles, which also received high quality and local relaxivities [23, 24].

Nanotubes are another members of the fullerene structure family, which are categorized as single-walled nanotubes (SWCNTs), double-walled nanotubes (DWCNTs) and multi-walled nanotubes (MWCNTs). The application of carbon nanotubes (CNTs) was limited because of the toxicity issue to biological systems [25]. Currently, strategies for chemically functionalizing of CNTs have been reported and summarized as, (i) Covalent sidewall chemistry; (ii) Covalent chemistry at defects or open ends; (iii) Non-covalent surfactant encapsulation; (iv) Non-covalent polymer wrapping; (v) Molecular insertion into the CNTs interior [26, 27]. Based on these modifications, carbon nanotubes rank among the major, newly developed nanomaterials are of interest for a board range of biomedical applications. For the past few years, high increased studies are emerged focusing on their biomedicinal properties, including high tensile strength, flexibility, absorptivity, durability, and light weight, have led to the anticipation of a high production volume. Lee et al. investigated the application of amine-modified single-walled carbon nanotubes as a scaffold for stem cell therapy and protection of neuron from injury in a rat stroke model, which associated to improve the tolerance of neurons to ischaemic injury and decreased infarction area caused by transient middle cerebral artery occlusion (MCAO) surgery [28]. This study lays the foundation for further studies to elucidate the relationship between nanomaterials and pathology action. In addition, carbon nanotubes have strong ability to induce malignant transformation and tumorigenesis [29, 30], decrease ROS-mediated toxicological response [31], and reduce cytotoxicity by binding blood proteins [32], etc.

Additionally, gold nanomaterials, such as spherical NPs, nanorods and gold nanoshells have been developed as multifunctional probes due to their diagnostic effects and unique optical properties [33, 34], including size controllability [35], good biocompatibility [36], surface modification or shape control [37], and photothermal effects [35, 38]. Huo et al. [39] indicated that 50 nm Au@tiopronin NPs can accumulate effectively in tumor xenografts in vivo and have superior penetration and retention behavior (Fig. 6.1). Another study also demonstrated 50 nm gold nanoparticles can easily be taken into cells through endocytosis, which eventually blocked autophagic influx and induced lysosome impairment [40].

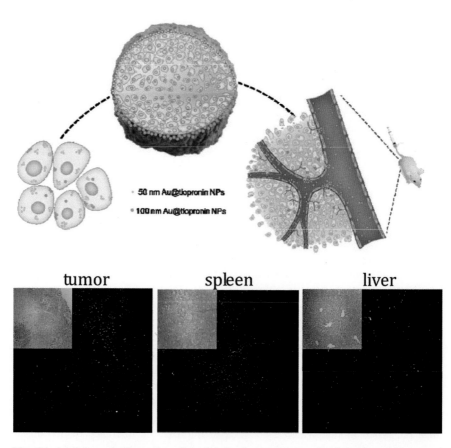

Fig. 6.1 (**a**) Schematic illustration of the localization and penetration behavior of 50 nm and 100 nm Au@tiopronin nanoparticles in monolayer cancer cells *in vitro*, multicellular tumor spheroid *ex vivo*, and xenograft tumor model *in vivo*. (**b**) Au content in the tissues including tumor, spleen, liver at 24 h after i.v. injections of gold nanoparticles (Reproduced with permission [39])

6.3   Designs of Multifunctional Nanoparticle for Medical Imaging

6.3.1   Accumulation and Absorbance

Almost all imaging motifs can be used in pharmaceutical studies to provide anatomic, pharmacokinetic and pharmacodynamic information to address specific properties of complex disease. Currently, nuclear-imaging methods (PET and single photon emission computed tomography, etc.) offer superior sensitivity compared with other modalities, such as ultrasound, computed tomography (CT) and magnetic resonance imaging (MRI). Yet, CT and particularly MRI can provide high-resolution images with great pathological resolution and soft-tissue contrast for diagnostic or prognostic of diseases. Consequently, multimodality imaging is emerging as a powerful combination to provide complementary information for pharmaceutical applications. Small gold NPs (e.g. 5–30 nm), which are considered to provide lower background in blood and tissue and more effectively target biomarkers, can be delivered more easily to cancer cells with different methods, compared to larger NPs [41]. For example, gold nanoparticles provide photothermal ablation therapy due to their unique surface plasmon resonance effects. The radiosensitisation of gold nanoparticles (GNPs) with an average diameter of 5 nm coated with the gadolinium chelating agent dithiolated diethylenetriaminepentaacetic gadolinium (Au@DTDTPA:Gd), which preferentially performed superior accumulation in vitro and in mice bearing tumors, exhibited a chemotherapeutic action [42]. The use of nanostructure based imaging technology requires imaging agents having high accumulating density and specificity for targeting specific cells and tissues. There are also other therapeutic nanoparticles designed based on fluorescent methods. Cabral H. et al. [43] focus on the penetration and efficacy of DACHPt-loaded polymer micellar nanomedicine with different diameters and concluded only the 30 nm micelles could penetrate poorly permeable pancreatic tumors to achieve an antitumor effect in mice (Fig. 6.2).

6.3.2   Distribution

Nanoparticle, which binds with high affinity to specific receptors, could be used to monitor receptor density and distribution in real time. Nanomedical application is designed to track their pharmacokinetics and distribution in real time or monitored at the drug therapeutic target site. To facilitate biodistributional analysis, the versatility of different imaging modalities for nanotheranostics is utilized if the circulation time and the imaged-guided characterization could be advanced in vivo. Various new tools for biomedical research and clinical applications have emerged with recent advances in nanotechnology. In the majority of cases, radionuclides have been used in clinic for improve the understanding of drug delivery process both in animal models and in patients. Types of radionuclide-labeled antibodies, micelles, polymers, liposomes have been subjected to biodistributional analysis for many years.

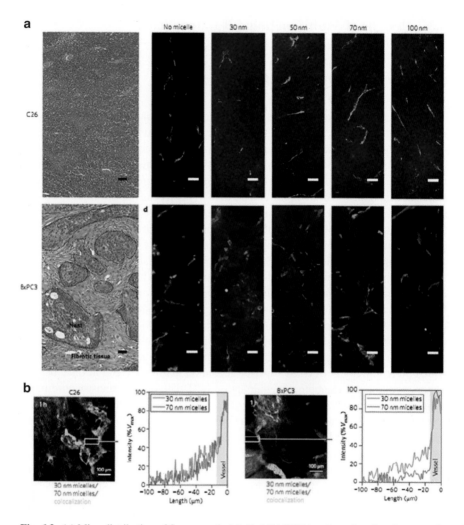

Fig. 6.2 (**a**) Microdistribution of fluorescently labelled DACHPt/m of varying sizes in tumors by histological examination of C26 tumour and BxPC3. (**b**) Real-time microdistribution of DACHPt/m with fluorescently labelled 30 nm (*green*) and 70 nm (*red*) micelles 1 h after injection into C26 and BxPC3 tumours (Reproduced with permission [43])

The high capacity for nanoparticle modification has led to their use as amplifiers for in vivo imaging. Magnetic materials also provide multiple functions in nanotheranostics by the magnetic attraction sensing and the heat generation, as well as serving manipulation by magnetic fields to improve drug delivery or allowing for localized heat therapy [44, 45]. As examples of nanoparticle based products already approved for clinical use, liposomes can encapsulate drug molecules to maintain drug integrity, shield toxicity and minimize side effects to other tissues during drug delivery, while superparamagnetic iron oxide nanoparticles (SIONPs) have superparamagnetic properties that make them and further evaluation of their efficacy as a powerful magnetic resonance imaging (MRI) contrast agent for imaging and diagnosis of lesions [46]. Torres Martin de Rosales et al. [47] used ^{64}Cu(II)-bis(dithiocarbamate bisphosphonate)-conjugated super paramagnetic iron nanoparticles as dual-modality PET-MRI agent to evaluation the imaging in vivo. The magnetic nanoparticles are ideal for accurate diagnostic use with the sensitive and quantifiable signal of PET and the high soft-tissue resolution of MRI. Namiki et al. [48] also described the utility of lipid-based magnetic nanoparticles as an efficient vector for the RNA interference of cancer. The systemic infusion of LipoMag/fluorescence-labeled siRNA was detected mainly in the magnetic field-irradiated tumor vessels, because EGFR contributed as a target motif. Under a magnetic field, a heavy accumulation of siRNA in tumor lesions was observed only in the LipoMag group (Fig. 6.3). Antitumor effect of the siRNA^{EGFR} was delivered by Lipo/Mag in gastric cancer model.

The ease of functionalizing the surface of nanoparticles with targeting moieties is anther clear advantage in designing molecular probes. Many cancer cells overexpress the folate receptor, and nanoparticles modified with surface folate have been shown to accumulate in tumor cells [49, 50]. Chen et al. covalent conjugated CdHgTe quantum dots and folic acid. These nanoparticles characterized with high targeting affinity and sensitivity were bound to folate transporter-positive cells both in vitro and in vivo, promising candidates for imaging, monitoring and early diagnosis of cancer [51]. The overexpression of folate receptor-α (FR-α) in 90–95 % of epithelial ovarian cancers prompted the investigation of intraoperative tumor-specific fluorescence imaging. In patients with ovarian cancer, intraoperative tumor-specific fluorescence agent folate-FITC imaging for real-time surgical visualization of tumor tissue showed cased the potential applications first-in-human use with ovarian cancer for improved intraoperative staging and more radical cytoreductive surgery [52].

6.3.3 Metabolism and Drug Release

Whereas identifying and monitoring tissues in vivo with various pathologies has received most attention to date, the use of molecular imaging to noninvasively quantify biological changes is potentially far more informative but also more challenging. For this, the imaging probe is a biosensor that not only requires specific targeting but also a reporter system that can be imaged. These probes may be useful for quantifying changes in a biological processes such as enzyme activity, drug metabolism, dynamic analysis, protein-protein interactions, or gene expression [53, 54]. A number of different dynamic analysis have been imaged in vivo using some innovative combinations of fluorescent probes [55]. A hybrid optical probe by incorporating nanoparticle and peptides with artificial tag molecules, which was utilized to detection of prostate cancer related serine protease activity with a high-fidelity and high signal-to-noise-ratio cancer nanoprobe, was applicable for dynamic detection of sensitive cancer biomarker enzymes [56]. Photoacoustic imaging is an emerging

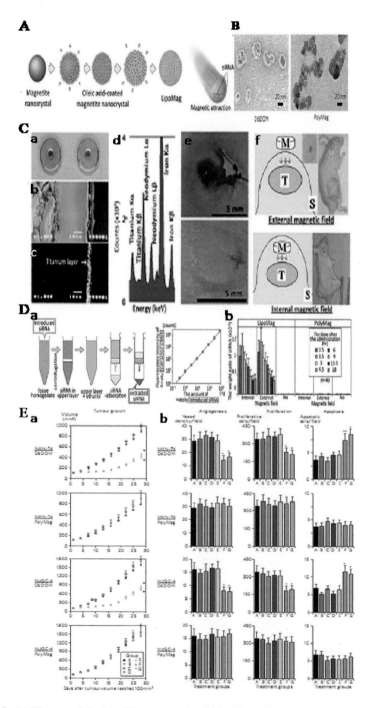

Fig. 6.3 (a) Diagram of the Lipo/Mag preparation with oleic acid-coated magnetic nanocrystal cores and lipid shells. (b) Transmission electron micrograph of French Press-treated D6DOM40 (*left*) and PolyMag (*right*). (c) Metal allergy-free magnets used to the subcutaneous tumor model in

modality that overcomes to a great extent the resolution and depth limitations of optical imaging while maintaining relatively high-contrast. The process of metastasis is the main cause of cancer death [57–59]. Nevertheless, detecting common markers for the development of metastasis, like circulating tumor cells (CTCs), is difficult to trace due to the deficiency of sensitivity diagnosis methods in vivo. To improve detection sensitivity and specificity, gold-plated carbon nanotubes conjugated with folic acid, which were functionalized to target a receptor commonly found in breast cancer cells, were used as a second contrast agent for photoacoustic imaging [60]. It was developed a new platform for in vivo magnetic enrichment and detection of rare targeted magnetic nanoparticles in combination with two-color photoacoustic flow cytometry and accomplished to capture circulating tumor cells under a magnet (Fig. 6.4). Highly sensitive and target specific multiplex detection with infections in static and dynamic conditions (e.g. in blood) that were otherwise previously difficulty to achieve using conventional methods.

6.3.4 Excretion and Clearance

AuNPs excretion can be detected in the kidneys or liver and further research is needed to clarify the properties that affect excretion pathways. A single intravenous injection of AuNPs was rapidly and consistently accumulated in both liver and spleen through 2 months. Significant accumulation in the kidney and testis was starting at 1 month and coincided with a decrease in gold levels in the urine and feces [61]. Regarding overcoming toxicities related to renal clearance of contrast agents, novel gold-silver alloy nanoshells as magnetic resonance imaging contrast agents as an alternative to typical gadolinium (Gd)-based contrast agents. These nanoparticles are very innovative and have the potential to utilize to nephrogenic systemic fibrosis as new candidates for nanotheranostics and imaging [62]. Boros et al. also reported that new Gd derivatives of the $H_{(2)}$ dedpa scaffold greatly improved blood, lung and

---

Fig. 6.3 (continued) external or internal magnetic field. (*a*) Magnets were coated with titanium nitride (gold-colored, *left*) and regular nickel-coated magnet (silver-colored, *right*), and (*b*) scanning electron micrograph and (*c*, *d*) X-ray analysis of the titanium nitride-coated magnet were shown. (*e*) A titanium-coating can avoid allergic inflammation caused by nickel-coated magnets, which was observed at 28 days after the implantation of titanium nitride-coated magnet. (*f*) External magnetic field for antitumor effect as titanium-nitride coated magnet (*M*) was attached to the tumor lesion (*T*) over the skin (*S*) using adhesive tape (*upper*), and internal magnetic field for tumor lesions as shown a magnet was implanted on tumor tissue under the skin (*lower*). (**d**) (*a*) The distribution of siRNA was quantified by Alexa Fluor-488-labelled siRNA, which were extracted and purified from siRNA in the introduced tissue. (*b*) Weight ratio of siRNA delivered by D6DOM or PolyMag for tumour lesion (*n*=6). In=internal; Ex=external; No=no magnets. (**e**) (*a*) A significant antitumor effect of the D6DOM/siRNA[EGFR] under a magnetic field was shown in the subcutaneously injected mice. (*b*) Two days after the last treatment, the degrees of angiogenesis, proliferation, and apoptosis were compared by immune-staining of vWF, Ki-67, and ssDNA, respectively. *$P<0.01$, compared with no treatment group (**a**) (Reproduced in part with permission [48])

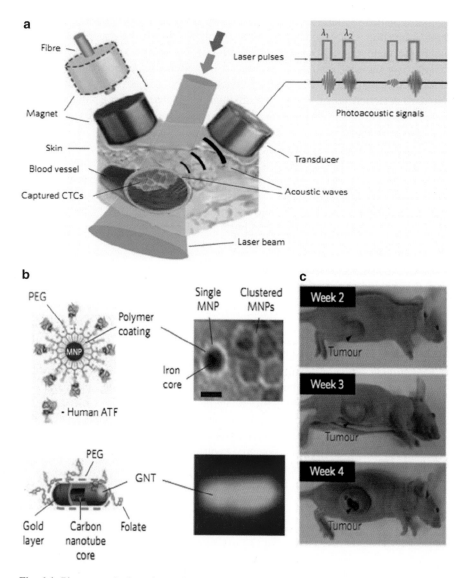

Fig. 6.4 Photoacoustic detection and magnetic enrichment of CTCs. (**a**) Schematic showing the detection setup. CTCs targeted by two-colour nanoparticles can be illuminated by laser pulses at wavelengths of 639 and 900 nm at 10-ms delay. (**b**) The 10 nm magnetic NPs (MNPs) coated with a thin layer (~2 nm) of amphiphilic triblock polymers, polyethylene glycol (PEG) and the amino-terminal fragment of urokinase plasminogen activator (ATF). (**c**) Photoacoustic detection and magnetic enrichment of CTCs in tumour-bearing mice. The size of the primary breast cancer xenografts at different stages of tumour development (Reproduced with permission [60])

kidney clearance compared to previously reported derivatives [63]. Multifunctional mesoporous silica nanospheres with cleavable Gd(III) chelates as MRI contrast agents was quickly cleaved by the blood pool thiols and eliminated through the renal excretion pathway [64]. These multifunctional imaging agents intensively guarantee the nanosafety during the proceed of imaging and theranostics.

Long period real-time imaging is another emergency request, which propose to overcome by nanotechnology. Bourrinet et al. [65] reported renal excretion of radiolabeled iron was retained minimal at day 84 and only 17–22 % injected doses could been excreted in the feces. Incorporation into the body's iron pool upon degradation of iron oxide core is one contributor leading to slow excretion. Briefly, it exists a homeostasis between iron oxide nanoparticles (IONPs) and iron-associated protein (e.g., transferritin, apoferritin, hemoglobin (Hb) molecules), which controls the degradation and clearance of INOPs [66–68]. Moreover, iron from degraded iron oxide nanoparticles could be found in these different forms in mice or rats and contribute to the long-term IONPs biodistribution and clearance kinetics [29, 30, 69, 70].

6.4 Conclusions

In this chapter, we reviewed the design of NPs for both theranoscis and medical imaging. The approach of using multifunctional nanoparticles featuring targets and other auxiliary moieties have exhibited as one of the most promising field to be used as theranostic agents, and trials are underway for the ultimate cure associated to human disease [71]. Based on the new concept of nano-theranostic, clinically approved nanoparticles have consistently shown value in reducing drug toxicity, improving the pharmacokinetics of drugs, even more and more feasible and personalized. It will be finally come in a foreseeable future that nanoscale-based nano-theranastics and imaging will truly help us providing the effective approaches of prevention, therapy and recovery for patients.

6.5 Future Perspective

Nanoparticle scaled agents, which possess unique physicochemical properties that allow integration of multiple functionalities in a single design, have shown tremendous promise in the development of imaging and possess the potential to greatly advance the theranosics and treatment of disease. Multifunctional nanoparticle holds considerable promise as the next generation of medicine that able to contain the early detection of disease, simultaneous monitoring and treatment and targeted therapy with minimal toxicity. The ability of visualization or monitor the absorption, distribution, metabolism and excretion in real time for an intact organism can reveal physiological and pathological changes that often cannot be studied in isolated cells or tissues. With the advent of highly sensitive and specific imaging technologies, more advanced approaches to image functional changes at the tissue and cellular level are now being explored. To meet the requirements for nanotheragnostic agents, nanoparticles must be required to high stability and controlling function in extreme conditions such as high salt concentrations and wide pH and temperature ranges. Therefore, a flexible and adaptive nanoparticle platform is highly desirable that allows interchangeable therapeutics with high biocompatibility and low toxicity. Future nanoparticle-based theranostics will offer new hope in an attractive way to combat severe or fatal diseases, such as cancer, cardiovascular and neurodegenerative diseases.

Acknowledgments This work was financially supported by the National Key Basic Research Program of China (MOST 973 projects 2009CB930200) and the program of National Natural Science Foundation of China (30970784 and 81171455). The authors are grateful for the support of the Chinese Academy of Sciences (CAS) "Hundred Talents Program" and the CAS Knowledge Innovation Program. Authors declare that there are no conflicts of interest in this study.

References

1. Ahmed N, Fessi H, Elaissari A (2012) Theranostic applications of nanoparticles in cancer. Drug Discov Today 17:928–934
2. Kuhlmeier D, Sandetskaya N, Allelein S (2012) Application of nanotechnology in miniaturized systems and its use in medical and food analysis. Recent Pat Food Nutr Agric 4:187–199
3. Osminkina L, Tamarov K, Sviridov A et al (2012) Photoluminescent biocompatible silicon nanoparticles for cancer theranostic applications. J Biophotonics 5:529–535
4. Wujcik E, Monty C (2013) Nanotechnology for implantable sensors: carbon nanotubes and graphene in medicine. Wiley Interdiscip Rev Nanomed Nanobiotechnol 5:233–249
5. Zhou X, Porter A, Robinson D et al (2014) Nano-enabled drug delivery: a research profile. Nanomed: Nanotechnol Biol Med 10(5):889–896
6. Ferrari M (2005) Cancer nanotechnology: opportunities and challenges. Nat Rev Cancer 5:161–171
7. Muthu M, Leong D, Mei L et al (2014) Nanotheranostics- application and further development of nanomedicine strategies for advanced theranostics. Theranostics 4(6):660–677
8. Fang C, Zhang M (2010) Nanoparticle-based theragnostics: integrating diagnostic and therapeutic potentials in nanomedicine. J Control Release 146:2–5
9. Lu Z, Dai T, Huang L et al (2010) Photodynamic therapy with a cationic functionalized fullerene rescues mice from fatal wound infections. Nanomedicine (Lond) 5:1525–1533
10. Mroz P, Xia Y, Asanuma D et al (2011) Intraperitoneal photodynamic therapy mediated by a fullerene in a mouse model of abdominal dissemination of colon adenocarcinoma. Nanomedicine 7:965–974
11. Cai X, Jia H, Liu Z et al (2008) Polyhydroxylated fullerene derivative C(60)(OH)(24) prevents mitochondrial dysfunction and oxidative damage in an MPP(+) -induced cellular model of Parkinson's disease. J Neurosci Res 86:3622–3634
12. Lao F, Chen L, Li W et al (2009) Fullerene nanoparticles selectively enter oxidation-damaged cerebral microvessel endothelial cells and inhibit JNK-related apoptosis. ACS Nano 3: 3358–3368

13. Nishizawa C, Hashimoto N, Yokoo S et al (2009) Pyrrolidinium-type fullerene derivative-induced apoptosis by the generation of reactive oxygen species in HL-60 cells. Free Radic Res 43:1240–1247
14. Palyvoda K, Grynyuk II, Prylutska S et al (2010) Apoptosis photoinduction by C60 fullerene in human leukemic T cells. Ukr Biokhim Zh 82:121–127
15. Saitoh Y, Xiao L, Mizuno H et al (2010) Novel polyhydroxylated fullerene suppresses intracellular oxidative stress together with repression of intracellular lipid accumulation during the differentiation of OP9 preadipocytes into adipocytes. Free Radic Res 44:1072–1081
16. Yamakoshi Y, Aroua S, Nguyen T (2014) FD173: water-soluble fullerene materials for bioapplication. Faraday Discuss. doi:10.1039/C4FD00076E
17. Sitharaman B, Zakharian T, Saraf A et al (2008) Water-soluble fullerene (C60) derivatives as nonviral gene-delivery vectors. Mol Pharm 5:567–578
18. Montellano A, Da Ros T, Bianco A et al (2011) Fullerene C(60) as a multifunctional system for drug and gene delivery. Nanoscale 3:4035–4041
19. Sigwalt D, Holler M, Iehl J et al (2011) Gene delivery with polycationic fullerene hexakis-adducts. Chem Commun (Camb) 47:4640–4642
20. Xing G, Yuan H, He R et al (2008) The strong MRI relaxivity of paramagnetic nanoparticles. J Phys Chem B 112:6288–6291
21. Liang X, Meng H, Wang Y et al (2010) Metallofullerene nanoparticles circumvent tumor resistance to cisplatin by reactivating endocytosis. Proc Natl Acad Sci U S A 107:7449–7454
22. Yang D, Zhao Y, Guo H et al (2010) [Gd@C(82)(OH)(22)](n) nanoparticles induce dendritic cell maturation and activate Th1 immune responses. ACS Nano 4:1178–1186
23. Bui T, Stevenson J, Hoekman J et al (2010) Novel Gd nanoparticles enhance vascular contrast for high-resolution magnetic resonance imaging. PLoS One 5:e13082
24. Feldmann V, Engelmann J, Gottschalk S et al (2012) Synthesis, characterization and examination of Gd[DO3A-hexylamine]-functionalized silica nanoparticles as contrast agent for MRI-applications. J Colloid Interface Sci 366:70–79
25. Firme C 3rd, Bandaru P (2010) Toxicity issues in the application of carbon nanotubes to biological systems. Nanomedicine 6:245–256
26. Hirsch A (2002) Functionalization of single-walled carbon nanotubes. Angew Chem Int Ed Engl 41:1853–1859
27. Hersam M (2008) Progress towards monodisperse single-walled carbon nanotubes. Nat Nanotechnol 3:387–394
28. Lee H, Park J, Yoon O et al (2011) Amine-modified single-walled carbon nanotubes protect neurons from injury in a rat stroke model. Nat Nanotechnol 6:121–125
29. Wang F, Kim D, Yoshitake T et al (2011) Diffusion and clearance of superparamagnetic iron oxide nanoparticles infused into the rat striatum studied by MRI and histochemical techniques. Nanotechnology 22:015103
30. Wang L, Luanpitpong S, Castranova V et al (2011) Carbon nanotubes induce malignant transformation and tumorigenesis of human lung epithelial cells. Nano Lett 11:2796–2803
31. Zhang Y, Xu Y, Li Z et al (2011) Mechanistic toxicity evaluation of uncoated and PEGylated single-walled carbon nanotubes in neuronal PC12 cells. ACS Nano 5:7020–7033
32. Ge C, Du J, Zhao L et al (2011) Binding of blood proteins to carbon nanotubes reduces cytotoxicity. Proc Natl Acad Sci U S A 108:16968–16973
33. Zhou X, Xu W, Liu G et al (2010) Size-dependent catalytic activity and dynamics of gold nanoparticles at the single-molecule level. J Am Chem Soc 132:138–146
34. Lee D, Koo H, Sun I et al (2012) Multifunctional nanoparticles for multimodal imaging and theragnosis. Chem Soc Rev 41:2656–2672
35. Juve V, Cardinal M, Lombardi A et al (2013) Size-dependent surface plasmon resonance broadening in nonspherical nanoparticles: single gold nanorods. Nano Lett 13:2234–2240
36. De Jong W, Hagens W, Krystek P et al (2008) Particle size-dependent organ distribution of gold nanoparticles after intravenous administration. Biomaterials 29:1912–1919
37. Elbakry A, Wurster E, Zaky A et al (2012) Layer-by-layer coated gold nanoparticles: size-dependent delivery of DNA into cells. Small 8:3847–3856

38. Lkhagvadulam B, Kim J, Yoon I et al (2013) Size-dependent photodynamic activity of gold nanoparticles conjugate of water soluble purpurin-18-N-methyl-d-glucamine. Biomed Res Int 2013:720579

39. Huo S, Ma H, Huang K et al (2013) Superior penetration and retention behavior of 50 nm gold nanoparticles in tumors. Cancer Res 73:319–330

40. Ma X, Wu Y, Jin S et al (2011) Gold nanoparticles induce autophagosome accumulation through size-dependent nanoparticle uptake and lysosome impairment. ACS Nano 5:8629–8639

41. Zharov V, Kim J, Curiel D et al (2005) Self-assembling nanoclusters in living systems: application for integrated photothermal nanodiagnostics and nanotherapy. Nanomedicine 1:326–345

42. Hebert E, Debouttiere P, Lepage M et al (2010) Preferential tumour accumulation of gold nanoparticles, visualised by Magnetic Resonance Imaging: radiosensitisation studies in vivo and in vitro. Int J Radiat Biol 86:692–700

43. Cabral H, Matsumoto Y, Mizuno K et al (2011) Accumulation of sub-100 nm polymeric micelles in poorly permeable tumours depends on size. Nat Nanotechnol 6:815–823

44. Liu-Snyder P, Webster T (2006) Designing drug-delivery systems for the nervous system using nanotechnology: opportunities and challenges. Expert Rev Med Devices 3:683–687

45. Liu J, Tabata Y (2010) Photodynamic therapy of fullerene modified with pullulan on hepatoma cells. J Drug Target 18:602–610

46. Wang Y, Wang K, Sun Q et al (2009) Novel manganese(II) and cobalt(II) 3D polymers with mixed cyanate and carboxylate bridges: crystal structure and magnetic properties. Dalton Trans 28:9854–9859

47. Torres Martin de Rosales R, Tavare R, Paul R et al (2011) Synthesis of 64Cu(II)-bis(dithiocarbamatebisphosphonate) and its conjugation with superparamagnetic iron oxide nanoparticles: in vivo evaluation as dual-modality PET-MRI agent. Angew Chem Int Ed Engl 50:5509–5513

48. Namiki Y, Namiki T, Yoshida H et al (2009) A novel magnetic crystal-lipid nanostructure for magnetically guided in vivo gene delivery. Nat Nanotechnol 4:598–606

49. Miki K, Oride K, Inoue S et al (2010) Ring-opening metathesis polymerization-based synthesis of polymeric nanoparticles for enhanced tumor imaging in vivo: synergistic effect of folate-receptor targeting and PEGylation. Biomaterials 31:934–942

50. Muller C, Schibli R et al (2011) Folic acid conjugates for nuclear imaging of folate receptor-positive cancer. J Nucl Med 52:1–4

51. Chen H, Li L, Cui S et al (2011) Folate conjugated CdHgTe quantum dots with high targeting affinity and sensitivity for in vivo early tumor diagnosis. J Fluoresc 21:793–801

52. van Dam G, Themelis G, Crane L et al (2011) Intraoperative tumor-specific fluorescence imaging in ovarian cancer by folate receptor-alpha targeting: first in-human results. Nat Med 17:1315–1319

53. Larson D, Zipfel W, Williams R et al (2003) Water-soluble quantum dots for multiphoton fluorescence imaging in vivo. Science 300:1434–1436

54. Rao J, Dragulescu-Andrasi A, Yao H (2007) Fluorescence imaging in vivo: recent advances. Curr Opin Biotechnol 18:17–25

55. Abulrob A, Brunette E, Slinn J et al (2008) Dynamic analysis of the blood-brain barrier disruption in experimental stroke using time domain in vivo fluorescence imaging. Mol Imaging 7:248–262

56. Liu G, Chen F, Ellman J et al (2006) Peptide-nanoparticle hybrid SERS probe for dynamic detection of active cancer biomarker enzymes. Conf Proc IEEE Eng Med Biol Soc 1:795–798

57. Christofori G (2006) New signals from the invasive front. Nature 441:444–450

58. Pantel K, Brakenhoff R, Brandt B (2008) Detection, clinical relevance and specific biological properties of disseminating tumour cells. Nat Rev Cancer 8:329–340

59. Liang X (2011) EMT: new signals from the invasive front. Oral Oncol 47:686–687

60. Galanzha E, Shashkov E, Kelly T et al (2009) In vivo magnetic enrichment and multiplex photoacoustic detection of circulating tumour cells. Nat Nanotechnol 4:855–860

61. Balasubramanian S, Jittiwat J, Manikandan J et al (2010) Biodistribution of gold nanoparticles and gene expression changes in the liver and spleen after intravenous administration in rats. Biomaterials 31:2034–2042
62. Gheorghe D, Cui L, Karmonik C et al (2011) Gold-silver alloy nanoshells: a new candidate for nanotherapeutics and diagnostics. Nanoscale Res Lett 6:554
63. Boros E, Ferreira C, Patrick B et al (2011) New Ga derivatives of the H(2)dedpa scaffold with improved clearance and persistent heart uptake. Nucl Med Biol 38:1165–1174
64. Vivero-Escoto J, Taylor-Pashow K, Huxford R et al (2011) Multifunctional mesoporous silica nanospheres with cleavable Gd(III) chelates as MRI contrast agents: synthesis, characterization, target-specificity, and renal clearance. Small 7:3519–3528
65. Liu H, Hua M, Yang H et al (2010) Magnetic resonance monitoring of focused ultrasound/magnetic nanoparticle targeting delivery of therapeutic agents to the brain. Proc Natl Acad Sci U S A 107:15205–15210
66. Kaldor I (1955) Studies on intermediary iron metabolism. VIII. The fate in mice of injected saccharated oxide of iron. Aust J Exp Biol Med Sci 33:645–649
67. Pouliquen D, Le Jeune J, Perdrisot R et al (1991) Iron oxide nanoparticles for use as an MRI contrast agent: pharmacokinetics and metabolism. Magn Reson Imaging 9:275–283
68. Oria R, Sanchez L, Houston T et al (1995) Effect of nitric oxide on expression of transferrin receptor and ferritin and on cellular iron metabolism in K562 human erythroleukemia cells. Blood 85:2962–2966
69. Hass F, Lee P, Lourenco R (1976) Tagging of iron oxide particles with 99mTc for use in the study of deposition and clearance of inhaled particles. J Nucl Med 17:122–125
70. Jain T, Reddy M, Morales M et al (2008) Biodistribution, clearance, and biocompatibility of iron oxide magnetic nanoparticles in rats. Mol Pharm 5:316–327
71. Cheng Z, Al Zaki A, Hui J et al (2012) Multifunctional nanoparticles: cost versus benefit of adding targeting and imaging capabilities. Science 338:903–910

Chapter 7
Medical Nanobiosensors

Eden Morales-Narváez and Arben Merkoçi

Abbreviations

Abs	Antibodies
AD	Alzheimer's disease
AuNPs	Gold nanoparticles
CA125	Cancer antigen 125
CA15-3	Cancer antigen 15-3
CEA	Carcionoembryonic antigen
CJD	Creutzfeldt-Jakob disease
CNTs	Carbon nanotubes
cTnT	Cardiac troponin-T
EGFR	Epidermal growth factor receptor
ELISA	Enzyme linked immunosorbent assay
FRET	Fluorescence resonance energy transfer
HER2	Human epidermal growth factor receptor 2
HRP	Horseradish peroxidase
IUPAC	International Union of Pure and Applied Chemistry
MMP-9	Matrix metalloproteinase 9

E. Morales-Narváez
Nanobioelectronics & Biosensors Group, Catalan Institute of Nanoscience and
Nanotechnology, Barcelona 08193, Spain

ESAII Department, Polytechnic University of Catalonia, Barcelona 08028, Spain

A. Merkoçi (✉)
Nanobioelectronics & Biosensors Group, Catalan Institute of Nanoscience
and Nanotechnology, Barcelona 08193, Spain

ICREA, Barcelona, Spain
e-mail: arben.merkoci@icn.cat

© Springer Science+Business Media New York 2014   117
Y. Ge et al. (eds.), *Nanomedicine*, Nanostructure Science and Technology,
DOI 10.1007/978-1-4614-2140-5_7

MNPs	Magnetic nanoparticles
MWCNTs	Multi-walled carbon nanotubes
oxLDL	Oxidized low density lipoprotein
PCR	Polymerase chain reaction
PD	Parkinson's disease
PrP	Prion proteins
PSA	Prostate specific antigen
QDs	Quantum dots
SWCNT	Single-walled carbon nanotubes

7.1 Introduction

...the preservation of health is ... without doubt the first good and the foundation of all the others goods of this life...

(René Descartes and Discours de la method, 1637)

The concept of diagnosis based on biological samples dated back several thousand years ago documented from the ancient China, Egypt to the Middle Ages of Europe [1]. Nevertheless, it was not until the 1960s when Professor Leland C Clark Jnr., as the father of the biosensor concept, described how to perform reliable and robust measurements of analytes (molecules of interest) presents in the body [2]. Presently, cancer can be diagnosed by screening the levels of the appropriate analytes existing in blood and likewise diabetes is inspected by measuring glucose concentrations. Moreover, the most conventional techniques of diagnostic technologies are the enzyme-linked immunosorbent Assay (ELISA) and the polymerase chain reaction (PCR). Nevertheless, these techniques report different handicaps such as high cost and time required, significant sample preparation, intensive sample handling, and can become troublesome to patients. Accordingly, novel advances in diagnostic technology are highly desired.

Diagnostic technology is an important field for the progress of healthcare and medicine, specifically in early diagnosis and treatment of diseases (which can deeply reduce the expense of patient care related to advanced stages of several diseases). Since molecular diagnostics can profile the pathological state of the patient, nanobiosensors for molecular diagnostics represent a factual and interesting application of the nanotechnology in medicine.

A biosensor is defined by the International Union of Pure and Applied Chemistry (IUPAC) as a "device that uses specific biochemical reactions mediated by isolated enzymes, immunosystems, tissues, organelles, or whole cells to detect chemical compounds usually by electrical, thermal, or optical signals". Generally, biosensors include biorecognition probes (responsible for the specific detection of the analytes) and a transducer element (that converts a biorecognition event into a suitable signal) [3, 4]. In the twenty-first century, nanotechnology has been revolutionizing many fields including medicine, biology, chemistry, physics, and electronics. In this way, biosensors have been also benefited by nanotechnology, which is an emerging

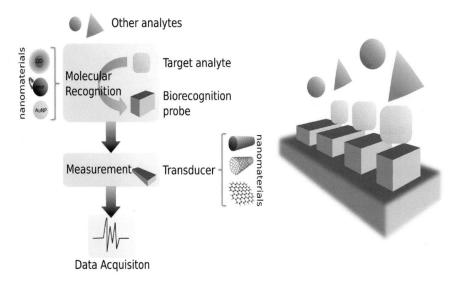

Fig. 7.1 Schematic representation of a nanobiosensor. Normally, a nanobiosensor relies on nanomaterials as transducer elements or reporters of biorecognition events

multidisciplinary field that entails the synthesis and use of materials or systems at the nanoscale (normally 1–100 nm). The rationale behind this technology is that nanomaterials possess optical, electronic, magnetic or structural properties that are unavailable for bulk materials. Since nanomaterials range in the same scale of the diagnostic molecules, when linked to biorecognition probes (such as antibodies, DNA and enzymes), nanostructures allow the control, manipulation and detection of molecules with diagnostic interest, even at the single molecule level. Normally, nanobiosensors are based on nanomaterials or nanostructures as transducer elements or reporters of biorecognition events [5, 6]. Figure 7.1 displays the schematic representation of a nanobiosensor.

This chapter aims at providing a description on the basic principles of the nanobiosensors and a brief survey on nanobiosensing strategies towards medical applications, specifically in the following disease categories: neurodegenerative diseases, cardiovascular diseases and cancer chosen as the most reported application fields.

7.2 Biorecognition Probes

Biorecognition probes, or molecular bioreceptors, are the key in the specificity of biosensors (a non-specific biorecognition event can yield a false result). Biomolecular recognition generally entails different interactions such as hydrogen bonding, metal coordination, hydrophobic forces, van der Waals forces, pi-pi interactions and electrostatic interactions. In this section the most common biorecognition probes in nanaobiosensors are briefly discussed (see Fig. 7.2a).

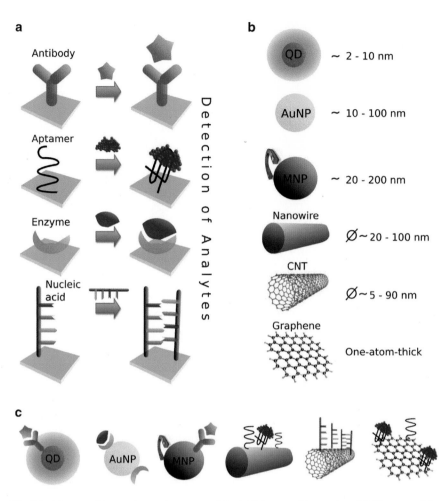

Fig. 7.2 Biorecognition probes and nanomaterials. (**a**) Biorecognition probes. (**b**) Nanomaterials. (**c**) Nanomaterials decorated with biorecognition probes. *QD* Quantum Dot, *AuNP* gold nanoparticle, *MNP* magnetic nanoparticle. Sketches are not at scale

7.2.1 Antibodies

Antibodies are soluble forms of immunoglobulin containing hundreds of individual amino acids arranged in a highly ordered sequence. These polypeptides are produced by immune system cells (B lymphocytes) when exposed to antigenic substances or molecules. Proteins with molecular weights greater than 5,000 Da are generally immunogenic. Antibodies contain in their structure recognition/binding sites for specific molecular structures of the antigen. Since an antibody interacts in a highly specific way with its unique antigen, antibodies are widely employed in biosensors.

7.2.2 *Aptamers*

Aptamers are novel artificial oligonucleic acid molecules that are selected (in vitro) for high affinity binding to several targets such as proteins, peptides, amino acids, drugs, metal ions and even whole cells [7–9].

7.2.3 *Enzymes*

Enzymes are protein catalysts of remarkable efficiency involved in chemical reactions fundamental to the life and proliferation of cells. Enzymes also possess specific binding capabilities and were the pioneer molecular recognition elements used in biosensors and continue still used in biosensing applications [10, 11, 126].

7.2.4 *Nucleic Acids*

Since the interaction between adenosine and thymine and cytosine and guanosine in DNA is complementary, specific probes of nucleic acids offer sensitive and selective detection of target genes in biosensors [12].

7.3 Transduction Modes

In order to detect biorecognition events, biosensors require a transduction mode. Transduction modes are generally classified according to the nature of their signal into the following types: (1) optical detection, (2) electrochemical detection, (3) electrical detection, (4) mass sensitive detection and (5) thermal detection.

7.3.1 *Optical Detection*

Optical biosensing is based on several types of spectroscopic measurements (such as absorption, dispersion spectrometry, fluorescence, phosphorescence, Raman, refraction, surface enhanced Raman spectroscopy, and surface plasmon resonance) with different spectrochemical parameters acquired (amplitude, energy, polarization, decay time and/or phase). Among these spectrochemical parameters, amplitude is the most commonly measured, as it can generally be correlated with the concentration of the target analyte [13].

7.3.2 Electrochemical Detection

Electrochemical detection entails the measurement of electrochemical parameters (such as current, potential difference or impedance) of either oxidation or reduction reactions. These electrochemical parameters can be correlated to either the concentration of the electroactive probe assayed or its rate of production/consumption [13].

7.3.3 Electrical Detection

Electrical detection is often based on semiconductor technology by replacing the gate of a metal oxide semiconductor field effect transistor with a nanostructure (usually nanowires or graphitic nanomaterials). This nanostructure is capped with biorecognition probes and a electrical signal is triggered by biorecognition events [14, 15].

7.3.4 Mass Sensitive Detection

Mass sensitive detection can be performed by either piezoelectric crystals or microcantilevers. The former relies on small alterations in mass of piezoelectric crystals due to biorecognition events. These events are correlated with the crystals oscillation frequency allowing the indirect measurement of the analyte binding [16]. Microcantilever biosensing principle is based on mechanical stresses produced in a sensor upon molecular binding. Such stress bends the sensor mechanically and can be easily detected [17].

7.3.5 Thermal Detection

Thermal biosensors are often based on exothermic reactions between an enzyme and the proper analyte. The heat released from the reaction can be correlated to the amount of reactants consumed or products formed [18].

7.4 Nanomaterials: The Nanobiosensors Toolbox

Recent advances of the nanotechnology focused on the synthesis of materials with innovative properties have led to the fabrication of several nanomaterials such as nanowires, quantum dots, magnetic nanoparticles, gold nanoparticles, carbon nanotubes and graphene. These nanomaterials linked to biorecognition probes are generally the basic components of nanobiosensors. In order to attach nanomaterials with biorecognition probes, nanomaterials are either electrostatically charged or functionalized with the suitable chemically active group [19–23] (see Fig. 7.2c).

In the following section the most widely used nanomaterials in biosensing are briefly described and they are sketched in Fig. 7.2b.

7.4.1 Zero-dimensional Nanomaterials

7.4.1.1 Quantum Dots (QDs)

QDs are semiconductors nanocrystals composed of periodic groups of II–VI (e.g., CdSe) or III–V (InP) materials. QDs range from 2 to 10 nm in diameter (10–50 atoms). They are robust fluorescence emitters with size-dependent emission wavelengths. For example, small nanocrystals (2 nm) made of CdSe emit in the range between 495 and 515 nm, whereas larger CdSe nanocrystals (5 nm) emit between 605 and 630 nm [24]. QDs are extremely bright (1 QD $\approx$ 10–20 organic fluorophores) [25]. They have high resistance to photobleaching, narrow spectral linewidths, large stokes shift and even different QDs emitters can be excited using a single wavelength, i.e. they have a wide excitation spectra [26, 27]. Because of their properties QDs are used in biosensing as either fluorescent probes [28, 29] or labels for electrochemical detection (Wang et al. 2011a).

7.4.1.1.1 Gold Nanoparticles (AuNPs)

Synthesis of AuNPs often entails the chemical reduction of gold salt in citrate solution. Their scale is less than about 100 nm. AuNPs have interesting electronic, optical, thermal and catalytic properties [30, 31]. AuNPs enable direct electron transfer between redox proteins and bulk electrode materials and are widely used in electrochemical biosensors, as well as biomolecular labels ([32]; Ambrosi et al. 2009b).

7.4.1.1.2 Magnetic Nanoparticles (MNPs)

MNP are often composed by iron oxide and due to their size (20–200 nm) they can possess superparamagnetic properties. MNP are used as contrast agents for magnetic resonance imaging and for molecular separation in biosensors devices [33–35].

7.4.2 One-Dimensional Nanomaterials

7.4.2.1 Carbon Nanotubes (CNTs)

CNTs consist of sheets (multi-walled carbon nanotubes, MWCNTs) or a single sheet (single-walled carbon nanotubes SWCNTs) of graphite rolled-up into a tube. Their diameters range about from 5 to 90 nm. The lengths of the graphitic tubes are normally in the micrometer scale. CNTs seem a remarkable scheme of excellent mechanical, electrical and electrochemical properties [36, 37] and even can display metallic, semiconducting and superconducting electron transport [38]. The properties of carbon nanotubes are highly attractive for electrochemical biosensors and also has been used as transducer in bio-field-effect transistors [39, 40].

7.4.2.2 Nanowires

Nanowires are planar semiconductors with a diameter ranging from 20 to 100 nm and length from submicrometer to few micrometer dimensions. They are fabricated with materials including but not limited to silicon, gold, silver, lead, conducting polymer and oxide [41, 42]. They have tunable conducting properties and can be used as transducers of chemical and biological binding events in electrically based sensors such as bio-field-effect transistors [43–45] or even as nanomotors [46].

7.4.3   The Innovative Two-Dimensional Material: Graphene

Graphene is a recently discovered one-atom-thick planar sheet of sp^2 bonded carbon atoms ordered in a two-dimensional honeycomb lattice and is the basic building block for carbon allotropes (e.g., fullerens, CNTs and graphite). Graphene has displayed fascinating properties such as electronic flexibility, high planar surface, superlative mechanical strength, ultrahigh thermal conductivity and novel electronic properties [47]. Owing to its properties, graphene has been employed as transducer in bio-field-effect transistors, electrochemical biosensors, impedance biosensors, electrochemiluminescence, and fluorescence biosensors, as well as biomolecular label [48, 49, 133].

7.5   Nanobiosensing Strategies Toward Medical Applications in Health Priorities: Biomarkers Detection

Biomarkers can be altered genes, RNA products, proteins, or other metabolites that profile biological processes in normal, pathogenic or pathological states and even during pharmacologic or therapy responses of the patient [50–53]. Molecular diagnostics relies on the detection of these biomarkers sourced from biological samples such as serum, urine, saliva and cerebrospinal fluid.

Nanobiosensors can perform a key role in biomarker detection. As innovative devices, nanobiosensors often comprise the following requirements: a tiny amount of sample and assay reagents, fast response, highly sensitive, high accuracy and reproducibility, portability, multiplexing capabilities, user friendly and low cost. This section contains an overview of the powerful advantages of different nanobiosensing strategies for biomarker detection focused on the following disease categories: neurodegenerative diseases, cardiovascular diseases and cancer. Figures 7.3, 7.4 and 7.5 display different biosensors for neurodegenerative diseases, cardiovascular diseases and cancer (respectively). All the exposed applications are summarized in Table 7.1. Since this chapter cannot cover all applications and technical details, the interested reader is referred to some recent literature referenced across the content.

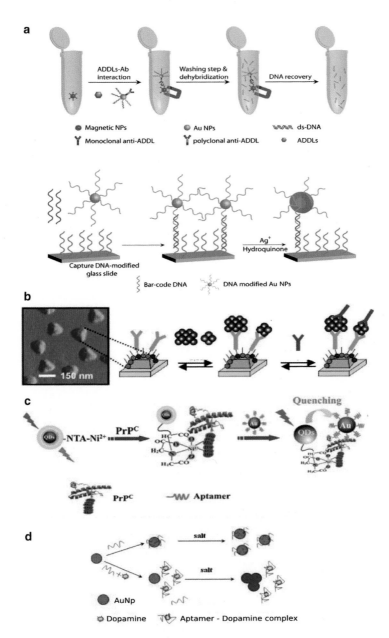

Fig. 7.3 Nanobiosensors for neurodegenerative diseases. (**a**) Bio-bar-code assay for amyloid β (ADDL) detection based on AuNP and MNP (Reprinted with permission from [121]. Copyright 2011, Elsevier). (**b**) Detection of amyloid β by using a nanopatterned optical biosensor based on silver nanoparticles and localized surface plasmon resonance (Modified with permission from Nam et al. 2005. Copyright 2005, American Chemical Society). (**c**) Detection of prion protein (PrP^C) with a long-range resonance energy transfer strategy based on quenching of the light of QDs by AuNPs (Modified with permission from [76]. Copyright 2010, Royal Society of Chemistry). (**d**) Aptamer-based colorimetric biosensing of dopamine using unmodified AuNPs (Modified with permission from [71]. Copyright 2011, Elsevier)

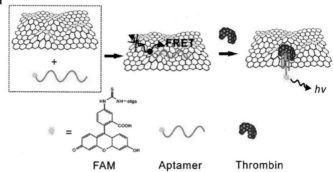

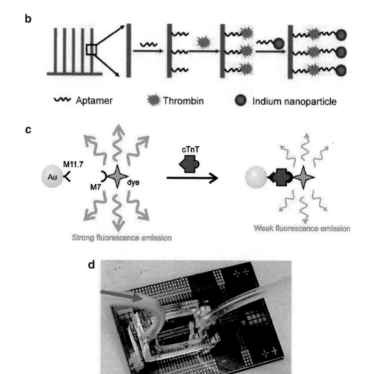

Fig. 7.4 Nanobiosensors for cardiovascular diseases. (**a**) Thrombin detection based on fluorescence resonance energy transfer through graphene as acceptor of organic dyes (FAM) donors connected with specific aptamers (Reprinted with permission from [88]. Copyright 2010, American Chemical Society). (**b**) Aptamer-functionalized indium nanoparticles as thermal probes for thrombin detection on silicon nanopillars using thermal detection (Reprinted with permission from [90]. Copyright 2010, Elsevier). (**c**) Long-range fluorescence quenching by gold nanoparticles in a sandwich immunoassay for cardiac troponin T (M11.7, specific antibody; M7, specific fragment antibody; cTnT, cardiac troponin T) (Reprinted with permission from [93]. Copyright 2009, American Chemical Society). (**d**) Setup for the detection of cardiac troponin T composed of arrays of highly ordered silicon nanowire clusters (Modified with permission from [92]. Copyright 2009, American Chemical Society)

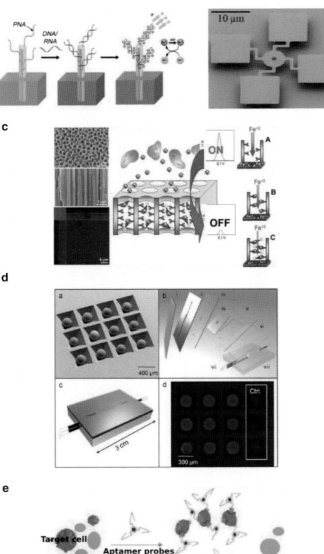

Fig. 7.5 Nanobiosensors for cancer. (**a**) Nanoscale electrode platform for the direct electrocatalytic mRNA (related with prostate cancer) detection using peptide nucleic acid-nanowire sensors (Reprinted with permission from [75]. Copyright 2009, American Chemical Society). (**b**) Trampoline shaped nanomechanical resonator made of silicon for the detection of prostate specific antigen (Reprinted with permission from [2]. Copyright 2009, Royal Society of Chemistry) (**c**) A nanochannel (porous alumina)/gold nanoparticle-based filtering and sensing platform for direct detection of cancer antigen 15-3 using Fe(2/3) as electrochemical signaling indicator (Reprinted with permission from [16]. Copyright 2011, John Wiley and Sons). (**d**) Silicon Nano-bio-chips for multiplexed protein detection: determinations of cancer biomarkers in serum and saliva using quantum dot bioconjugate labels (as fluorescent probes) (Reprinted with permission from [36]. Copyright 2009, Elsevier). (**e**) A sensitive fluorescence anisotropy method for the direct detection of cancer cells in whole blood based on aptamer-conjugated near-infrared fluorescent methylene blue nanoparticles (Modified with permission from [72]. Copyright 2010, Elsevier)

Table 7.1 Nanobiosensors for biomarker detection towards medical applications in different disease categories

Application/detection	Nanomaterials/platform	LOD	Characteristics	Reference
Neurodegenerative diseases				
Amyloid β	Magnetic beads and AuNPs, fluorescent detection (OD)	10 fM	Ultra-high sensitivity. A hopeful beginning towards a diagnosis tool	Khan et al. [57]
	Scanometric detection (OD)	10 aM		
	Nanoscale biosensor based on localized surface plasmon resonance (OD)	7.3 pM	Provides new information relevant to the understanding and possible diagnosis of Alzheimer	Haes et al. [59]
	AuNPs, surface plasmon resonance based sensor (OD)	1 fg/mL	Ultrasensitive detection, focused on early detection	Lee et al. [60]
Amyloid β aggregation inhibitor	Precipitation of AuNP. Optical density of supernatants (OD)	ND	Potential diagnostic tool for diseases involving abnormal protein aggregation as their key pathogenesis processes	Han et al. [62]
Apolipoprotein E	Quantum dots. Fluorescent microarray (OD)	62 pg/mL	High sensitivity. Can be extended to several AD biomarkers	Morales-Narváez et al. [122]
Dopamine	Electrodes modified with CNTs and other features (ECD)	1 nM	High sensitivity and selectivity	Ali et al. [69]
	SWCNTs (ECD)	15 nM	Selective, analyte is detected in the presence of ascorbic acid and uric acid at physiological pH	Alwarappan et al. [70]
	Graphene-modified electrodes (ECD)	ND	Selective detection of dopamine in a large excess of ascorbic acid	Wang et al. [47]
	AuNPs (OD)	0.36 μM	Simple, sensitive and selective aptamer based colorimetric detection	Zheng et al. [73]
	MWCNT as enhancer of electron transfer (ECD)	0.01 mM	Response enhanced by using CNT	Alarcón-Angeles et al. [71]
Prion proteins	Gold-coated magnetic nanoparticles (OD)	N/A	Can provide a useful insight into the affinity of PrP to nanoparticle-functionalized aptamers for diagnosis applications	Kouassi et al. [76]
	Nanomechanical resonator arrays and nanoparticles (MSD)	20 pg/mL	Can be suitably applied to develop ante mortem tests to directly detect prion proteins in body fluids	Varshney et al. [77]
	Long range resonance energy transfer from QDs to the surface of AuNPs (OD)	33 aM	Ultra-sensitive detection. Can be successfully applied in biological media	Hu et al. [78]
	QDs as a highly selective probe (OD)	3 nM	Sensitive, rapid and simple detection	Zhang et al. [79]

Cardiovascular diseases

Matrix metalloproteinase 9	Peptide decorated MNP (MRI)	N/A	Good candidate for an enzyme reporter probe for in-vivo and for whole-body imaging of protease activity in humans	Schellenberger et al. [85]
Thrombin	Catalytic enlargement of aptamer-functionalized AuNPs (OD)	2 nM	Sensitive detection. One of the first nanobiosensors for thrombin detection	Pavlov et al. [89]
	FRET between graphene (acceptor) and organic dye (donor) (OD)	31 pM	High sensitivity in blood serum samples	Chang et al. [90]
	Aptamer-functionalized indium nanoparticles as thermal probes (TD)	22 nM	Sensitive, selective and low-cost method	Wang et al. [72]
	Label-free colorimetric detection using fribirnogen and AuNPs (OD)	0.04 pM	Highly sensitive, selective, rapid and simple method employed in blood plasma	Chen et al. [92]
Cardiac troponin T	Arrays of highly ordered silicon nanowire clusters (ED)	0.8 fM	Ultra-high sensitivity. Potential point-of-care application	Chua et al. [94]
	FRET between AuNP (acceptor) organic dye (donor) (OD)	0.02 nM	High sensitivity	Mayilo et al. [95]
Oxidized low density lipoprotein	Antibody decorated nanowires in field effect transistors (EC)	N/A	Selectivity	Rouhanizadeh et al. [97]
Myoglobin	Surface electrode modified with AuNPs (ECD)	0.56 nM	The whole procedure takes 30 min. Do not require plasma pretreatment. Small sample amount (1 μL).	Suprun et al. [99]

Cancer

Probes made of peptide nucleic acid used to detect a gene fusion recently associated with prostate cancer	Nanowires, electrochemical detection	100 fM	One of the first electrochemical sensors to directly detect specific mRNAs in unamplified patient samples	Fang and Kelley. [105]
Prostate specific antigen	Arrays of nanomechanical resonators and nanoparticles (MSD)	1.5 fM	Ultra-high sensitivity in real samples	Waggoner et al. [106]
	Silver enhancement of AuNPs (OD)	15 fM	Diagnostic at earlier timepoint comparing with other procedures	Storhoff et al. [107]

(continued)

Table 7.1 (continued)

Application/detection	Nanomaterials/platform	LOD	Characteristics	Reference
Cancer antigen 15-3	Nanochannels and antibody decorated AuNPs (ECD)	52 U/mL	Simple procedure applied to whole blood that avoids tedious and time consuming labors	De la Escosura-Muñiz and Merkoçi. [108]
	Antibody decorated AuNPs (OD)	15 U/mL	High sensitivity	Ambrosi et al. 2009a
Carcinoembryonic antigen	Nanogold-enwrapped graphene nanocomposites (ECD)	10 pg/mL	Sensitive detection through novel nanocomposites that have a promising potential in clinical diagnosis	Zhong et al. [110]
	Quantum dots capped with silica onto gold substrates (ECD)	50 pg/mL	Satisfactory stability, acceptable veracity, proper for dual-tumor markers detection in clinic serum samples	Qian et al. [112]
Carcinoemrbionic antigen	Antibody decorated QDs and capture antibody capped agarose beads in microfluidic chip (OD)	20 pg/mL	Multiplexing capabilities. High sensitivity in saliva and serum samples	Jokerst et al. [113]
Cancer antigen 125		N/A		
Human epidermal growth factor receptor 2		0.27 ng/mL		
Human epidermal growth factor receptor 2	Antibody decorated graphene encapsulated nanoparticles are patterned as gate of a field effect transistor (ED)	1 pM	High specificity and sensitivity. Suitable for clinical applications	Myung et al. [111]
Epidermal growth factor receptor		100 pM		
Tumor cells	Antibody decorated AuNPs (ECD)	4×10^3 cells/ 0.7 mL	Rapid and simple assay	De la Escosura-Muñiz et al. [116]
Leukemia infected cells	Aptamer decorated near infrared fluorescence nanoparticles (OD)	4×10^3 cells/mL	Quick method, without the need of the complicated separation steps in whole blood samples	Deng et al. [117]
Tumor imaging	Antibody decorated MNP as contrast agent (MRI)	N/A	Could be considered for further research as an MRI contrast agent for the detection of tumors in human	Oghabian et al. [117], Rasaneh et al. [120]

LOD limit of detection, *AuNPs* gold nanoparticles, *QDs* quantum dots, *OD* optical detection, *CNT* carbon nanotube, *MWCNT* multi-walled carbon nanotube, *ECD* Electrochemical detection, *ED* electrical detection, *MSD* mass sensitive detection, *TD* thermal detection, *FRET* fluorescence resonance energy transfer, *MRI* magnetic resonance imaging, *N/A* not available

7.5.1 Neurodegenerative Diseases

Dementia is a meaningful health problem in developed countries with over 25 million people affected worldwide and probably over 75 million people at risk during the next 20 years [54]. Alzheimer's Disease (AD) is the most frequent cause of dementia and results in a progressive loss of cognitive function affecting one in eight people by the time they reach 65 years of age [55, 123]. Diverse sources of evidence suggest that amyloid-β (Aβ) have a causal role in its pathogenesis [56]. Therefore, Aβ is a potential AD biomarker. An overview on recent nanobiosensing approaches for Aβ detection is presented below.

Khan et al. [57] have employed the "bio-bar-code assay", previously developed by Mirkin and co-workers [58], to detect Aβ at femto molar level (10 fM by using fluorescent detection) and atomolar level (10 aM, by using scanometric detection). Plasma samples from control and Alzheimer's patients were assayed. This technic is based on oligonucleotide-modified gold nanoparticles. A complex: magnetic bead – capture antibody – Aβ – detection antibody – AuNP – DNA strands is performed. And finally, each analyte binding is reported by the presence of thousands of DNA strands; i.e. the more complexes are created, the more DNA released (see Fig. 7.3a). Despite the results were hard to reproduce, is a hopeful beginning to a clinical diagnostic tool. Haes and colleagues [59] have reported an optical nanobiosensing platform based on localized surface plasmon resonance spectroscopy, so as to monitor the interaction Aβ/specific antibodies. Clinical samples, extracted from cerebrospinal fluid, were assayed in this sensor based on nanopatterned surfaces achieving a picomolar sensitivity (7.3 pM) (see Fig. 7.3b). This technology provides new information relevant to the understanding and possible diagnosis of AD. Lee et al. [60] have developed a surface plasmon resonance based biosensor [61] for Aβ detection. The procedure enhances the surface plasmon resonance signal by using antibody decorated AuNP. This approach is focused on ultrasensitive detection towards early detection achieving a detection limit of 1 fg/mL. Han et al. [62] have proposed a screen for Aβ aggregation inhibitor by using Aβ-conjugated AuNPs. Aβ aggregation of AD patients serum was visualized through Aβ aggregation-induced AuNP precipitation. This approach is a potential diagnostic tool for diseases involving abnormal protein aggregation as their key pathogenesis processes. The authors of this chapter have recently studied a microarray for another potential Alzheimer Disease's biomarker screening; i.e., Apolipoprotein E. This microarray is reported by QDs nanocrystals, exploiting their advantageous photonics properties, a limit of detection up to 62 pg mL^{-1} can be achieved [63].

Parkinson's disease (PD) is a common chronic neurodegenerative disorder. The classical clinical features are progressive tremor, rigidity and bradykinesia [64] Dopamine (DA) is a neurotransmitter with a variety of functions in the central nervous system. It affects the brain's control of learning, feeding and neurocognition. Disorders in DA levels have been associated with Parkinson's disease, among others psychiatric disorders such as schizophrenia, and depression [65–66]. Thus, DA has an active research as biomarker and is a promising analyte toward molecular diagnosis of Parkinson's disease.

Since DA is electrochemically active, electrochemical detection of DA by oxidative methods is a preferred procedure. Nevertheless, DA is always subsisting with ascorbic acid in real samples and the products of DA oxidation can react with ascorbic and regenerate dopamine again impacting the accuracy of detection gravely. Different approaches have been reported to avoid this shortcoming. Ali et al. [67] demonstrated that DA can be electrochemically detected by altering the electrode surface with a thin layer of an in situ polymerized poly(anilineboronic acid)/CNT composite and a thin layer of the highly permselective Nafion film. The detection limit is of ~1 nM. Alwarappan and co-workers [68] explored the performance of electrochemically pretreated SWCNTs for the electrochemical detection of DA in the presence of ascorbic acid and uric acid at physiological pH. This approach showed a successful selective response with a detection limit of about 15 nM. A MWCNT as enhancer of electron transfer combined with β-Cyclodextrin (β-CD) as molecular receptor is also reported as a DA electrochemical sensor system. The proposed molecular host–guest recognition based sensor shows a electrochemical sensitivity for amperometric detection of DA over the range 0.01–0.08 mM [69].

Wang et al. [70] proposed a graphene-modified electrode that was applied in the selective detection of DA with a linear range from 5 to 200 µM in a large excess of ascorbic acid. Zheng and co-workers [73] explored the interaction of DA-binding aptamer and DA by using unmodified citrate-coated AuNPs as colorimetric signal readout, where AuNP aggregation is induced in the presence of the analyte and the stability of the solution is preserved in the absence of the analyte (see Fig. 7.3d). A selective nanobiosensor was achieved with a detection limit of 0.36 µM. The sensitivity and selectivity of these approaches enable their potential use in diagnosis of PD.

Creutzfeldt-Jakob disease (CJD) is a neurodegenerative disease characterized by rapidly progressive dementia, myoclonus, ataxia and visual disturbances, extrapyramidal and pyramidal involvement, as well as a kinetic mutism [74]. Prion proteins (PrP) once transformed from their normal cellular counterparts (PrP^c) into infectious form (PrP^{res}) are transmissible infectious particles, destitute of nucleic acid, that are believed to be responsible for causing the fatal CJD in humans [75]. Since disorders in PrP levels are expected to be present in samples derived from CJD patients, there is an enormous interest in PrP screening technologies; some nanobiosening approaches are presented below.

Kouassi et al. [76] have developed an assay for PrP assessment based on aptamer-mediated magnetic and gold-coated magnetic nanoparticles. Analyte detection was reported by using Fourier transform infrared spectroscopy. The proposed assay can provide a useful insight into the affinity of PrP to nanoparticle-functionalized aptamers for diagnosis applications. Varshney and co-workers [77] have reported PrP detection in serum by exploiting micromechanical resonator arrays. Secondary antibodies and nanoparticles were used as mass amplifiers to detect the presence of small amounts of PrP onto mechanical resonators. This device showed a limit of detection of about 20 pg/mL and the authors are currently working so as to enhance the detection limit. Hu et al. [78], have explored an ultra-sensitive detection strategy for PrP based on the long range resonance energy transfer from QDs to the surface of AuNPs. Energy transfer from QDs to the surface of AuNPs occurs with high efficiency and the fluorescent signal of QDs was quenched as a consequence of the molecular recognition between PrP (bound with high specificity to QDs) and an aptamer specific for the PrP (conjugated to AuNPs) (see Fig. 7.3c). This procedure achieves a very low detection limit of 33 aM and might be successfully applied in biological media. Zhang and co-workers [79] have reported the use of QDs as a highly selective probe for PrP detection. When the suitable treated QDs were mixed with PrP, the concentration of QDs in supernatant decreased due to the precipitation resulting on a reduced fluorescence intensity of the supernatant. This phenomenon was used for quantitative detection with a detection limit of 3 nM. The reported method shows a sensitive, rapid and simple performance.

7.5.2   Cardiovascular Diseases

Cardiovascular diseases is the cause of nearly half of all deaths in the Western world [80] and is also a major cause of death, morbidity, and disability in Asia and Africa [81, 129–131]. Cardiovascular diseases include hypertension with or without renal disease, stroke, atherosclerosis, other diseases of arteries, arterioles, and capillaries, and diseases of veins and lymphatics. In addition, there are different forms of heart disease such as rheumatic fever/rheumatic heart disease, hypertensive heart disease, heart and renal disease, ischemic heart disease, diseases of pulmonary circulation and so on [82]

Matrix metalloproteinase 9 (MMP-9), also known as gelatinase B, is an enzyme important in inflammation, atherosclerosis and tumor progression processes; furthermore MMP-9 is a potential biomarker involved in cardiovascular diseases [83, 84]. Schellenberger and colleagues [85] have developed a technology for MMP-9 screening. They have reported a nanobiosensor based on peptide decorated MNPs that is proper for in vivo imaging of MMP-9 activity by magnetic resonance imaging. This system has a great potential as reporter probes for assessing enzyme activity of proteases by in vivo magnetic resonance imaging and can help towards diagnosis and monitoring applications [86].

Thrombin (also known as factor IIa) is the last enzyme protease involved in the coagulation cascade and it converts fibrinogen to insoluble fibrin, causing blood clotting [87]. Therefore, thrombin plays a central role in cardiovascular diseases [88]. Thrombin detection is under active research; for example, currently, an inquiry on the Web of Knowledge displays more than 50 approaches based on nanotechnology related with thrombin detection. About 30 of them exploit the use of AuNPs and aptamers are the most common biorecognition probes in these sensors.

As far as is known, one of the first nanobiosensors for thrombin detection was reported by Pavlov and colleagues [89]. They developed an amplified optical detection of thrombin onto solid phase by the catalytic enlargement of thrombin aptamer-functionalized AuNPs. This system exhibited a detection limit of ca. 2 nM. Chang and co-workers [90], have proposed a biosensor based on fluorescence (or Förster) resonance energy transfer (FRET) [91] between graphene sheets as acceptors and fluorescent dyes as donors. Fluorescence of dye labeled aptamer is quenched when aptamer attach to graphene due to energy transference between dyes and graphene.

Dyes emission is recovered by biorecognition events, i.e. when thrombin binds with aptamers to perform quadruplex-thrombin complexes (FRET is not present). Fluorescence emission is proportional to thrombin concentration (see Fig. 7.4a). The procedure was applied to blood serum samples achieving a high sensitivity with a detection limit of ca. 31 pM. Wang et al. [47] have reported an innovative thermal biosensor for thrombin detection in serum samples by using aptamer-functionalized indium nanoparticles as thermal probes onto silicon nanopillars (see Fig. 7.4b). This device is characterized by a sensitive (detection limit of ca. 22 nM), selective and low-cost method. Chen and co-workers [92], have published a label-free colorimetric thrombin detection by using fibrinogen decorated AuNPs. Mixing of thrombin into the fibrinogen decorated AuNPs solutions in the attendance of excess fibrinogen yields the agglutination of the capped AuNPs. The absorbance of the supernatant of the assayed solution is inversely proportional to the thrombin concentration. This procedure is highly sensitive, selective, rapid and simple and can be suitably employed in blood plasma with a detection limit of 0.04 pM.

Human cardiac troponin-T (cTnT) is a key protein biomarker related specifically to myocardial damage [93]. Since cTnT prevails in elevated concentrations in the bloodstream of a patient suffering from heart attack, this biomarker can be employed as a helpful linker between a patient suffering from unstable angina or a more dangerous case of myocardial infarction [94]. Mayilo and colleagues [95] have developed a biosensing platform based on FRET effect with AuNP as acceptors of organic dyes donors. The fluorescence of the antibody fragment – organic dye conjugated is quenched by performing an immunocomplex with an antibody decorated AuNP. The fluorescence quenching is proportional to the detected cTnT (see Fig. 7.4c). This platform accomplished a detection limit of 0.02 nM in serum. Chua and co-workers [94] have designed a label-free electrical detection of cTnT by using complementary metal-oxide semiconductor-compatible silicon nanowire sensor arrays. Nanowires are decorated with specific antibodies against cTnT and biorecognition events are reported by electrical signals. Setup is displayed in Fig. 7.4d. This system achieved a limit of detection as low as ca. 0.85 fM fg/mL in undiluted serum and it possesses portability characteristics for point-of-care application.

Oxidized low density lipoprotein (oxLDL) is recognized as a biomarker for acute myocardial infarction in patients with coronary artery disease. Specifically, oxLDL at elevated levels can forecast acute heart attack or coronary syndromes [96, 128]. Rouhanizadeh and colleagues [97] have reported a biosensor based on electrical detection by using antibody decorated nanowires in field effect transistors. This biosensor enables to differentiate the LDL cholesterol between the reduced (native LDL) and the oxidized state (oxLDL). Acute myocardial infarction can also be diagnosed preventively by measuring myoglobin protein levels [98]. Suprun et al. [99] have designed a electrochemical nanobiosensor for cardiac myoglobin screening based on direct electron transfer between Fe(III)-heme and electrode surface modified with antibody decorated AuNPs. Notably, 1 μL of undiluted plasma of healthy donors and patients with acute myocardial infarction was analyzed with a limit of detection as low as 0.56 nM. Since the procedure takes 30 min, it can be used for rapid diagnosis.

7.5.3 Cancer

Cancer is the predominant cause of death in economically developed countries and the second major cause of death in developing countries [100]. Cancer is a pathology mainly characterized by the chaotic growth of cells with an altered cell cycle control. Since the name of a specific cancer depends upon the tissue or body cells in which it originated, there are many different types of cancers and the most frequent are: breast, lung, prostate, skin, cervical, colon and ovarian cancer among others.

Cancer results in complex molecular alterations. These alterations can be unveiled by using technologies that assess changes in the content or sequence of DNA, its transcription into messenger RNA or microRNA, the production of proteins or the synthesis of several metabolic products [101]. Nevertheless, validation of accurate cancer biomarkers has been slow and is under active research [101, 102]. Table 7.2 shows a list of some potential cancer biomarkers. More details about cancer markers can be founded in the literature [103, 104,127, 129,134, 135].

Biosensors for cancer detection are highly attractive and they are under vigorous development. An overview of some recent nanobiosensing approaches is provided below (Table 7.2).

Fang et al. (2009) [105] have developed an electrochemical sensor to directly detect specific mRNAs in unamplified patient samples (without PCR amplification). The sensor relies on peptide nucleic acid decorated nanowires. Probes made of peptide nucleic acid have been used to detect a gene fusion recently associated with prostate cancer (see Fig. 7.5a). This system exhibits a sensitivity of 100 fM. Waggoner and colleagues [106] have designed a nanobiosensing platform based on arrays of nanomechanical resonators for PSA detection. The surfaces of the proposed trampoline-like devices (see Fig. 7.5b) are capped with specific antibodies and the mass of bound analyte is detected as a reduction in the resonant frequency. Antibody decorated nanoparticles are used in order to enhance sensitivity. Real samples were assayed with a detection limit of 1.5 fM. Storhoff et al. [107] have applied a procedure for the detection of prostate cancer recurrence by using AuNP in monitoring PSA levels of clinical samples. This strategy employs functionalized AuNPs as a probe of PSA captured onto antibody capped glass slides. PSA amount was quantified through silver enhancement of AuNPs [108]. Since this method show a detection limit of 15 fM, they achieved a diagnostic of prostate cancer recurrence in

Table 7.2 Potential cancer biomarkers and their applications

Biomarker	Abbreviation	Type of cancer	Application
Human epidermal growth factor receptor 2	HER2	Breast	Predictive
Epidermal growth factor receptor	EGFR	Breast	Predictive
Cancer antigen 15-3	CA15-3	Breast	Monitoring
Carcionoembryonic antigen	CEA	Colon	Monitoring
Cancer antigen 125	CA125	Ovarian	Monitoring
Prostate specific antigen	PSA	Prostate	Monitoring

clinical samples at earlier timepoint comparing with other procedures. In this regard, the device designed by Waggoner et al., probably might exhibit a similar clinical performance.

De la Escosura-Muñiz and Merkoçi [109] have designed a nanochannel/nanoparticle-based filtering and sensing platform for direct electrochemical detection of CA15-3 in blood. A membrane with nanochannels capped with specific antibodies enable both the filtering of the whole blood assayed without previous treatment and the specific detection of the target analyte. Captured analytes are reported by adding antibody decorated AuNPs as electrochemical reporters (see Fig. 7.5c). This process avoids tedious and time consuming labors and has shown a limit of detection of 52 U/mL. The same group have developed an optical ELISA for the analysis of the same antigen useful for the follow-up of the medical treatment of breast cancer. AuNPs were used as carriers of the signaling antibody anti-CA15-3 – HRP (horseradish peroxidase) in order to achieve an amplification of the optical signal. The developed assay resulted in higher sensitivity and shorter assay time when compared to classical ELISA procedures while working between 0 and 60 U CA15-3/mL (Ambrosi et al. 2009a).

Zhong et al. [110] have reported the detection of CEA through nanogold-enwrapped graphene nanocomposites as enhancer probes of electrochemical immunodetection. An immunocomplex capture antibody – CEA is reported by antibody decorated nanocomposites yielding an amplified signal with detection limit as low as 10 pg/mL. Myung et al. [111] have performed a graphene-encapsulated nanoparticle-based biosensor for the selective detection of HER2 and EGFR. Antibody decorated graphene encapsulated nanoparticles are patterned as gate of a field effect transistor where biorecognition events are detected electrically. This biosensor has shown high specificity and sensitivity (1 pM for HER2 and 100 pM for EGFR). These novel procedures have a promising potential in clinical diagnosis.

Qian and colleagues [112] have proposed a simultaneous detection of CEA and IgG by using QDs coated silica. Two different types of QDs (CdSe and PbS) are capped with two different types of detection antibodies respectively. Capture antibodies against the two target analytes are deposited onto gold substrates so as to perform immunocomplexes. After assay steps, since the selected QDs have a different electrochemical response, the target analytes are measured by voltammetry with a detection limit of 50 pg/mL. Jokerst and colleagues [113] have developed a microfluidic biosensor [114] for the multiplexed screening of CEA, CA125 and HER2 based on detection antibody decorated QDs and capture antibody capped agarose beads. QDs are used as fluorescent probes of specific biorecognition events performed onto an array of localized agarose beads (see Fig. 7.5d). Notably, ELISA method and the employment of organic dyes as fluorescent probes in the same platform were compared with this approach. The best limit of detection was achieved by taking advantage of the powerful optical properties of the QDs. For example, real samples of saliva and serum were assayed with a detection limit of 0.02 ng/mL CEA and 0.27 ng/mL HER2. These systems could be applied to multiple tumor markers screening in clinic samples.

Since cancer cells can quickly infect their surrounding cells and the disease can spread subsequently, the detection of a tiny amount of infected cells is vital towards early diagnosis [115]. De la Escosura-Muñiz et al. [116], have designed a rapid and simple tumor cell detection device based on the specific binding between cell surface proteins and antibody decorated AuNPs. AuNPs are employed as electrochemical probes (based on the enhancement of hydrogen catalysis) achieving a detection limit of ca. 4×10^3 cells/0.7 mL. Deng and colleagues [117] have developed a system based on aptamer decorated near infrared fluorescence methylene blue nanoparticles for leukemia infected cells detection (see Fig. 7.5e). Cancer cells have been detected quickly without the need of the complicated separation steps in whole blood samples by using the fluorescence properties of the nanoparticles. They have been able to detect from 4×10^3 to 7×10^4 cells/mL in a linear range. Gold nanoparticles have also been used, through a simple electrochemical approach, for cancer cell monitoring by Maltez-da Costa et al. [118]. This platform has achieved a limit of detection around 8.3×10^3 cells/mL. Oghabian and co-workers [119] and Rasaneh et al. [120] have proposed strategies based on antibody decorated MNP as contrast agent for tumor screening by magnetic resonance imaging. Antibodies against HER2 were used for the specific detection of tumor mice cells. These strategies could be considered for further research as an MRI contrast agent for the detection of tumors in human.

7.6   Conclusions and Future Perspectives

We have described the basic principles of the nanobiosenors and we have discussed several nanobiosensing approaches for biomarker detection towards medical applications in different disease categories. Nanobiosensors offer powerful capabilities to diagnostic technology. They can enable to reduce cost, sample amount and assay time. These novel sensors can exhibit high selectivity and unprecedented sensitivity. Despite these advantages, nanobiosensors possess some potential weaknesses; for example, some of the nanomaterials production technologies are still expensive and the inherent toxicity of these materials overall while being applied for in-vivo analysis is little known yet. The exposed nanobiosensors are successfully applied as research devices and they are far from being applied in the public domain; nevertheless, close consensus with regulatory agencies (such as the European Medicines Agency or the U.S. Food and Drugs Administration) to develop comprehensive standards for nanobiosensors and procedures will ensure the operative and realistic transition of nanobiosensors to common medical devices.

Acknowledgements We acknowledge funding from the fellowship program grant given by CONACYT (Mexico) to Eden Morales-Narváez. MCINN (Madrid) through project MAT2011-25870 and E.U. through FP7 "NADINE" project (contract number 246513) have sponsored this work.

References

1. O'Farrel B (2009) Evolution in lateral flow–based immunoassay systems. In: Wong RC, Tse HY (eds) Lateral flow immunoassay. Humana Press, New York, pp 1–33
2. Turner APF (2013) Biosensors: then and now. Trends Biotechnol 31:119–120
3. Mascini M, Palchetti I (2014) Biosensors, electrochemical. Encycl Appl Electrochem 136–140
4. Schallmey M, Frunzke J, Eggeling L et al (2014) Looking for the pick of the bunch: high-throughput screening of producing microorganisms with biosensors. Curr Opin Biotechnol 26:148–154
5. Gdowski A, Ranjan A, Mukerjee A et al (2014) Nanobiosensors: role in cancer detection and diagnosis. Infect Dis Nanomed I 807:33–58, Springer India
6. Sagadevan S, Periasamy M (2014) Recent trends in nanobiosensors and their applications-a review. Rev Adv Mater Sci 36:62–69
7. Song S, Wang L, Li J et al (2008) Aptamer-based biosensors. TrAC Trend Anal Chem 27:108–117
8. Mairal T, Ozalp VC, Lozano Sánchez P et al (2008) Aptamers: molecular tools for analytical applications. Anal Bioanal Chem 390:989–1007
9. Ruigrok VJB, Levisson M, Eppink MHM et al (2011) Alternative affinity tools: more attractive than antibodies? Biochem J 436:1–13
10. Li H, Liu S, Dai Z, Bao J, Yang X (2009) Applications of nanomaterials in electrochemical enzyme biosensors. Sensors 9:8547–8561
11. Leca-Bouvier BD, Blum LJ (2010) Enzyme for biosensing applications. In: Zourob M (ed) Recognition receptors in biosensors. Springer, New York, pp 177–220
12. Suman AK (2008) Recent advances in DNA biosensor. Sens Transducers J 92:122–133
13. Kubik T, Bogunia-Kubik K, Sugisaka M (2005) Nanotechnology on duty in medical applications. Curr Pharm Biotechnol 6:17–33
14. Gruner G (2006) Carbon nanotube transistors for biosensing applications. Anal Bioanal Chem 384:322–335
15. Lee J-H, Oh B-K, Choi B (2010) Electrical detection-based analytic biodevice technology. BioChip J 4:1–8
16. Vo-Dinh T, Cullum B (2000) Biosensors and biochips: advances in biological and medical diagnostics. Fresenius J Anal Chem 366:540–551
17. Fritz J (2008) Cantilever biosensors. Analyst 133:855–863
18. Ramanathan K, Danielsson B (2011) Principles and applications of thermal biosensors. Biosens Bioelectron 16:417–423
19. Hu X, Dong S (2008) Metal nanomaterials and carbon nanotubes – synthesis, functionalization and potential applications towards electrochemistry. J Mater Chem 18:1279–1295
20. Martin AL, Li B, Gillies ER (2009) Surface functionalization of nanomaterials with dendritic groups: toward enhanced binding to biological targets. J Am Chem Soc 131:734–741
21. Prencipe G, Tabakman SM, Welsher K (2009) PEG branched polymer for functionalization of nanomaterials with ultralong blood circulation. J Am Chem Soc 131:4783–4787
22. Liu Z, Kiessling F, Gaetjens J (2010) Advanced nanomaterials in multimodal imaging: design, functionalization, and biomedical applications. J Nanomater 894303

23. Mehdi A, Reye C, Corriu R (2011) From molecular chemistry to hybrid nanomaterials. Design and functionalization. Chem Soc Rev 40:563–574

24. Alivisatos AP, Gu W, Larabell C (2005) Quantum dots as cellular probes. Annu Rev Biomed Eng 7:55–76

25. Jiang W, Singhal A, Fischer H (2006) Engineering biocompatible quantum dots for ultrasensitive, real-time biological imaging and detection. In: Ferrari M, Desai T, Bhatia S (eds) BioMEMS and biomedical nanotechnology. Springer, New York City, pp 137–156

26. Bruchez M, Morronne M, Gin P et al (1998) Semiconductor nanocrystals as fluorescent biological labels. Science 281:2013–2016

27. Chan W, Nie S (1998) Quantum dot bioconjugates for ultrasensitive nonisotopic detection. Science 281:2016–2018

28. Algar WR, Tavares AJ, Krull UJ (2010) Beyond labels: a review of the application of quantum dots as integrated components of assays, bioprobes, and biosensors utilizing optical transduction. Anal Chim Acta 673:1–25

29. Rosenthal SJ, Chang JC, Kovtun O et al (2011) Biocompatible quantum dots for biological applications. Chem Biol 18:10–24

30. Liu S, Leech D, Ju H (2003) Application of colloidal gold in protein immobilization. Anal Lett 36:1–19

31. Katz E, Willner I, Wang J (2004) Electroanalytical and bioelectroanalytical systems based on metal and semiconductor nanoparticles. Electroanalysis 16:19–44

32. Yáñez-Sedeño P, Pingarrón JM (2005) Gold nanoparticle-based electrochemical biosensors. Anal Bioanal Chem 382:884–886

33. Pankhurst Q (2006) Nanomagnetic medical sensors and treatment methodologies. BT Technol J 24:33–38

34. Roca AG, Costo R, Rebolledo AF et al (2009) Progress in the preparation of magnetic nanoparticles for applications in biomedicine. J Phys D Appl Phys 42:224002

35. Sandhu A, Handa H, Abe M (2010) Synthesis and applications of magnetic nanoparticles for biorecognition and point of care medical diagnostics. Nanotechnology 21:442001

36. Wang J (2005) Carbon-nanotube based electrochemical biosensors: a review. Electroanalysis 17:7–14

37. Balasubramanian K, Burghard M (2006) Biosensors based on carbon nanotubes. Anal Bioanal Chem 385:452–468

38. Davis JJ, Coleman KS, Azamian BR (2003) Chemical and biochemical sensing with modified single walled carbon nanotubes. Chemistry 9:3732–3739

39. Zhao Q, Gan Z, Zhuang Q (2002) Electrochemical sensors based on carbon nanotubes. Electroanalysis 14:1609–1613

40. Compton RG, Wildgoose GG, Wong ELS (2009) In carbon nanotube–based sensors and biosensors. In: Merkoçi AE (ed) Biosensing using nanomaterials. Wiley, Hoboken, NJ, USA, pp 1–38

41. Huo Z, Tsung C-kuang, Huang W et al (2008) Sub-two nanometer single crystal Au. Nano Lett 8:2041–2044

42. Wang J, Liu C, Lin Y (2007) Nanotubes, nanowires, and nanocantilevers in biosensor development. In: Kumar CSSR (ed) Nanomaterials for biosensors. WILEY, Hoboken, NJ, USA, pp 56–100

43. Cui Y, Wei Q, Park H, Lieber CM (2001) Nanowire nanosensors for highly sensitive and selective detection of biological and chemical species. Science 293:1289–1292

44. Patolsky F, Zheng G, Lieber CM et al (2006) Nanowire sensors for medicine and the life sciences. Nanomedicine 1:51–65

45. Hangarter CM, Bangar M, Mulchandani A, Myung NV (2010) Conducting polymer nanowires for chemiresistive and FET-based bio/chemical sensors. J Mater Chem 20:3131–3140

46. Wang J, Manesh KM (2010) Motion control at the nanoscale. Small 6:338–345

47. Wang C, Hossain M, Ma L et al (2010) Highly sensitive thermal detection of thrombin using aptamer-functionalized phase change, nanoparticles. Biosens Bioelectron 26:437–443

48. Shao Y, Wang J, Wu H (2010) Graphene based electrochemical sensors and biosensors: a review. Electroanalysis 22:1027–1036

49. Pumera M (2011) Graphene in biosensing. Mater Today 14:308–315
50. Chambers G, Lawrie L, Cash P, Murray G (2000) Proteomics: a new approach to the study of disease. J Pathology 192:280–288
51. Mattson MP (2004) Pathways towards and away from Alzheimer's disease. Nature 430:631–639
52. Bild AH, Yao G, Chang JT et al (2006) Oncogenic pathway signatures in human cancers as a guide to targeted therapies. Nature 439:353–357
53. Alizadeh AA, Eisen MB, Davis RE et al (2000) Distinct types of diffuse large B-cell lymphoma identified by gene expression profiling. Nature 403:503–511
54. Takeda M, Martínez R, Kudo T (2010) Apolipoprotein E and central nervous system disorders: reviews of clinical findings. Psychiatry Clin Neurosci 64:592–607
55. Ray S, Britschgi M, Herbert C (2007) Classification and prediction of clinical Alzheimer's diagnosis based on plasma signaling proteins. Nat Med 13:1359–1362
56. Palop JJ, Mucke L (2010) Amyloid-[beta]-induced neuronal dysfunction in Alzheimer's disease: from synapses toward neural networks. Nat Neurosci 13:812–818
57. Khan S, Klein W, Mirkin C et al (2005) Fluorescent and scanometric ultrasensitive detection technologies with the bio-bar code assay for Alzheimer's disease diagnosis. Nanoscape 2:7–15
58. Nam J-M, Park S-J, Mirkin CA (2002) Bio-barcodes based on oligonucleotide-modified nanoparticles. J Am Chem Soc 124:3820–3821
59. Haes AJ, Chang L, Klein WL, Duyne RP (2005) Detection of a biomarker for Alzheimer's disease from synthetic and clinical samples using a nanoscale optical biosensor. J Am Chem Soc 127:2264–2271
60. Lee J, Kang D, Lee T (2009) Signal enhancement of surface plasmon resonance based immunosensor using gold nanoparticle-antibody complex for beta-amyloid (1–40) detection. J Nanosci Nanotechnol 9:7155–7160
61. Piliarik M, Vaisocherová H, Homola J (2009) Surface plasmon resonance biosensing. In: Rasooly A, Herold KE (eds) Biosensors and biodetection. Humana Press, New York, pp 65–88
62. Han S-H, Chang YJ, Jung ES et al (2011) Effective screen for amyloid β aggregation inhibitor using amyloid β-conjugated gold nanoparticles. Int J Nanomedicine 6:1–12
63. Morales-Narv.ez E, Mont.n H, Fomicheva A, Merko.i A (2012) Signal enhancement in antibody microarrays using quantum dots nanocrystals: application to potential Alzheimer's disease biomarker screening. Anal Chem 84:6821–6827
64. Michell AW, Lewis SJG, Foltynie T, Barker R (2004) Biomarkers and Parkinson's disease. Brain 127:1693–1705
65. Volkow ND, Fowler JS, Wang GJ, Swanson JM (2004) Dopamine in drug abuse and addiction: results from imaging studies and treatment implications. Mol Psychiatry 9:557–569
66. Naranjo CA, Tremblay LK, Busto UE (2011) The role of the brain reward system in depression. Prog Neuropsychopharmacol Biol Psychiatry 25:781–823
67. Lang AE, Lozano AM (1998) Parkinson's disease. N Engl J Med 339:1044
68. Kapur S, Mamo D (2003) Half a century of antipsychotics and still a central role for dopamine D2 receptors. Prog Neuropsychopharmacol Biol Psychiatry 27:1081–1090
69. Ali SR, Ma Y, Parajuli RR et al (2007) A nonoxidative sensor based on a self-doped polyaniline/carbon nanotube composite for sensitive and selective detection of the neurotransmitter dopamine. Anal Chem 79:2583–2587
70. Alwarappan S, Liu GD, Li CZ (2010) Simultaneous detection of dopamine, ascorbic acid, and uric acid at electrochemically pretreated carbon nanotube biosensors. Nanomedicine 6:52–57
71. Alarcón-Angeles G, Pérez-López B, Palomar-Pardave M et al (2008) Enhanced host–guest electrochemical recognition of dopamine using cyclodextrin in the presence of carbon nanotubes. Carbon 46:898–906
72. Wang Y, Li Y, Tang L (2009) Application of graphene-modified electrode for selective detection of dopamine. Electrochem Commun 11:889–892

73. Zheng Y, Wang Y, Yang X (2011) Aptamer-based colorimetric biosensing of dopamine using unmodified gold nanoparticles. Sens Actuators B 156:95–99
74. Maltête D, Guyant-Maréchal L, Mihout B, Hannequin D (2006) Movement disorders and Creutzfeldt-Jakob disease: a review. Parkinsonism Relat Disord 12:65–71
75. Triantaphyllidou IE, Sklaviadis T, Vynios DH (2006) Detection, quantification, and glyco-typing of prion protein in specifically activated enzyme-linked immunosorbent assay plates. Anal Biochem 359:176–182
76. Kouassi GK, Wan P, Sreevatan S, Irudayaraj J (2007) Aptamer-mediated magnetic and gold-coated magnetic nanoparticles as detection assay for prion protein assessment. Biotechnol Prog 23:1239–1244
77. Varshney M, Waggoner PS, Montagna RA, Craighead HG (2009) Prion protein detection in serum using micromechanical resonator arrays. Talanta 80:593–599
78. Hu PP, Chen LQ, Liu C et al (2010) Ultra-sensitive detection of prion protein with a long range resonance energy transfer strategy. Chem Commun 46:8285–8287
79. Zhang L-Y, Zheng H-Z, Long Y-J et al (2011) CdTe quantum dots as a highly selective probe for prion protein detection: colorimetric qualitative, semi-quantitative and quantitative detection. Talanta 83:1716–1720
80. Allender S, Scarborough P, Peto V, Rayner M, Leal J, Luengo-Fernandez R et al (2008) European cardiovascular disease statistics, 2008 edition, 3rd edn. European Heart Network, Brussels
81. Chang K, Chiu J (2005) Clinical applications of nanotechnology in atherosclerotic diseases. Curr Nanosci 1:107–115
82. Nature Biotechnology (2000) Prognostic/diagnostic testing and a raft of new drug targets from genomics promise to transform cardiovascular medicine. Nat Biotechnol 18:IT15–IT17
83. Blankenberg S, Rupprecht HJ, Poirier O et al (2003) Plasma concentrations and genetic variation of matrixmetalloproteinase 9 and prognosis of patients with cardiovascular disease. Circulation 107:1579–1585
84. Martín-ventura JL, Blanco-colio LM, Tuñón J et al (2009) Biomarkers in cardiovascular medicine. Revista Española de Cardiología 62:677–688
85. Schellenberger E, Rudloff F, Warmuth C (2008) Protease-specific nanosensors for magnetic resonance imaging. Bioconjug Chem 19:2440–2445
86. McCarthy JR (2011) Nanomedicine and cardiovascular disease. Curr Cardiovasc Imaging Rep 3:42–49
87. Holland CA, Henry AT, Whinna HC, Church FC (2010) Effect of oligodeoxynucleotide thrombin aptamer on thrombin inhibition by heparin cofactor II and antithrombin. FEBS Lett 484:87–91
88. Tracy RP (2003) Thrombin inflammation, and cardiovascular disease: an epidemiologic perspective cardiovascular disease. Chest 124:49s–57s
89. Pavlov V, Xiao Y, Shlyahovsky B et al (2004) Aptamer-functionalized Au nanoparticles for the amplified optical detection of thrombin. J Am Chem Soc 126:11768–11769
90. Chang H, Tang L, Wang Y et al (2010) Graphene fluorescence resonance energy transfer aptasensor for the thrombin detection. Anal Chem 82:2341–2346
91. Schäferling M, Nagl S (2011) Förster resonance energy transfer for quantification of protein–protein interactions on microarrays. In: Wu CJ (ed) Protein microarray for disease analysis: methods and protocols. Springer, London, pp 303–320
92. Chen C-K, Huang C-C, Chang H-T (2010) Label-free colorimetric detection of picomolar thrombin in blood plasma using a gold nanoparticle-based assay. Biosens Bioelectron 25:1922–1927
93. Apple FS (1999) Tissue specificity of cardiac troponin I, cardiac troponin T and creatine kinase-MB. Clin Chim Acta 284:151–159
94. Chua JH, Chee R-E, Agarwal A et al (2009) Label-free electrical detection of cardiac bio-marker with complementary metal-oxide semiconductor-compatible silicon nanowire sensor arrays. Anal Chem 81:6266–6271
95. Mayilo S, Kloster MA, Wunderlich M et al (2009) Long-range fluorescence quenching by gold nanoparticles in a sandwich immunoassay for cardiac troponin T. Nano Lett 9:4558–4563

96. Austin M (1994) Small, dense low-density lipoprotein as a risk factor for coronary heart disease. Int J Clin Lab Res 24:187–192
97. Rouhanizadeh M, Tang T, Li C (2006) Differentiation of oxidized low density lipoproteins by nanosensors. Sens Actuators B 114:788–798
98. McDonnell B, Hearty S, Leonard P, O'Kennedy R (2009) Cardiac biomarkers and the case for point-of-care testing. Clin Biochem 42:549–561
99. Suprun E, Bulko T, Lisitsa A (2010) Electrochemical nanobiosensor for express diagnosis of acute myocardial infarction in undiluted plasma. Biosens Bioelectron 25:1694–1698
100. Jemal A, Bray F, Ferlay J (2011) Global cancer statistics. CA Cancer J Clin 61:69–90
101. Sawyers CL (2008) The cancer biomarker problem. Nature 452:548–552
102. Novak K (2006) Biomarkers: taking out the trash. Nat Rev Cancer 6:92
103. Perfézou M, Turner A, Merkoçi A (2011) Cancer detection using nanoparticle-based sensors. Chem Soc Rev 7:2606–2622
104. Bohunicky B, Mousa S (2011) Biosensors: the new wave in cancer diagnosis. Nanotechnol Sci Appl 4:1–10
105. Fang Z, Kelley SO (2009) Direct electrocatalytic mRNA detection using PNA-nanowire sensors. *Analytical chemistry* 81:612–617
106. Waggoner PS, Varshney M, Craighead HG (2009) Detection of prostate specific antigen with nanomechanical resonators. Lab Chip 9:3095–3099
107. Storhoff J, Lubben T, Lefebvre P et al (2008) Detection of prostate cancer recurrence using an ultrasensitive nanoparticle-based PSA assay In: Journal of Clinical Oncology. 45th annual meeting of the American-Society-of-Clinical-Oncology, Orlando, p e16146
108. Oliver C (1994) Use of immunogold with silver enhancement. In: Javois LC (ed) Immunocytochemical methods and protocols, vol 34. Springer, New York, pp 211–216
109. de la Escosura-Muñiz A, Merkoçi A (2011) A nanochannel/nanoparticle-based filtering and sensing platform for direct detection of a cancer biomarker in blood. Small 7:675–682
110. Zhong Z, Wu W, Wang D (2010) Nanogold-enwrapped graphene nanocomposites as trace labels for sensitivity enhancement of electrochemical immunosensors in clinical immunoassays: carcinoembryonic antigen as a model. Biosens Bioelectron 25:2379–2383
111. Myung S, Solanki A, Kim C et al (2011) Graphene-encapsulated nanoparticle-based biosensor for the selective detection of cancer biomarkers. Adv Mater (Deerfield Beach, Fla) 23:2221–2225
112. Qian J, Dai H, Pan X, Liu S (2011) Simultaneous detection of dual proteins using quantum dots coated silica nanoparticles as labels. Biosens Bioelectron 28:314–319
113. Jokerst JV, Raamanathan A, Christodoulides N et al (2009) Nano-bio-chips for high performance multiplexed protein detection: determinations of cancer biomarkers in serum and saliva using quantum dot bioconjugate labels. Biosens Bioelectron 24:3622–3629
114. Ng AHC, Uddayasankar U, Wheeler AR (2010) Immunoassays in microfluidic systems. Anal Bioanal Chem 397:991–1007
115. Jain KK (2010) Advances in the field of nanooncology. BMC Med 8:83
116. de la Escosura-Muñiz A, Sánchez-Espinel C, Díaz-Freitas B et al (2009) Rapid identification and quantification of tumor cells using an electrocatalytic method based on gold nanoparticles. Anal Chem 81:10268–10274
117. Deng T, Li J, Zhang L-L et al (2010) A sensitive fluorescence anisotropy method for the direct detection of cancer cells in whole blood based on aptamer-conjugated near-infrared fluorescent nanoparticles. Biosens Bioelectron 25:1587–1591
118. Maltez-da Costa M, de la Escosura-Muñiz A, Nogués C et al (2012) Simple monitoring of cancer cells using nanoparticles. Nano Lett 12:4164–4171
119. Oghabian M, Jeddi-Tehrani M, Zolfaghari A (2011) Detectability of Her2 positive tumors using monoclonal antibody conjugated iron oxide. J Nanosci Nanotechnol 11:5340–5344
120. Rasaneh S, Rajabi H, Babaei M (2011) MRI contrast agent for molecular imaging of the HER2/neu receptor using targeted magnetic nanoparticles. J Nanopart Res 13:2285–2293
121. Perry M, Li Q, Kennedy RT (2009) Review of recent advances in analytical techniques for the determination of neurotransmitters. Anal Chim Acta 653:1–22

122. Morales-Narváez E, Montón H, Fomicheva A, Merkoçi A (2012) Signal enhancement in antibody microarrays using quantum dots nanocrystals: application to potential Alzheimer's disease biomarker screening. Anal Chem 84:6821–6827
123. Ambrosi A, Airò F, Merkoçi A (2009) Enhanced gold nanoparticle based ELISA for breast cancer biomarker. Anal Chem 82:1151–1156
124. Ambrosi A, De La Escosura-Muñiz A, Castañeda MT, Merkoçi A (2009) Gold nanoparticles: a versatile label for affinity. In: Merkoçi A (ed) Biosensing using nanomaterials. Wiley, Hoboken, NJ, USA, pp 177–197
125. Brambilla D, Le Droumaguet B, Nicolas J et al (2011) Nanotechnologies for Alzheimer's disease: diagnosis, therapy, and safety issues. Nanomedicine 7(5):521–540
126. Clark LC, Lyons C (1962) Electrode systems for continuous monitoring in cardiovascular surgery. Ann N Y Acad Sci 102:29–45
127. Donzella V, Crea F (2011) Optical biosensors to analyze novel biomarkers in oncology. J Biophotonics 4:442–452
128. Ehara S, Ueda M, Naruko T et al (2001) Elevated levels of oxidized low density lipoprotein show a positive relationship with the severity of acute coronary syndromes. Circulation 103:1955–1960
129. Ferrari M (2005) Cancer nanotechnology: opportunities and challenges. Nat Rev Cancer 5:161–171
130. Godin B, Sakamoto J, Serda R (2010) Emerging applications of nanomedicine for therapy and diagnosis of cardiovascular diseases. Trends Pharmacol Sci 31:199–205
131. Jayagopal A, Linton MRF, Fazio S, Haselton FR (2010) Insights into atherosclerosis using nanotechnology. Curr Atheroscler Rep 12:209–215
132. Kong LX, Peng Z, Sugumar D (2006) Management of cardiovascular diseases with micro systems and nanotechnology. J Nanosci Nanotechnol 6:2754–2761
133. Morales-Narváez E, Merkoçi A (2012) Graphene oxide as an optical biosensing platform. Adv Mater 24:3298–3308
134. Soper SA, Brown K, Ellington A et al (2006) Point-of-care biosensor systems for cancer diagnostics/prognostics. Biosens Bioelectron 21:1932–1942
135. Veiseh O, Kievit FM, Ellenbogen RG, Zhang M (2011) Cancer cell invasion: treatment and monitoring opportunities in nanomedicine. Adv Drug Deliv Rev 63:582–596
136. Wang C, Knudsen B, Zhang X (2011a) Semiconductor quantum dots for electrochemical biosensors. In: Li S, Singh J, Li H, Banerjee IA (eds) Biosensor nanomaterials. Wiley, Hoboken, NJ, USA, pp 199–219
137. Wang Y, Li Z, Wang J et al (2011b) Graphene and graphene oxide: biofunctionalization and applications in biotechnology. Trends Biotechnol 29:205–212

Chapter 8
Nanotechnology in Advanced Medical Devices

Sabeeh Habib-Ullah, Dan Fei, and Yi Ge

Abbreviations

AFM	Atomic Force Microscopy
AMD	Advanced Medical Device
CCMV	Cowpea Chlorotic Mottle Virus
CT	Computed tomography
EBID	Electron Beam Induced Deposition
ECM	Extracellular Matrix
ESF	European Science Foundation
HSE	Health and Safety Executive
LOC	Lab on a Chip
MD	Medical Device
MEMs	Microelectromechanical System
MNP	Magnetic Nanoparticles
MPA	Mercaptopropionic Acid (MPA)
MRI	Magnetic Resonance Imaging
NEMs	Nanoelectromechanical Systems
NMs	Nanomaterials
NP	Nanoparticles
OSHA	Occupational Safety and Health Act

S. Habib-Ullah • Y. Ge (✉)
Centre for Biomedical Engineering, Cranfield University,
Cranfield, Bedfordshire MK43 0AL, UK
e-mail: y.ge@cranfield.ac.uk

D. Fei
Leicester School of Pharmacy, De Montfort University,
The Gateway, Leicester LE1 9BH, UK

© Springer Science+Business Media New York 2014
Y. Ge et al. (eds.), *Nanomedicine*, Nanostructure Science and Technology,
DOI 10.1007/978-1-4614-2140-5_8

PAMAM   Poly(amidoamine)
PB   Prussian Blue
PEG   Polyethylene Glycol
PMMA   Polymethyl Methacrylate
POC   Point of Care
QD   Quantum Dots
REACH   Registration Evaluation, Authorisation and restriction of Chemicals
SPION   Superparamagnetic Iron Oxide Nanoparticles
SPR   Surface Plasmon Resonance
SWCNT   Single Walled Carbon Nanotube
USPION   Ultra-small Superparamagnetic Iron Oxide Nanoparticles
VNP   Viral Nanoparticles
WHO   World Health Organisation

8.1   Introduction

Nano refers to materials and systems being in the proportion of a billionth of a metre, i.e. 10^{-9} m and it originates from the Greek word 'nano' meaning dwarf. The term of 'nano' was first formally introduced by Tokyo Science University Professor Norio Taniguchi in a paper on ionsputter machining 1974 [5]. Generally these systems are agreed to be between 1–100 nm.

The potential of nanoscience was first expounded by Richard P. Feynman in 1959 [4], in his famous speech "Plenty of Room at the Bottom". Nanotechnology is now one of the world's leading industrial, academic, and political drivers. In a short time its emergence as an innovative and multidisciplinary platform has secured funding and investment on an unheard of scale [1]. Nanotechnology is projected to be worth $2.6 trillion in manufactured goods by 2014 [2]. Moreover the market impact of nano-based applications is estimated to be $300 billion within the next 12 years for the United States alone [3].

Nanomaterials (NMs) and nanotechnology hold the potential to initiate a new industrial revolution. The characteristics that make these materials commercially suitable are their size, aspect ratio and possibility to functionalise their surfaces. Most experts agree that this new phase of technology could have an impact on every aspect of life. Their uses range from drug delivery, food hygiene, to astrobiology and ocular implants.

Recently NMs have caused a flurry of activity in the medical field. A large effort has been made to use the technology itself as an analytical tool, as well as using nano-sized material to enhance the current paradigm. This was codified in 2004 [6], the definition for nanomedicine that the Medical Standing Committee of the European Science Foundation (ESF) compiled is:

> the science and technology of diagnosing, treating, and preventing disease and traumatic injury, of relieving pain, and of preserving and improving human health, using molecular tools and molecular knowledge of the human body

Further, the ESF demarcated five main disciplines of nanomedicine:

1. Analytical tools
2. Nanoimaging
3. Nanomaterials and nanodevices
4. Novel therapeutics and drug delivery systems
5. Clinical, regulatory, and toxicological issues

These five categories are generating a considerable amount of research [7]. However, it is difficult to interpret which of these fields are classed under medical devices, because they can be a range of articles. The World Health Organisation (WHO) has developed its own definition for medical devices [8]:

"Medical device" means any instrument, apparatus, implement, machine, appliance, implant, *in vitro* reagent or calibrator, software, material or other similar or related article, intended by the manufacturer to be used, alone or in combination, for human beings for one or more of the specific purposes of:

• diagnosis, prevention, monitoring, treatment or alleviation of disease
• diagnosis, monitoring, treatment, alleviation of or compensation for an injury

This clearly illustrates the breadth of products that could be considered as medical devices (MDs). Tools such as tongue depressors, injections and even spoons can be considered as MDs. Advanced MDs (AMDs) would clearly be a step above these mundane tools, clearly the use of NMs could potentially cause a device to be labelled in that bracket [9]. Instead of looking at a broad range of devices, it is beneficial to focus on the exceptional that are making a noticeable impact in the current climate of AMDs.

Nanoimaging and diagnostics are showing potential to revolutionise the gold standard techniques of today. Imaging systems can be grouped by the energy used to construct pictorial information, the type of information that is acquired or the spatial resolution that is obtained. Macroscopic imaging equipment that provide data for anatomical and physiological systems are regularly used in clinics across the world, the most common being computed tomography (CT), magnetic resonance imaging (MRI) and ultrasound. Whereas the emergence of molecular imaging techniques, that can be attributed to nanotechnology, are beginning to be used in clinical and pre-clinical settings [10]. This advancement has opened the possibilities of actually imaging biological functions [11].

Figure 8.1 shows the possibilities of nanoimaging and diagnostics; its potential to measure biological processes [12] is similar to a biopsy, but better. This can be achieved without surgery and possible long term monitoring of specific conditions. Applications comprise early detection and diagnosis of state and stage of disease, assessing response to treatment, and studying biological processes in real time [13]. Compared to CT scans, MRI scans and others of this ilk, they are advantageous in providing information for evidence based medicine.

Similar to this are biosensors and nanobiosensors [14, 15], which are comprised of two elements, a detection device and a transducer. The first provides sensitivity and specificity, while the latter gives an indication of the presence of the desired substrate [16]. These devices are being primed to make point of care (POC)

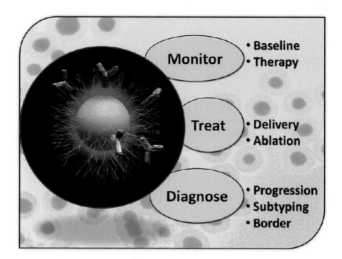

Fig. 8.1 The potential and consequences of nanoimaging and diagnostics [12]

technology feasible, the purpose of which is to deliver real time information on the state of the body. Hence barriers to overcome are to produce something easily fabricated and inexpensive with recognition capabilities, focusing on increasing sensitivity, specificity and enhanced response time [17].

There is already evidence of this on the market with glucose sensors for diabetics. However with nanotechnology in the ascendency, a fast paced diagnostic tool that can run hundreds of tests in less than a minute is conceivable. This lab on a chip (LOC) will utilise the miniaturisation capabilities of nanotechnology to revolutionise patient care. The idea is to produce a small chip that has multiple detection points and transducers to produce a signal, which can then be converted into meaningful information [18]. A key example is the work of Sung et al., in 2010 they demonstrated a microfluidic approach to analyse the pharmacokinetic-pharmacodynamic profile of an anti-cancer drug [19]. The results indicated that their approach was a better alternative to assess drug profiles on cell cultures. Although the work was completed on a micro level, the idea is to exemplify the direction the field is heading towards.

The impact of technology on surgery is often discussed but not truly examined Surgery is also being advanced through the rise of nanotechnology. It can range from the simplicity of the removal of a mole to the complexity of an open heart quadruple bypass. However, surgeons can now use an array of nano-driven materials and devices to aid their work. This extends from advances in intraoperative imaging, tissue healing and wound care to the procedures of nanosurgery involving nanoneedles, nanotweezers and precision lasers [20]. This is leading to more accurate and effective clinical practice combined with surgical precision to produce better invasive procedures.

For completeness, it would also be beneficial to examine the risk to reward ratio. There is scope within nanotechnology for misuse and exploitation. Therefore any regulatory and ethical quandaries should come under consideration in the subsequent discussion.

8.1.1 Nanoimaging

Imaging is a very important aspect within industry. Its uses vary from the x-rays in airports, X-raycrystallography, and MRI to determine chemical compositions. It also plays a critical part in medicine and can be classed as a MD. The WHO clearly states that devices can be used to investigate, diagnose, prevent and alleviate injuries, diseases or a physiological process [8]. The "advanced" can clearly be attributed to the introduction of nanotechnology. This can range from quantum dots, to high-tech processes such as Atomic Force Microscopy (AFM), or a combination of current machinery with a nanotechnology component; such as contrast agents combined with MRI. These tools can be used in prevention, monitoring and treatment of diseases. There is a lag time of information received and can hinder clinician decisions. However, the arrival of new technologies provides real time data along with a unheard of depth of information leading to a more complete and informed diagnosis.

8.1.1.1 Quantum Dots

Quantum dots (QDs) are nanoparticles (NPs) of a few hundred atoms. They absorb light at a wide range of wavelengths, but radiate near monochromatic light of a wavelength that depends on the size of the crystals. Of particular usefulness is the amount of control exerted over QD characteristics; temperature, duration and ligand molecules used during production of QDs can govern size and shape [21]. Hence, the radiation absorbed and light emitted can also be controlled [22]. Furthermore because the materials employed are inorganic, they are more durable in the body than their organic counterparts, they are highly observable using electron microscopy [23].

There are some toxicity issues with QDs, but functionalising them with multiple moieties has led to some favourable outcomes. In 2002, Akerman et al. used polyethylene glycol (PEG) as a surface cap for cadmium selenide QDs [24]. They examined their penetration of tumours in immune-deficient mice and compared it to peptide tagging (Fig. 8.2). Their results demonstrated that QD uptake potential and fluorescent characteristics were enhanced by PEGylation.

Furthermore endocytosis uptake of QDs and labelling of cell surface proteins with QDs conjugated to antibodies were investigated and demonstrated [25]. This procedure has since been improved to show that cells can be imaged and observed in real time. This technology has progressed towards a pilot study of QDs in non-human primates. The use of phospholipid micelle-encapsulated CdSe/CdS/ZnS QDs were observed over a 90 day period. Our results show that acute toxicity of these quantum dots in vivo can be minimal. Blood and biochemical markers remained within normal ranges, but chemical analysis revealed that majority of the initial dose of cadmium remained in the liver, spleen and kidneys. This suggests that the breakdown and clearance warrants further investigation [26]. Furthermore, this technology has been used to investigate specific processes that range from binding of QDs conjugated to the nerve growth factor to membrane specific receptors and intracellular uptake, tracking of membrane protein at the single molecule level, and recognition of ligand bound QDs by T cell receptors [27].

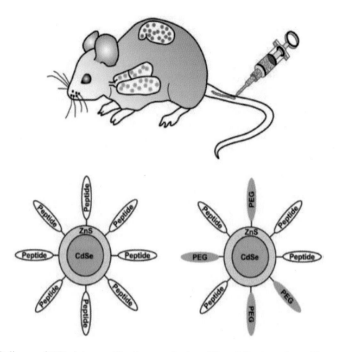

Fig. 8.2 Delivery of QDs into specific tissues of mice. (*Upper*) Design of peptide-coated QDs. (*Lower*) QDs was coated with either peptide only or with peptides and PEG, increasing solubility and bioavailability [24]

8.1.1.2   Magnetic Resonance Imaging

Magnetic resonance imagings (MRIs) are the most widely used diagnostic tools across the world. It is the instrument of choice due to its range, non-invasiveness and minimal side effects. It can be used to give high resolution images of soft tissues; specifically for tumours, brain, central nervous system and cardiac function [28]. MRIs exploit the difference in relaxation times of protons. When protons are exposed to a magnetic field, they become excited and then return to their equilibrium state at different rates T_1 and T_2. It is this information that is used to construct images to help diagnosis and treatment of patients.

Viral nanoparticles (VNPs) are virus-based NP preparations that can be used as a scaffold with a variety of properties. VNPs can be found in the form of bacteriophages and plant or animal viruses. Additionally they are biocompatible and biodegradable; there are organic and non-organic forms. However there is potential to combine these formulations to produce a more functional NP. Yet there have been no explorations of their toxicity, either as singular (organic or non-organic) or in a conjugated format [29].

In 2005, Allen et al. investigated Cowpea chlorotic mottle virus (CCMV) as MRI contrast agents [30]. It has 180 metal binding sites and *in vivo* CCMV binds calcium ions at specific sites; however gadolinium ions (Gd^{3+}) can also do this. Gadolinium is more useful because it increases proton relaxation during MRI, which gives sharper signals. The unusually high proton relaxation values of the Gd^{3+}-CCMV complex can be attributed to the VNP size and number of ions bound. These findings are encouraging, but at the moment VNPs remain proof of concepts until further investigations are completed.

Like VNPs magnetic nanoparticles (MNPs) have gained attention due to their properties. Their magnetic characteristics enable tracking through MRI. MNPs include metallic, bimetallic, Ultra-small and super-paramagnetic iron oxide nanoparticles (USPIONs and SPIONs) [31]. The USPIONs and SPIONs are preferable because of their lack of cytotoxicity, and surface being amenable to functionalization [32]. Hence, there is greater penetration in biological systems, and increased proton relaxation times present sharper signals for better imaging.

New vehicles are being developed that encapsulate multiple imaging molecules onto the MNP for use in integrated imaging systems. These agents can assist investigators to produce images across varied techniques. This signifies that one contrast agent could be used for MRI as well as CT and positron emission tomography scans, offering clinicians the ability to gain a selection of information efficiently [10]. It has been reported that gold NPs coated with gadolinium chelates produces MR contrast [33]. Such MRI active inorganic can be used for imaging of the vasculature, liver and other organs, as well as molecular imaging and cell tracking.

Of course these are not the only particles used for MRI contrast agents. Nanotechnology is developing so quickly that multimodal NPs are being developed not only for MRIs but other imaging techniques. Table 8.1 clearly shows that NPs of different types and sizes can target different organs, and all of them are in the preclinical stages of trials.

8.1.1.3 Atomic Force Microscopy

AFMs are a multipurpose instrument, they can be employed for imaging, determining and manipulating materials at the nanoscale. It can be used to work out surface structure, using forces and interactions such hydrogen bonding, Van der waal, and electrostatic forces. It consists of a small cantilever (of nano-dimensions) that produces a signal when deflected. This is extremely useful in biological imaging, mapping interactions at the cell surface, using high resolution images.

AFM has been used for a multitude of purposes. It can be classed as an AMD due to the fact that it is an analytical tool that is used for monitoring/imaging. Most recently it has proved an effective tool in diagnosis the root causes of protein folding diseases such as Parkinson's, Alzheimer's, and Huntington's diseases. The use of AFM and other nanoimaging techniques has been instrumental in understanding the structure and aggregation of key proteins [35]. Its ability to work on

Table 8.1 A comparison of different nanoparticles, their functions and the cancers they effect [34]

Cancer site	Nanoparticle	Clinical application	Current status
Breast	Iron oxide nanoparticles + Herceptin	Detection of small tumors on MRI	Preclinical
	Iron oxide nanoparticles + uMUC- tumor antigen	MRI and monitoring tumor response to chemotherapy via antigen expression and change in size	Preclinical
	Dendrimer	Contrast agent for micro-MR lymphangiography	Preclinical
	Iron oxide nanoparticle	Detection of sentinel lymph node	Clinical
Colon	Iron oxide particles	MRI of CRC and métastasés	N/A
	QDs	Visualization of cancer using fiber optics	Preclinical
Prostate	Iron oxide nanoparticles	Detection of metastasis with high resolution MRI	Preclinical
	Dendrimer + Prostate specific antibody	Targeting of antigen expressing cells	Preclinical
Brain	Iron oxide nanoparticles	Dual function particles to help define tumor margins accurately intra-operatively	Preclinical
Pancreas	Iron oxide particles	Enhances normal pancreatic tissue on MRI enabling easy visualization of PDAC	Preclinical

singular molecules and isolating its interactions has led to progress in developing understanding and therapies for such diseases. Neurodegeneration in Alzheimer's disease has been linked to β-amyloid (Aβ) peptide build up. AFM has shown that Aβ interacts with the plasma membrane by changing the structure, which leads to the disruption of ionic homeostasis [36]. This was further investigated to conclude that amyloid proteins disrupt distinct regions of the bilayer membranes, which may represent a common mechanism of membrane disruption for protein folding diseases [37].

Nanotechnology has provided the means to make breakthrough machinery more efficient. AFM imaging is generally slow; it takes between 1 and 2 min per frame, making it difficult to record biological processes that generally occur in seconds. Collagen (it is one of the main components of the extracellular matrix) imaging was improved by an inversions feedback/feedforward process. Furthermore it reduced positioning errors as well as increased scan time at high frequencies [38].

In 2006, Voitchovsky and colleagues were able to examine Purple membranes formed by bacteriorhodopsin, it is a combination of crystals and lipids [39]. Using AFM the group were able to observe stiffness and lipid mobility. This work was extended by using AFM to actually study the dynamics of the purple membrane. Yamashita and co-workers were able to film the dynamics of the membrane using high speed AFM, determining how the crystals within it were assembled [40].

8.1.1.4   Optical Tweezers

Optical tweezers are another method of imaging, they can control objects with light. They offer a spatial resolution of 1 nm, a force resolution of a piconewton, and a time resolution of a millisecond. This makes them ideally suited to examine biological processes ranging from the size of a single cell to the minuteness of isolating a single molecule [41].

They were first introduced by AT&T Bell Laboratories. An optical tweezer uses force exerted by the electric field of a focused beam of light. In response to this, a small object develops a dipole, which allows it to be focused at a particular point. The focal point is formed on the optical axis, and the radiation pressure interacts with the electric force, allowing the isolation of a nanosized object. The object of interest can then be examined.

Ermilov and colleagues used optical tweezers to study the interaction of plasma membranes with their own cytoskeletons [42]. Their results showed that combining this technique with fluorescence imaging allowed them to create a profile of forces and stresses acting on the structures of interest. Pine and Chow went further by not only immobilising a neuron but isolating it as well [43]. They found they could move these neurons at a speed of 200 μm/s without actually causing damage to it. The duo also managed to cage these neurons for cell culture. The implication of this study is vast, this technique could be potentially used to grow neurons and graft them for victims of spinal injuries.

Combining this technique with confocal microscopy has allowed the multiplanar imaging of intercellular immune synapses [44]. This collective technology enabled the characterization of complex behaviour of highly dynamic clusters of T cell receptors at the T cell/antigen-presenting cell synapse. This work also identified the presence of receptor rich molecules.

The ability of such techniques to manipulate and measure forces has found several applications such as understanding the dynamics of biological molecules, cell-cell interactions and the micro-rheology of both cells and fluids [45]. A prime example is using optical tweezers to measure the interaction between NPs and prostate cancer cells [46]. The study revealed that the functionalised NP and cell binding was improved by the presence of folic acid. The works presented here show how nanotechnology has made a positive impact on medical devices and their capabilities.

8.1.2   Nanobiosensors

The biosensors market has become very lucrative, it is estimated that annual global investment in this technology for research and development was thought to be $300 million. The introduction and fruition of nanotechnology has produced a huge amount of investigation and innovation; over 6,000 articles and 1,100 patents were issued and pending between 1998 and 2004 [16]. This indicates that effort within this field is commercially worthwhile and is rapidly progressing. It is important to note that the addition of NMs is what changes a biosensor to a nanobiosensor.

The uses of biosensors have increased rapidly and have become applicable across a range of fields. Most commonly they are used for clinical, environmental and food purposes, to name but a few [47]. As mentioned before there are two main parts to a biosensor, a recognition element that combines with a transducer (Fig. 8.3). The signal produced is then amplified against some baseline to produce a significant reading. The most dominants types of biosensors can be categorised into three classes, electrochemical, optical and followed by piezoelectric, the latter being the least common. Each of these different categories is differentiated by their transducer technology, as opposed to the molecular recognition element, which can be the same for each of them.

The structure of biosensors has an obvious effect on their functions. The bio-recognition element is based on the high affinity of receptor and analyte, such as an enzyme, a strand of nucleic acid or even an antibody [48].

8.1.2.1   Electrochemical Biosensors

This particular type of biosensor is the most common. It has been used for a variety of reasons. Of interest is the fact that they have been used to detect emerging infectious diseases, however despite the abundance of technology this aspect is still underdeveloped. Environmental detection has been based on monitoring toxic effects in cells, genes and possible endocrine disruptions. This technology is also being heavily mooted to make an impact in POC diagnosis potentially providing information for early detection of disease [17]. Where electrochemical sensors differ is in how signalling is achieved in the transducer element. It can be based on a measured voltage (potentiometric), a current (amperometric), or the transport of charge (conductometric) [16].

8.1.2.2   Optical Biosensors

These particular sensors have been developing rapidly, even more so considering the impact of nanotechnology. Optical sensors employ optical fibres orplanar waveguides to direct light. This is then directed to the sample and interactions produce a signal which can then be compared to its baseline. The signals produced include absorbance, fluorescence, chemiluminescence, surface plasmon resonance to probe refractive index, or changes in light reflectivity and can be read at the sensing film [16, 49].

The advantages of optical biosensors are their rapidity, the insusceptibility of the signal to interference and the potential for greater data, as a result of the changes in the electromagnetic spectrum. Optical methods can be employed to use multiple wavelengths on a sample without interfering with one another. This arrangement can lead to direct or indirect detection [49]. Deposition techniques can be used to produce optical biosensors. Screen printing and ink-jet printing are extremely fast and precise, producing great quantities of low-cost and reproducible biosensors [50].

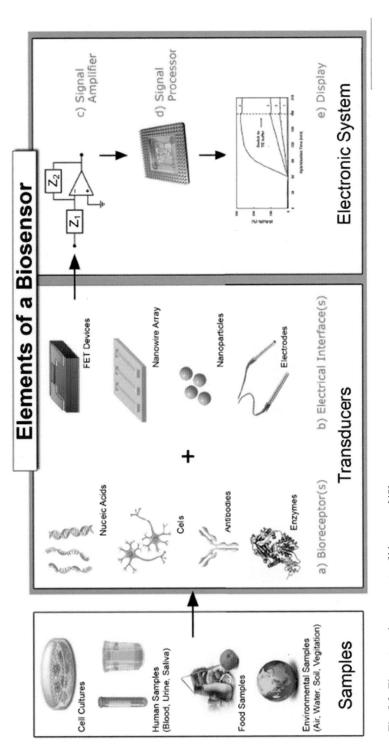

Fig. 8.3 Elements and components of biosensors [48]

8.1.2.3 Piezoelectric Biosensors

As the name suggests this particular biosensor is based on the piezoelectric phenomenon. This was first investigated in 1880 by Jacques and Pierre Curie, who observed that mechanical stress applied to the surfaces of various crystals, caused a correspondingly proportional electrical potential across the crystal, the converse was also true [51]. A multitude of crystalline materials (tourmaline, lithium niobate, aluminium nitride to name but a few) exhibit the piezoelectric effect, however the characteristics of quartz make it the most common crystal employed [52]. These crystals have been used as microbalances owing to their small size, high sensitivity, simplicity of construction and operation, light weight, and the low power required [92]. Similar to the other classes of biosensor, this particular methodology has been employed across a wide range of settings, including environmental and clinical.

8.1.2.4 Nanotechnological Impact

Due to the ubiquitous properties of NMs their use to construct these AMDs is no surprise. In 2003, Davis and co-workers reported on the investigations carried out in their laboratory, which culminated in the construction of a glucose biosensor [53]. The novelty of this sensor was that its construction took place on a single-walled carbon nanotube (SWCNT). The results showed that SWCNTs are highly biocompatible, and their unique structure has added benefits, including having a large surface area that makes it ideal for biological loading. This functionalised surface can exchange electrons and this mechanism can be employed in the fabrication of SWCNT-based biosensing devices.

QDs have also been used in biosensors. Cadmium Telluride was employed because its surface area was conducive to functionalization [54]. Furthermore its conductive properties allowed facilitated the oxidation of the thiocholine-acetylcholinesterase complex, thereby increasing sensitivity. Considering that QDs are so widely used and investigated as biological probes and contrast agents, this is a novel way of using them.

In relation to this green and orange Cadmium Telluride QDs were used as part of a fluorescent biosensor [55]. They were employed to observe proton flux driven by ATP synthesis in viruses. The key findings showed that different fluorescence occurred simultaneously and independently, further they were pH-dependant and showed no interference. An added benefit was that the fabrication of these QDs was inexpensive and convenient, making this commercially viable.

Gold NPs have many uses, in biosensors they are employed because their functions are heavily linked to their structural conformation. Depending on their shape, the surface plasmon resonance (SPR) is affected [56]. This is the oscillation of conduction electrons resonating with the wavelength of light used for excitation. By changing of the gold NPs shape from spheres to rods, the aspect ratio changes, thereby affecting the altering the SPR from the visible region to that of near infrared.

Gold NPs provide better biocompatibility, it also significantly increases the receptor area, thus improving sensitivity. In 2008, Li et al. conjugated gold NPs with

3-mercaptopropionic acid (MPA), poly(amidoamine) (PAMAM) dendrimer to obtain films on which Prussian blue (PB) was electrochemically deposited [57]. The purpose of this was to examine the response to the reduction of hydrogen peroxide. The investigation found that the sensitivity and limit of detection was enhanced, pH range and electrochemical stability and response were also improved. Furthermore, gold NP assemblies were used to trap proteins [58]. Using localised surface plasmon resonance nanotransducers delivers new leverages in hot spot-based nanosensing. The intensity of this electromagnetic hot-spot can be fine-tuned to gain picomolar sensitivity.

Recently, silicon nanowires were found to be able to detect microRNA, at a limit of <100 1 fM [59]. It was found that optimising the surface functionalisation and fabrication protocol, a theoretical limit of detection of 1 fM could be achieved.

It is clear that NMs have enhanced the field of biosensors to new heights. Never before have academics been able to explore biological and cellular processes at such close quarter. Moreover, the addition of NMs can actually enhance specificity, sensitivity and reproducibility. The miniaturisation of this technology is leading to POC and LOC diagnostics.

8.1.2.5 Progress to Lab On a Chip (LOC)

It is imperative to demonstrate the impact that nanotechnology has on the progression from a biosensor to a LOC. In 2009 Lakshmi et al. presented N-phenylethylenediamine methacrylamide (NPEDMA), which contains both aniline and a methacrylamide group, capable of independent polymerisation via free radical or electrochemical methods [60]. This work was then furthered by the group by developing an electrochemical sensor for catechol and dopamine using NPEDMA and a molecularly imprinted polymer (MIP) [61]. The MIP employed in this case is a tyronisase mimicking polymer. It has two copper binding sites and is more stable than its natural counterpart.

MIPs can be considered as polymeric NPs that are able to recognize biological and chemical molecules. The synthesis is based on the formation of a moulding complex between an analyte (template) and a functional monomer. The template is removed leaving impressions of precise recognition sites matching in characteristics to the analyte, intermolecular interactions determine the molecular recognition [62]. Hence it is particularly desirable to use a MIP in conjunction with NPEDMA.

Electropolymerisation of the NPEDMA resulted in the formation of a polyaniline (PANI) layer, a conducting polymer, on a gold electrode surface. The catechol specific MIP was then photochemically grafted on to this layer, in effect creating a network of molecular wires, to conduct a signal for molecule capture. This was a novel method of producing a biosensor with recognition and transducer capabilities.

This approach was then further modified again to examine what impact PANI nanostructures had on the catechol sensor. Berti et al. sputtered gold on an alumina membrane and used it as a mould for poly-NPEDMA [63]. The tryonisase-mimicking MIP was then attached and resulted in the improvement of catechol detection, the lower limit was determined to be 29 nm, a thousand fold increase.

This emphasises the capability of nanostructures to improve diagnostic performances of sensors.

This approach was expanded upon by Akbulut and colleagues. In 2011 they used polychemical grafting to coat NPEDMA to polystyrene microtitre plates [64]. The purpose of this was to create a library of different polymers that could be attached to the plates. The idea of was to assess the feasibility of this to identify a small catalogue of biological and organic compounds. The analytes were a herbicide-atrazine, organic dyes (eosin, bernthsen and meldola), and bovine serum albumin. The results showed that this method of screening is suitable for a myriad of applications and is cost effective, reproducible and easily manufactured.

This work was furthered improved upon by synthesising a polymer to improve and enhance the characteristics shown by NPEDMA [65]. The new polymer is called N-(N_0,N_0-diethyldithiocarbamoylethylamidoethyl) aniline (NDDEAEA). This structure is similar to NPEDMA in that it is based on PANI, but has the ability to be functionalised twice.

This could have far reaching implications over some future applications. In effect, this case study has developed two new polymers with a myriad of different characteristics, the first of which has already been explored to some extent. NPEDMA and NDDEAEA seem to offer potential not only in analytical technology but in other fields. The impact of nanotechnology is vast on AMDs, as demonstrated by this series of work. A polymer and MIP were constructed to form a biosensor, this was essentially modified and optimised to form precursor to a lab on a chip with enhanced limits of detection. This work is only one step short of forming an LOC, consider the key components, there is a molecular circuit linked to a possibility of a library of polymers that can be manipulated to detect any analyte.

The possibilities are endless; furthermore the construct is reproducible and inexpensive. An example of this, is use of nanolithography to create nanoarrays of gold [66]. To demonstrate its multiplex analyses, horseradish peroxidase and anti-horseradish peroxidase antibody was used as a model for a recognition system. The enzyme-linked immunosorbent assay performed had a detection limit 100 pg/mL. It was found that these chips could be stored for 50 days when stored at 4 °C without any significant loss of activity.

8.1.3   Surgery and Clinical Applications

Surgery and clinical applications have been on an upward curve since the exploration of nanotechnology. This was also predicted by Richard P. Feynmann in 'Plenty of room at the bottom' talks about the impact of nanotechnology on surgery and its applications:

> It would be very interesting in surgery … if you could swallow the surgeon. You could put the little mechanical surgeon inside the blood vessel and he goes into the heart and looks around (of course the information has to be fed out). He finds out which valve is the faulty one and slices it out. Other small machines may be permanently incorporated into the body to assist some inadequately functioning organ.

It is clear that some of the advances in AMDs are leading to the fulfilment of this vision. This involves the applications of nanotechnology in the surgery. The idea of molecular machines, high precision tools, and nanoimaging are leading to the beginnings of a new field, nanosurgery.

Moreover, giant strides are being made in clinical applications of nanotechnology. This consists of tissue engineering, regenerative medicine and implantation of these AMDs.

8.1.3.1   Nanosurgery

Surgery has always been a macro-scale project, involving cutting, controlling and splicing of organs, muscles and bones. However, as technology has progressed, so have the requirements of this particular field in medicine, leading to more precision and less invasive measures. In the latter twentieth century, the emphasis was on miniaturisation, small incisions, laparoscopic procedures by fibre-optic visualisation, vascular surgery by catheters and microsurgery under microscopes to refine protocols and diminish trauma [67].

One of the major tools that advocate nanosurgery is the use of high precision lasers. A landmark work by Shen and colleagues demonstrated the selectivity and suitability of a femtosecond laser for the ablation of subcellular structures. They focused laser pulses beneath the cell membrane to ablate cellular material. Shen and colleagues were able to selectively remove regions of the cytoskeleton and individual mitochondria without affecting nearby structures or compromising cell viability. Using this approach demonstrated that mitochondria are structurally independent functional units [68].

This work has led to a plethora of development in the use of this 'nanoscissors' technique. Most recently, experiments conducted on human metaphase chromosomes and fixed cell nuclei show it is possible to induce sub-100 nm effects with near-infrared femtosecond laser pulses [69]. Another study used a multimodal imaging system with nanosurgery capabilities, for the selective ablation of sub-cellular components in cancer cells [70]. The work resulted in precise destruction of structures in the cancer cells while leaving them fully functional.

As mentioned before AFM is an instrument that is used in nanomedicine and is highly precise. A study was done to modify the tips of AFM for using electron beam induced deposition (EBIM). This technique was used to make a 'nanoscalpel', 'nanotome' and 'nanoneedle' [71].

Figure 8.4a–c shows the structures of the nanoscalpel being constructing. This was fabricated using an electron beam of low velocity depositing carbon atoms along the tip. This extended along the tip in a self-supporting structure that can then form a blade in the nano scale.

Nanotomes use a similar technique. The process that follows is identical to nanoscalpels, but there are two blades deposited and extended. The difference is that a single filament is extended between the blades, giving it a similar functionality to a vegetable peeler.

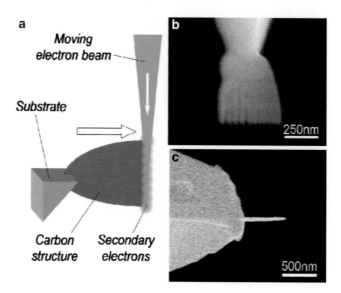

Fig. 8.4 (**a**) Fabrication of nanotool structures using electron beam induced deposition. (**b, c**) Nanoscalpel blades imaged from the side (**b**) and from the top (**c**) using SEM [71]

Nanoneedle is processed in a similar way to a scalpel, but the deposition is lateral as well. This allows the needle to be thickened and provide more support, as the thin structure limits the force that can be applied to the nanoneedle. The nanoneedle has several uses, it can be used for profiling structures, and if the needle is thickened then indentations can be made and studied. Beard and colleagues showed that this particular tip could also be used to remove a cluster of proteins from a cell [71]. While the nanoscalpel function can be used for cellular dissections, making controlled incisions. The nanotome can be used to peel back layers of a cell, which can then be imaged.

These nano-surgical tools provide a new route to the manipulation and dissection of cellular structures. They can be used to ablate and manipulate subcellular structures. Although in their infancy, these methods could be developed and used to complete complex surgeries at the nanoscale. Cellular level surgery has been proposed using a nano robot based on AFM technology [72]. It has multiple functions including imaging, characterizing mechanical properties, and tracking. Furthermore, the technique of tip functionalisation facilitates the robot ability for precisely delivering a drug. Therefore, the nano robot can be used for conducting complicated nano surgery on samples such as live cells and bacteria. Additionally, the software in this nano robot provides a "videolized" visual feedback for simultaneously monitoring the operation of nano surgery and observation of the surgery results.

8.1.3.2   Implantable Devices

Nanotechnology is in the process of modernising implantable devices. AMDs that were of macro proportions are now being miniaturised and made more efficient. This is partly due to increased knowledge in the field of biomimetics, the process of using the way in which nature successfully produces something to create a material or device [20].

Implantable devices can cover many aspects and can deal with sensory issues as well as regulatory issues such as glucose sensors, retinal implants, prostheses with nerve control to name but a few. The ideal scenario is to reach a stage of miniaturisation and biocompatibility that can then produce devices capable of autonomous power, self-diagnosis, remote control and external transmission of data [73].

The auditory nerve contains near to 30,000 axons which cochlear implants stimulate [74]. From 1972 near a 100,000 people have been fitted with the device, however the work being developed is focusing on the nanoscale as the ear contains such structures [75]. The research that produced in this field is focused on developing smaller implants that can aid the work of axons to produce and relay an audible signal for the brain to process.

The optical nerve contains near a million fibres, and must deal with complex data and its processing [74]. In 2007, Alteheld and colleagues compiled a review to report on the developments of visual implants [76]. Electrical stimulation of the retina of blind subjects resulted in ambulatory vision (allowed movements without collisions and stumbling) and some character recognition. This was done by developing a wireless intraocular prosthesis, and having external feedback. Most recently a trial was carried out on 20 patients with varying degrees of blindness, to determine the electric charges needed to stimulate visual perception [77]. In 15 patients this was concluded to be in the range of 20–768 nC. This work illustrates the progress being made within this field, from exacting some visualisation to working out the exact range of charge needed to induce a visual event.

Another key AMD that is being developed for clinical use, is an implantable drug reservoir. This can be used in many scenarios but primarily for diabetics. This concept is to have a system in the body that has sensor that monitors a particular metabolite and releases the needed drug. Microelectromechnical systems (MEMs) and nanoelectromechanical system (NEMs) are ideal for this purpose. Some of these systems include microneedle-based transdermal devices, and micropump-based devices. Most notably the latter has been used for insulin delivery, glucose injection for diabetes, and administration of neurotransmitters to neurons [74].

It is apparent that progress within implantable devices is being made rapidly. This can be attributed to the technological advances and ease of access to information. The range of work being done is broad, focusing on producing an applicable product or technique, as well as optimising it.

8.1.3.3 Tissue Regeneration and Prosthetics

The goal of tissue engineering and prosthetics are very similar. The idea is to provide a platform for the body to either encourage repair or an opportunity to assimilate a new material (the surface of a joint replacement) and possibly to graft new material to a damaged organ. With an ageing population, these options are becoming necessary, especially when you consider the damage done by lifestyles and everyday use of organs and joints.

Considering the lifetime associated with orthopaedic implants and the rigours it goes through, nanomaterials are a viable alternative for current materials. The lifetime of implants are severely hampered by the eventual loosening between the joint and prosthesis [78]. This implies the bonding cement, polymethyl methacrylate (PMMA) that is widely used, is inefficient. Using nano-manipulated titanium showed an improvement in adhesion and promotion of cell growth. By introducing sub-micron features on titanium compared to flat surfaces, led an increase of endothelial cell adhesion density by 200 %. Whereas nanometre surface features promoted a 50 % increase in endothelial cell adhesion density compared to flat titanium surfaces. Using aligned patterns of such features on titanium, results highlighted that both endothelial and bone cells preferentially adhered onto sub-micron and nanometre enhanced surface features when compared to flat regions [79]. The use of titanium oxide nanotubes in dental implants were shown to be arrayed like collagen fibres. Furthermore the nanotubes increased roughness and surface area providing superior performances in multiple areas. It had greater hydrophilicity, greater cell adhesion and growth for osteoblasts and a better bioactivity and compatibility than current materials [80].

Tissue engineering is similar to this aspect in that it wants to promote the body to repair regenerate new cells. This can be done by introducing an extracellular matrix (ECM) that the body recognises as 'self'. Signals are transmitted between the cell and the ECM would facilitate communication for "cell adhesion, migration, growth, differentiation, programmed cell death, modulation of cytokine and growth factor activity, and activations of intracellular signalling" [74].

For example, hybrid biomaterials are being employed to better emulate natural ECM. These materials can be used to encourage healing while reducing the formation of scarring [93]. RGD peptides (R: arginine; G: glycine; D: aspartic acid) have been found to promote cell adhesion, RGD have also been amalgamated into man made ECMs to promote cell proliferation [81].

Some of the topics mentioned in this section are varied; however they do fall under the category of medical devices under the definition of the WHO. The connection may not be obvious but the "replacement, modification, or support of the anatomy or of a physiological process" combined with the monitoring and diagnosis of using a technology, renders that a medical device. The added caveat of the device being of nano-dimensions or functionality not only drives this branch of medicine but provides the epithet of 'advanced'.

8.1.4   Hindrances and Effects to be Considered

So far the focus has been on the positive impacts of nanotechnology and AMDs. However there are some minor and/or serious hindrances and effects that pertain to the success, ethical considerations, regulatory issues and the inherent toxicity of some NMs need to be considered before making an informed decisions.

8.1.4.1 Scientific Hindrances

The work discussed here are some of the more important breakthroughs within AMDs. However this does not even begin to compare to the volume and standard of work begun. In 2007, the amount of papers being published that were nano-related numbered near 15,000, this influx of publications is also reflected within the movements in intellectual property. In 2003, 90,000 patents were submitted for consideration. This illustrates the sheer amount of activity this field has generated.

However, the issues surrounding the use of this technology are manifold. A prime example of this is the use of NPs as imaging agents that are introduced within the body. Usually only 5 % of NPs administered to the body remain after 12 h, while 80 % of the initial dose are eliminated from the body. Further to this it is very difficult to ascertain how swiftly these particles would go to the intended target, NPs would be in systemic circulation for an extended period of time once they are released in a subject [82]. QDs are especially perilous as their outer shells can be worn away and can cause the production of free radicals and DNA nicking [83]

Another issue within progression and realisation of AMDs are the kind of studies that are being done. There is a distinct lack of comparability between investigations which can be attributed to one cause: the drive of getting published. Researchers and their funding are heavily linked to their work being heralded in journals of repute. Consequently, the techniques they use are always new, protocols and models are novel and inventive. Therefore, promising work is sometimes not explored and improved upon by other scientists, which leaves avenues incomplete.

Another issue with publications is that the lack of "no effect" studies is hindering the body of knowledge accrued [84]. This would provide researcher with protocols and methods of investigations that are a waste of resources. Moreover, having access to such data would provide a more complete picture, this would be crucial in designing possible resolutions for projects. In light of this the editors of three scientific journals have agreed to also publish the results of "no-effect studies" [85]. This would be instrumental in the use of resources of academia, governments and industries to translate scientific research into consumable products such as AMDs.

8.1.4.2 Regulatory Hindrances

As has already been discussed NMs size heavily contributes to their unique properties. However, this same feature that allow NMs to interact with biological entities but can also have deleterious effects on cellular mechanisms, and cell viability itself. It must also be considered that as of 2010, NMs have been used or present within over 800 consumer products [86]. There is a growing concern among cynics and advocates of NMs about their toxic potential, and there are calls from some quarters to have a global moratorium on all research and release of products, until regulations are in place to ensure protection from hazards [87].

Toxicity issues range from the points of entry to the range of effects themselves as shown in Table 8.1. Size and biopersistence play a large role in the extent of damage incurred. NMs can enter the body upon contact of the skin, inhalation and ingestion. This can cause a variety of issues ranging from free radical causing DNA damage, cell death and the formations of cancer [85].

In 2007, the European health and safety executive (HSE) initiated the Registration, Evaluation, Authorisation and restriction of Chemicals (REACH). Previous to this regulation, it was supposed that NMs act identically to their bulk equivalents, whereas studies show this is far from the truth. The necessity to weigh the impact of NMs, and also set some guidelines in how they should be approached commercially is key to the propagation of nanotechnology in AMDs. This however is not enough, and to remedy this, a motion has been passed by the European parliament to reassess the regulation of handling of NMs. Guidance has also been set for companies and they are recommended to adopt the precautionary principle [88].

In the US the Occupational Safety and Health Act (OSHA) of 1970 protects employees from illness at work. The slow response of OSHA standards-setting compared to the swiftly developing nanotechnology sector has left many staffs unprotected [89].

However in March 2008, the FDA and nano-health Alliance organised a workshop to identify the most pressing barriers to the success of nanotechnology. They classified the followings as the most important hurdles:

• Determination of the distribution of nanoparticulate carriers in the body following systemic administration through any route;
• Development of imaging modalities for visualizing the biodistribution over time;
• Understanding mass transport across compartmental boundaries in the body;
• Development of new mathematical and computer models that will lead to predicting risk and benefit parameters;
• Establishment of standards or reference materials and consensus testing protocols that can provide benchmarks for the development of novel classes of materials; and
• Realization of an analytical toolkit for nanopharmaceutical manufacturing, accompanied by a specification sheet of toxicological, safety, and biodistribution properties obtained through standardized, validated methods" [74].

These initiatives that are being initialised by various taskforces are an indication of intent. The progress of nanotechnology is held in high regard by industry and academia and its success must be ensured.

8.2 Summary and Outlook

AMDs are part of a wave of health related technology that has been pervading the medical field for at least three decades. However, this aspect of medicine is being driven to new heights by the dawn of nanotechnology. The research presented here is not definitive; the purpose of this work was to highlight the present climate of AMDs and the impact of NMs, NPs and novel techniques.

AMDs range from imaging agents, to techniques such as optical tweezers. The purpose of which is to reach a resolution within cellular activity that actually identifies the structure, function and even movement of specific cell mechanisms. This is typified by the capabilities of AFM and optical tweezers. The issue with the research examined, is that none of these techniques are making it to the clinical setting.

However, the potential within the diagnostic techniques covered is vast. The most common future is the combination of QDs and MRI contrast agents. They have the potential to be combined into theranostics, devices/particles that not only give information about disease but deliver drugs to them as well. Ideally these particles would be loaded with therapeutic agents and functionalised with target and imaging ligands, thereby providing an intermediary that has bio-penetration, medication and surveillance capabilities.

Similarly biosensors have been revolutionised by the introduction of nanotechnology. Generally they all have the same type of technology; a recognition element and the transducer technology. They can be defined by their transducer element, and the three most common were discussed. The issue between them of course lies between the interface between the recognition and transducer, the more integrated it is, the better the sensor. The dynamic range, specificity and sensitivity have been improved beyond compare by NMs. It is yet to be seen whether a particular type of architecture will translate on to the commercial market, what is certain is the possibilities on the horizon [90]. Of course the discovery of NMs and conducting polymers has made this easier and is bringing POC and LOC technologies within grasp, as the case study of MIPs has conveyed.

Surgical applications of AMDs have also been accelerated by the nano-revolution. For example, as shown in Fig. 4, the tips of AFM could be modified using EBIM to create 'nano scalpels' and 'nano needles'. This technique has allowed surgery at the subcellular level. This could be combined with implantable devices and tissue grafts to make near flawless additions to the human body with minimal damage. However most of the literature only deals with in vivo and in vitro. This technology will only move forward if more robust studies can push this into the clinical arena.

The future looks quite interesting in the way of coupling nanosurgery with wireless robotic surgery such as the da vinci robot. In 2010, over 300,000 surgeries were completed using robotic techniques. It is motivating that miniaturisation is being heralded as one of key implications to drive this technology forward [91]. Using this with nanotools and combining them with gold standard clinical practice would drive this even further.

It has been demonstrated that nanotechnology has impacted the development and evolution of AMDs. The use of NMs can promote an ordinary MD to the advanced category. There are some barriers to overcome, but it is globally understood that this technology needs to be nurtured correctly for it to come to fruition. It has provided platforms for interdisciplinary collaborations and is commercially lucrative for all parties involved. Most importantly this nanotechnology and AMDs will change the face of global healthcare.

References

1. Zheng J (2014) Layout of nanotechnology patents in global market. Adv Mat Res 889: 1578–1584
2. Hansen SF, Maynard A, Baun A, Tickner JA (2008) Late lessons from early warnings for nanotechnology. Nat Nanotechnol 3:444–447
3. Flynn T, Wei C (2005) The pathway to commercialization for nanomedicine. Nanomedicine 1(1):47–51
4. Feynman R (1960) There's plenty of room at the bottom. Caltech's Eng Sci 23:22–36
5. Patel DN, Bailey SR (2007) Nanotechnology in cardiovascular medicine. Catheter Cardiovasc Interv 69:643–654
6. European Science Foundation (2004) Nanomedicine – an ESF–European Medical Research Councils (EMRC) forward look report, ESF, Strasbourg cedex
7. Webster TJ (2006) Nanomedicine: what's in a definition? Ch. 2. Int J Nanomed 1:115–116
8. World Health Organization (2003) Medical device regulations: global overview and guiding principles. World Health Organization, Geneva
9. Cheng M (2003) Medical device regulations global overview and guiding principles. WHO, Geneva
10. Weissleder R, Pittet MJ (2008) Imaging in the era of molecular oncology. Nature 452:580–589
11. Massoud TF, Gambhir SS (2003) Molecular imaging in living subjects: seeing fundamental biological processes in a new light. Genes Dev 17(5):545–580
12. Jokerst JV, Gambhir SS (2011) Molecular imaging with theranostic nanoparticles. Acc Chem Res 44:1050–1060
13. Weissleder R, Ross BD, Rehemtulla A, Gambhir SS (2010) Molecular imaging: principles and practice. People's Medical Publishing House, Shelton
14. Jamali AA, Pourhassan-Moghaddam M, Dolatabadi JEN, Omidi Y (2014) Nanomaterials on the road to microRNA detection with optical and electrochemical nanobiosensors. TrAC Trend Anal Chem 55:24–42
15. Liu TY, Lo CL, Huang CC, Lin SL, Chang CA (2014) Engineering nanomaterials for biosensors and therapeutics. In: Cai W (ed) Engineering in translational medicine. Springer, London, pp 513–534
16. Luong JHT, Male KB, Glennon JD (2008) Biosensor technology: technology push versus market pull. Biotechnol Adv 26:492–500
17. Privett BJ, Shin JH, Schoenfisch MH (2008) Electrochemical sensors. Anal Chem 80: 4499–4517
18. Pushkarsky I, Tseng P, Murray C, Di Carlo D (2014) Research highlights: microfluidics and magnets. Lab Chip 14:2882–2886
19. Sung JH, Kama C, Shuler ML (2010) A microfluidic device for a pharmacokinetic–pharmacodynamic (PK–PD) model on a chip. Lab Chip 10:446–455
20. Asiyanbola B, Soboyejo W (2008) For the surgeon: an introduction to nanotechnology. J Surg Educ 65:155–161
21. Rosenthal SJ, McBride J, Pennycook SJ, Feldman LC (2007) Synthesis surface studies, composition and structural characterization of CdSe, core/shell and biologically active nanocrystals. Surf Sci Rep 62:111–157
22. Michalet XF, Pinaud F, Bentolila LA, Tsay JM, Doose S, Li JJ, Sundaresan G, Wu AM, Gambhir SS, Weiss S (2005) Quantum dots for live cells in vivo imaging, and diagnostics. Science 307:538–544
23. Parak WJ, Gerion D, Pellegrino T, Zanche D, Micheel C, Williams SC, Boudreau R, Le Gros MA, Larabell CA, Alivisatos AP (2003) Biological applications of colloidal nanocrystals. Nanotechnology 14:15–27
24. Akerman ME, Chan WC, Laakkonen P, Bhatia SN, Ruoslahti E (2002) Nanocrystal targeting in vivo. Proc Natl Acad Sci 99:12617–12621
25. Jaiswal JK, Mattoussi H, Mauro JM, Simon SM (2002) Long-term multiple colour imaging of live cells using quantum dot bioconjugates. Nat Biotechnol 21:47–51

26. Ye L, Yong K-T, Liu L, Roy I, Hu R, Zhu J, Cai H, Law W-C, Liu J, Wang K, Liu J, Liu Y, Hu Y, Zhang X, Swihart MT, Prasad PN (2012) A pilot study in non-human primates shows no adverse response to intravenous injection of quantum dots. Nat Nanotechnol 7:453–458

27. Mattoussi H, Palui G, Na HB (2012) Luminescent quantum dots as platforms for probing in vitro and in vivo biological processes. Adv Drug Deliv Rev 64:138–166

28. Na HB, Song IC, Hyeon T (2009) Inorganic nanoparticles for MRI contrast agents. Adv Mater 21:2133–2148

29. Steinmetz NF (2010) Viral nanoparticles as platforms for next-generation therapeutics and imaging devices. Nanomedicine 6:634–641

30. Allen M, Bulte JWM, Liepold L, Basu G, Zywicke HA, Frank JA, Young M, Douglas T (2005) Paramagnetic viral nanoparticles as potential high-relaxivity magnetic resonance contrast agents. Magn Reson Med 54:807–812

31. Sun C, Lee JSH, Zhang M (2008) Magnetic nanoparticles in MR imaging and drug delivery. Adv Drug Deliv Rev 60:1252–1265

32. Veiseh O, Gunn JW, Zhang M (2010) Design and fabrication of magnetic nanoparticles for targeted drug delivery and imaging. Adv Drug Deliv Rev 62(3):284–304

33. Cormode DP, Sanchez-Gaytan BL, Mieszawska AJ, Fayad ZA, Mulder WJM (2013) Inorganic nanocrystals as contrast agents in MRI: synthesis coating and introduction of multifunctionality. NMR Biomed 26:766–780

34. Gunasakera UA, Pankhurst QA, Douek M (2009) Imaging applications of nanotechnology in cancer. Target Oncol 4:169–181

35. Lyubchenko YL, Kim BH, Krasnoslobodtsev AV, Yu J (2010) Nanoimaging for protein misfolding diseases. Nanomed Nanobiotechnol 2:526–543

36. Connelly L, Jang H, Teran Arce F, Capone R, Kotler SA, Ramachandran S, Kagan BL, Nussinov R, Lal R (2012) Atomic force microscopy and MD simulations reveal pore-like structures of all-d-enantiomer of Alzheimer's β-amyloid peptide: relevance to the ion channel mechanism of AD pathology. J Phys Chem B 116:1728–1735

37. Burke KA, Yates EA, Legleiter J (2013) Amyloid-forming proteins alter the local mechanical properties of lipid membranes. Biochemistry 52:808–817

38. Zou Q, Leang KK, Sadoun E, Reed MJ, Devasia S (2004) Control issues in high-speed AFM for biological applications: collagen imaging example. Asian J Control 6:164–178

39. Voitchovsky K, Contera SA, Kamahira M, Watts A, Ryan JF (2006) Differential stiffness and lipid mobility in the leaflets of purple membranes. Biophys J 90:2075–2085

40. Yamashita H, Voitchovsky K, Uchihashi T, Contera SA, Ryan JF, Toshio A (2009) Dynamics of bacteriorhodopsin 2D crystal observed by high-speed atomic force microscopy. J Struct Biol 167:153–158

41. Van Mamaren J, Wuite GJL, Heller I (2011) Introduction to optical tweezers: background system designs, and commercial solutions. Methods Mol Biol 783:1–20

42. Ermilov SA, Murdock DR, Qian F, Brownell WE, Anvari B (2007) Studies of plasma membrane mechanics and plasma membrane of plasma mem interactions using optical tweezers and fluorescence imaging. J Biomech 40:476–480

43. Pine J, Chow G (2009) Moving live dissociated neurons with an optical tweezer. Trans Biomed Eng 56:1184–1188

44. Oddos S, Dunsby C, Purbhoo MA, Chauveau A, Owen DM, Neil MAA, Davis DM, French PMW (2008) High-speed high-resolution imaging of intercellular immune synapses using optical tweezers. Biophys J 95:66–68

45. Ashok PC, Dholakia K (2012) Optical trapping for analytical biotechnology. Curr Opin Biotechnol 23:16–21

46. Blesener T, Mondal A, Menon JU, Nguyen KT, Mohanty S (2013) Optical tweezers based measurement of PLGA-NP interaction with prostate cancer cells. Prog Biomed Opt Imaging – Proc SPIE 8594, Article no. 859407

47. Nakamura H, Karube I (2003) Current research activity in biosensors. Anal Bioanal Chem 377:446–468

48. Grieshaber D, MacKenzie R, Voros J, Reimhult E (2008) Electrochemical biosensors – sensor principles and architectures. Sensors 8:1400–1458

49. Velasco-Garcia MN (2009) Optical biosensors for probing at the cellular level: a review of recent progress and future prospects. Sem Cell Dev Biol 20:27–33
50. Dey D, Goswami T (2011) Optical biosensors: a revolution towards quantum nanoscale electronics device fabrication. J Biomed Biotechnol, Article no. 348218, 7 pages, 2011. doi:10.1155/2011/348218
51. Tombelli S, Minunni M, Mascini M (2005) Piezoelectric biosensors: strategies for coupling nucleic acids to piezoelectric devices. Methods 37:48–56
52. Cooper MA (2003) Label-free screening of bio-molecular interactions. Anal Bioanal Chem 377:834–842
53. Davis JJ, Coleman KS, Azamian BR, Bagshaw CB, Green MLH (2003) Chemical and biochemical sensing with modified single walled carbon nanotubes. Chem Eur J 9:3732–3739
54. Du D, Chen S, Song D, Li H, Chen X (2008) Development of acetylcholinesterase biosensor based on CdTe quantum dots/gold nanoparticles modified chitosan microspheres interface. Biosens Bioelectron 24:475–479
55. Deng Z, Zhang Y, Yue J, Tang F, Wei Q (2007) Green and orange CdTe quantum dots as effective pH-sensitive fluorescent probes for dual simultaneous and independent detection of viruses. J Phys Chem 111:12024–12031
56. Zhang X, Guo Q, Cui D (2009) Recent advances in nanotechnology applied to biosensors. Sensor 9:1033–1053
57. Li NB, Park JH, Park K, Kwon SJ, Shin H, Kwak J (2008) Characterization and electrocatalytic properties of Prussian blue electrochemically deposited on nano-Au/PAMAM dendrimer-modified gold electrode. Biosens Bioelectron 23:1519–1526
58. Abbas A, Fei M, Tian L, Singamaneni S (2013) Trapping proteins within gold nanoparticle assemblies: dynamically tunable hot-spots for nanobiosensing. Plasmonics 8:537–544
59. Dorvel BR, Reddy B, Go J, Duarte Guevara C, Salm E, Alam MA, Bashir R (2012) Silicon nanowires with high-k hafnium oxide dielectrics for sensitive detection of small nucleic acid oligomers. ACS Nano 6:6150–6164
60. Lakshmi D, Bossi A, Whitcombe MJ, Chianella I, Fowler SA, Subrahmanyam S, Piletska EV, Piletsky SA (2009) Electrochemical sensor for catechol and dopamine based on a catalytic molecularly imprinted polymer-conducting polymer hybrid recognition element. Anal Chem 81:3576–3584
61. Piletsky SA, et al. (2005) Molecularly imprinted polymers-tyrosinase mimics. Ukrainskii biokhimicheskii zhurnal 77.6:63
62. Vasapollo G, Del Sole R, Mergola L, Lazzoi MR, Scardino A, Scorrano S, Mele G (2011) Molecularly imprinted polymers: present and future prospective. Int J Mol Sci 12:5908–5945
63. Berti F, Todrosb S, Lakshmi D, Whitcombec MJ, Chianellac I, Ferronib M, Piletskyc SA, Turnerc APF, Marrazzaa G (2010) Quasi-monodimensional polyaniline nanostructures for enhanced molecularly imprinted polymer-based sensing. Biosens Bioelectron 26:497–503
64. Akbulut M, Lakshmi D, Whitcombe MJ, Piletska EV, Chianella I, Guven O, Piletsky SA (2011) Microplates with adaptive surfaces. ACS Comb Sci 13:646–652
65. Ivanova-Mitseva PK, Fragkou V, Lakshmi D, Whitcombe MJ, Davis F, Guerreiro A, Crayston JA, Ivanova DK, Mitsev PA, Piletska EV, Piletsky SA (2011) Conjugated polymers with pendant iniferter units: versatile materials for grafting. Macromolecules 44:1856–1865
66. Dixit CK, Kaushik A (2012) Nano-structured arrays for multiplex analyses and lab-on-a-chip applications. Biochem Biophys Res Commun 419:316–320
67. Jain KK (2007) Nanotechnology in medical practice. Med Princ Pract 17:89–101
68. Shen N, Datta D, Schaffer CB, LeDuc P, Ingber DE, Mazur E (2005) Ablation of cytoskeletal filaments and mitochondria in live cells using a femtosecond laser nanoscissor. Mech Chem Biosys 2:17–25
69. Uchugonova A, Zhang H, Lemke C, König K (2011) Nanosurgery with near-infrared 12-femtosecond and picosecond laser pulses. Prog Biomed Opt Imaging –Proc SPIE 7903, Article no. 79031N
70. Tserevelakis GJ, Psycharakis S, Resan B, Brunner F, Gavgiotaki E, Weingarten K, Filippidis G (2012) Femtosecond laser nanosurgery of sub-cellular structures in HeLa cells by employing Third Harmonic Generation imaging modality as diagnostic too. J Biophotonics 5:200–220

71. Beard JD, Gordeev SN, Guy RH (2011) AFM nanotools for surgery of biological cells. J Phys Conf Ser 286, Article no. 012003
72. Song B, Yang R, Xi N, Patterson KC, Qu C, Lai KWC (2012) Cellular-level surgery using nano robots. J Lab Autom 17:425–434
73. Boisseau P, Loubaton B (2011) Nanomedicine, nanotechnology in medicine. C R Phys 12:620–636
74. Murday, James S., et al. (2009) Translational nanomedicine: status assessment and opportunities. Nanomedicine: Nanotechnology, Biology and Medicine 5.3:251–273
75. Duke T (2003) Hair bundles: nano-mechanosensors in the inner ear. J Phys Condens Matter 15:1747–1757
76. Alteheld N, Roessler G, Walter P (2007) Towards the bionic eye – the retina implant: surgical, opthalmological and histopathological perspectives. Acta Neurochir Suppl 97:487–493
77. Keserü M, Feucht M, Bornfeld N, Laube T, Walter P, Rössler G, Velikay-Parel M, Hornig R, Richard G (2012) Acute electrical stimulation of the human retina with an epiretinal electrode array. Acta Ophthalmol 90:1–8
78. Liu-Snyder P, Webster TJ (2008) Developing a new generation of bone cements with nanotechnology. Curr Nanosci 4:111–118
79. Khang D, Lu J, Yao C, Haberstroh KM, Webster TJ (2008) The role of nanometer and sub-micron surface features on vascular and bone cell adhesion on titanium. Biomaterials 29:970–983
80. Wang F, Shi L, He W-X, Han D, Yan Y, Niu Z-Y, Shi S-G (2013) Bioinspired micro/nano fabrication on dental implant-bone interface. Appl Surf Sci 265:480–488
81. Hersel U, Dahmen C, Kessler H (2003) RGD modified polymers: biomaterials for stimulated cell adhesion and beyond. Biomaterials 24:4385–4415
82. Bae YH, Park K (2011) Targeted drug delivery to tumors: myths reality and possibility. J Control Release 153:198–205
83. Green M, Howman E (2005) Semiconductor quantum dots and free radical induced DNA nicking. Chem Commun 1:121–123
84. Hankin S, Boraschi D, Dushci A, Lehr CM, Lechtenbeld H (2011) Towards nanotechnology regulation – publish the unpublishable. Nanotoday 6:228–231
85. Krug HF, Wick P (2011) Nanotoxicology: an interdisciplinary challenge. Angew Chem Int Ed 50:1260–1278
86. Savolainen K, Aleniusa H, Norppa H, Pylkkänen L, Tuomi T, Kasper G (2010) Risk assessment of engineered nanomaterials and nanotechnologies – a review. Toxicology 269:92–104
87. Nel A, Xi T, Madler L, Li N (2006) Toxic potential of materials at the nanolevel. Science 311:622–627
88. Seaton A, Tran L, Aitken R, Donaldson K (2010) Nanoparticles human health hazard and regulation. J R Soc Interface 7:119–129
89. Howard J (2011) Dynamic oversight: implementation gaps and challenges. J Nanopart Res 13:1427–1434
90. Bellan LM, Wu D, Langer RS (2011) Current trends in nanobiosensor technology. Nanomed Nanobiotechnol 3:229–246
91. Wedmid A, Llukani E, Lee DI (2011) Future perspectives in robotic surgery. BJU Int 108:1028–1103
92. Cooper, Matthew A, Victoria T. Singleton (2007) A survey of the 2001 to 2005 quartz crystal microbalance biosensor literature: applications of acoustic physics to the analysis of biomolecular interactions". Journal of Molecular Recognition 20.3:154–184
93. Caldorera-Moore, Mary, Nicholas A. Peppas (2009) Micro-and nanotechnologies for intelligent and responsive biomaterial-based medical systems. Advanced drug delivery reviews 61.15:1391–1401

Chapter 9
Wireless Actuation of Micro/Nanorobots for Medical Applications

Soichiro Tottori, Li Zhang, and Bradley J. Nelson

9.1 Introduction

People have envisioned tiny robots that can explore a human body, find and treat diseases since Richard Feynman's famous speech, "There's plenty of room at the bottom," in which the idea of a "swallowable surgeon" was proposed in the 1950s [1]. Even though we are at a state of infancy to achieve this vision, recent intense progress on nanotechnology and micro/nanorobotics has accelerated the pace toward the goal [2–6]. A number of research efforts have been recently published regarding the development from the basic principles and fabrication methods to practical applications [7–10]. Not limited in vivo applications, the integration of micro/nanorobots to lab-on-a-chip systems can also be foreseen because of the nature of their size and liquid operating environments [11, 12]. This interdisciplinary research of micro/nanorobot-based medical treatments or diagnosis has been investigated from many different aspects, such as locomotion, functionalization, imaging, biocompatibility, interface, etc (Fig. 9.1). In this chapter we focus on wireless actuation of micro/nanorobots, which plays an important role in locomotion and part of functionalization. The chapter starts from wireless locomotion by means of magnetic fields, bacteria, and chemical reaction, followed by wireless actuation of robotic tools that function to manipulate targets.

S. Tottori • B.J. Nelson (✉)
Institute of Robotics and Intelligent Systems, ETH Zurich, 8092 Zurich, Switzerland
e-mail: bnelson@ethz.ch

L. Zhang
Department of Mechanical and Automation Engineering, The Chinese University of Hong Kong, Shatin NT, Hong Kong, SAR, China

© Springer Science+Business Media New York 2014
Y. Ge et al. (eds.), *Nanomedicine*, Nanostructure Science and Technology,
DOI 10.1007/978-1-4614-2140-5_9

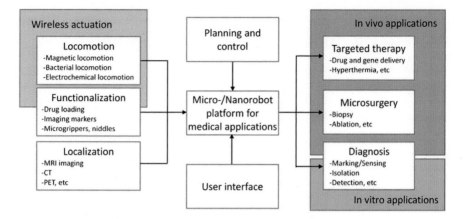

Fig. 9.1 A road map of micro/nanorobots for medical applications

9.2 Magnetic Actuation

Safety issues with respect to transmitting energy through body tissues and biocompatibility of the wirelessly driven device must be taken into account for in vivo applications. As MRI systems have been used extensively in clinical practice, it is accepted that the limited strength and frequency of these magnetic fields are safe. Therefore, magnetic actuation is one promising approach for powering and manipulating medical micro/nanorobots in vivo. Moreover, the human body is magnetically transparent, which implies that interference of magnetic fields by human bodies is practically negligible.

In general, magnetic actuation can be categorized into two types: force-driven and torque-driven. Magnetic attraction forces are generated by gradient fields, whereas magnetic torques are generated by misalignment of magnetizations of the devices and magnetic fields. Using the average magnetization $\mathbf{M}$ of the magnetic body in Amps per meter (A m^{-1}), the magnetic force and torque are described, respectively, as:

$$\mathbf{F} = V(\mathbf{M} \cdot \nabla)\mathbf{B} \qquad (9.1)$$

and

$$\mathbf{T} = V\mathbf{M} \times \mathbf{B} \qquad (9.2)$$

where V is the volume of the body, and $\mathbf{B}$ is the flux density of an external field in Tesla (T).

9.2.1 Gradient Field for Concentration and Steering of Magnetic Therapeutic Carriers

As described in Eq. 9.1, magnetic carriers are attracted to the direction of the high flux density of gradient fields. Magnetic nanoparticles are widely used to magnetically-tag therapeutic carriers [13, 14]. The approach of using gradient fields to move magnetic therapeutic carriers can be further divided, depending on the applied magnetic systems, into the following three groups: concentration using permanent magnets placed outside the body, direct concentration at implants, and steering of magnetic carriers in flows using the MRI system.

9.2.1.1 Concentrating Magnetic Carriers at Targeted Sites by External Magnets

Concentration of magnetic carriers using permanent magnets or electromagnets is probably the most straightforward method of using magnetic-field-based manipulation. Figure 9.2a shows that a strong magnet is placed on the surface or incised region of a patient body, and the magnetic therapeutic carriers flowing in the circulatory system are trapped in a local region by magnetic attraction [13, 16]. This approach has been investigated for treatment of tumors located at various sites in vivo, such as livers [17, 18], brains [19, 20] and lungs [15].

In Dames et al. [15], Rudolph and coworkers demonstrated that magnetic aerosols that contain superparamagnetic iron oxide nanoparticles (SPIONs) can be collected by external electromagnets in mouse lungs. The single SPIONs themselves do not have sufficient magnetization to be attracted by a gradient field, however, the magnetic aerosols (~3.5 μm) containing approximately 2,930 SPIONs can be guided in a high magnetic flux gradient (larger than 100 T m^{-1}). Figure 9.2b shows that SPIONs were collected in the right lung, as the magnet's tip was located directly above the right lung lobe, surgically exposed by thoracotomy. Figure 9.2c and d are histological images of right and left lungs, respectively, in which the brown areas show the accumulated SPIONs. The magnetic aerosol can also contain several different drugs, biodegradable nanoparticles, liposomes, nanocrystals and so on. The major challenge for scaling up from mouse to human size is generation of large magnetic gradient fields to a relatively far location from the tip of the electromagnet core.

9.2.1.2 Concentrating Magnetic Therapeutic Carriers Directly at Implants

An alternative approach to concentrate magnetic particles in vivo is proposed by using magnetic implants with an externally applied uniform field. Externally applied uniform fields induce magnetization of magnetic implants and flowing magnetic therapeutic carriers simultaneously. Gradient fields are generated surrounding the

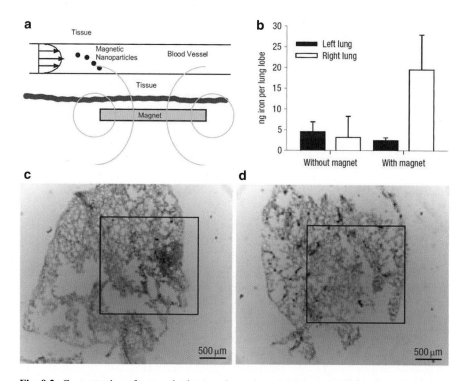

Fig. 9.2 Concentration of magnetic therapeutic carriers. (**a**) Schematic illustration of capturing magnetic therapeutic carriers: a magnet is place outside or an incised region in order to collect magnetic therapeutic carriers flowing in the circulatory system ((**a**) Reprinted with permission from Pankhurst et al. [14]. Copyright 2003 IOP Publishing Ltd.) (**b**) The bar graph shows the amount of SPION in the *left* (without a magnetic field) and *right* (with a magnetic field) lungs (thoracotomy, 400 breathing cycles, ~300 μl SPION solution, 12.5 mg ml^{-1} Fe, n=3±s.d.). (**c–d**) Lung histology after nanomagnetosol application. (**c**: right lung, **d**: left lung) ((**b–d**) Reprinted with permission from Dames et al. [15], Copyright 2007 Macmillan Publishing Ltd: Nature Nanotechnology)

ferromagnetic implants because the implants possess much higher magnetic permeabilities than that of air and bodily tissues. In comparison to the method described in the last section, i.e. direct use of gradient fields generated by external magnets, the method using magnetic implants can provide relatively strong gradient fields, because the source of gradient is installed at the targeted site inside the body [21]. Based on this concept, targeting of magnetic therapeutic carriers at the implanted sites has been demonstrated to deliver drugs or cells [22–26].

In Polyak et al. [26], Levy and co-workers demonstrated targeting delivery of magnetic nanoparticle-loaded endothelial cells to the stent surface, as illustrated in Fig. 9.3a. They applied the magnetic stents made from 304-grade stainless steel instead of conventional non-magnetic medical-grade stainless-steel 316 L. The bovine aortic endothelial cells (BAECs) functionalized with magnetic nanoparticles were accumulated onto the magnetic stents in the uniform field. In vivo acute rat

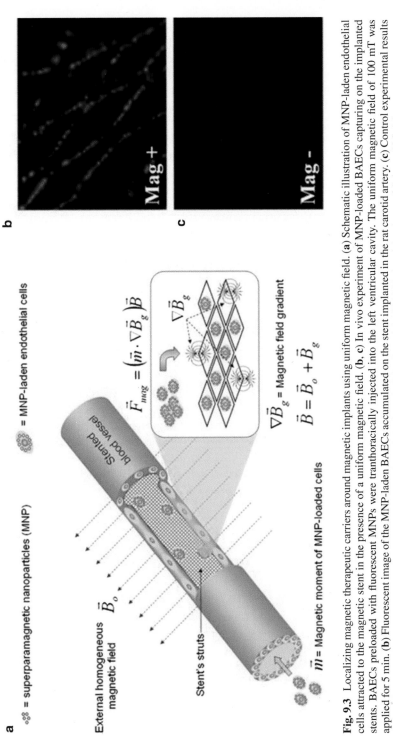

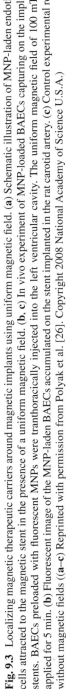

Fig. 9.3 Localizing magnetic therapeutic carriers around magnetic implants using uniform magnetic field. (**a**) Schematic illustration of MNP-laden endothelial cells attracted to the magnetic stent in the presence of a uniform magnetic field. (**b**, **c**) In vivo experiment of MNP-loaded BAECs capturing on the implanted stents. BAECs preloaded with fluorescent MNPs were tranthoracically injected into the left ventricular cavity. The uniform magnetic field of 100 mT was applied for 5 min. (**b**) Fluorescent image of the MNP-laden BAECs accumulated on the stent implanted in the rat carotid artery. (**c**) Control experimental results without magnetic fields ((**a–c**) Reprinted with permission from Polyak et al. [26]. Copyright 2008 National Academy of Science U.S.A.)

carotid stenting studies were performed by transthoracic injection of MNPs-loaded (MNP: magnetic nanoparticle) BAECs into the left ventricular cavity of the rat. Figure 9.3b and c show the fluorescent images of MNPs collected onto the 304-grade stainless steel stents with and without the uniform magnetic field, respectively. Clearly, the MNPs were collected on the surface of the stent in the presence of magnetic fields.

9.2.1.3   Steering Magnetic Therapeutic Carriers in Flows by MRI Systems

Using MRI systems to steer magnetic therapeutic carriers flowing in the circulatory systems is one of the emerging areas of magnetic control [27]. Localization of magnetic therapeutic carriers using permanent magnets is limited to the organs that are close to the skin because magnetic gradient fields decay rapidly as the distance between the target and the externally placed magnets increases. The gradient field provided by a MRI system is appealing for steering magnetic therapeutic carriers deep inside the body because of its gradient linearity, high-strength and real-time imaging feedback.

The MRI-system-based steering of magnetic therapeutic carriers in bifurcations in vitro and in vivo has been reported recently [28–31]. For example, in Pouponneau et al. [30] Martel and co-workers demonstrated in vivo guidance of magnetic therapeutic carriers for the treatment of hepatocellular carcinoma (HCC) via trans-arterial chemoembolization in the hepatic artery using rabbits. Figure 9.4a and b show that the magnetic therapeutic carriers: biodegradable polymer particles containing anti-tumor drug and magnetic nanoparticles were steered in the hepatic arterial bifurcation in the presence of a gradient field. In Fig. 9.4c, the result shows that the deposition of magnetic therapeutic carriers in the targeted and untargeted lobes were increased and decreased by MRI steering, respectively. In Fig. 9.4d, the magnetic therapeutic carriers, visualized as dark spots in MR images, were collected in the left lobe. The histological image in Fig. 9.4e shows that magnetic therapeutic carriers were deposited within a branch of the hepatic artery.

9.2.2   Magnetic-Torque-Driven Propulsion at Low Reynolds Number

Locomotion of micro/nanorobots in a fluidic environment at low Reynolds number is challenging, since viscosity dominates over inertia [32, 33]. A scallop-like motion, one of the typical examples of reciprocal motions, cannot get net forward displacement but just moves forward and backward repeatedly at the same place in a low Reynolds number regime because the flow is time-reversible. In nature, micro-organisms utilize various non-reciprocal locomotion strategies. For example, spermatozoa propel themselves by propagating a wave motion on their flexible tails, and *E. coli* bacteria swim by rotating their helical flagella.

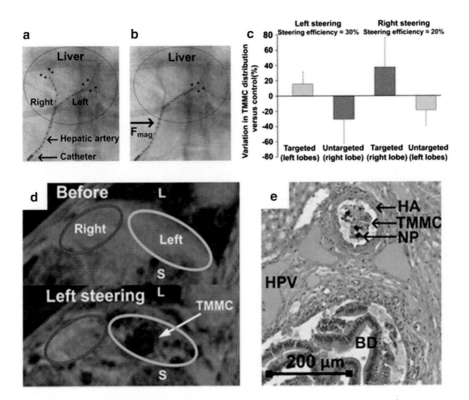

Fig. 9.4 Steering magnetic therapeutic carriers for liver chemoembolization using an upgraded MRI system. (**a, b**) Fluoroscopy images of the rabbit hepatic artery with superposed images of the magnetic therapeutic carriers without (**a**) and with (**b**) the MRI control. (**a**) The microparticles are released from the catheter in the artery and distributed to both lobes. (**b**) MRI steering of magnetic therapeutic carriers to left lobe to preserve the right lobe from the chemoembolization. (**c**) Variation in TMMC (therapeutic magnetic microcarriers) distribution versus control in the liver lobes with left and right steering based on Co analysis. Mean ± SD. (**d**) T2*-weighted MR images of the rabbit liver before and after the operation. The red and blue lines indicate the right and left lobe, respectively. "S" indicates the stomach and "L" the lung. (**e**) Histological image of liver parenchyma and the blood vessels dyed with hematoxylin and eosin. HA: branch of hepatic artery, HPV: hepatic portal vein, BD: bile duct, NP FeCo nanoparticles ((**a–e**) Reprinted with permission from Pouponneau et al. [30]. Copyright 2011 Elsevier Ltd)

Inspired by the above mentioned microorganisms, many microscopic artificial swimmers have been developed [34–44]. For example, artificial microswimmers with traveling wave-induced propulsion were presented by Dreyfus and co-workers in Dreyfus et al. [34]. Figure 9.5a and b show the illustration and optical microscope image of the microscopic swimmer consisting of a chain of colloidal magnetic particles connected by DNA linkages, respectively. Oscillating magnetic fields induce a wave motion into the flexible body, resulting in a net forward movement. Inspired by helical bacterial flagella, artificial bacterial flagella (ABFs) consisting of ferromagnetic Ni heads and rolled-up helical tails were reported by our group elsewhere

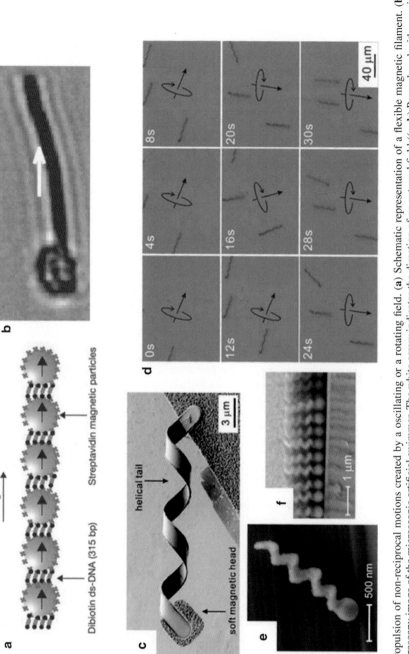

Fig. 9.5 Propulsion of non-reciprocal motions created by a oscillating or a rotating field. (**a**) Schematic representation of a flexible magnetic filament. (**b**) Optical microscopy image of the microscopic artificial swimmer. The white arrow indicates the direction of an external field ((**a**, **b**) Reproduced with permission from Dreyfus et al. [34]. Copyright 2005 Nature Publishing Group). (**c**) Helical microswimmer comprised of the magnetic head and the rolled-up helical tail. (**d**) Steering of the three ABFs as an entity ((**c**, **d**) Reproduced with permission from Zhang et al. [43]. Copyright 2009 American Chemical Society). (**e**) The microscopic helical swimmer fabricated with GLAD technique. (**f**) A wafer section with a nanostructured film after GLAD of SiO₂ helices ((**e**, **f**) Reproduced with permission from Ghosh and Fischer [37]. Copyright 2009 American Chemical Society)

[42, 43]. Figure 9.5c shows the SEM image of an as-fabricated ABF with a diameter of 2.8 µm and a length of approximate 25 µm. The top-down fabrication process of the ABFs is based on the self-rolling technique: curling generated by the stress releasing of multilayered nanoribbons [45, 46]. By applying a rotating magnetic field, the ABFs were rotated in sync with the external field rotation, resulting in a forward displacement. Swimming of the multi-agent ABFs was also demonstrated, in which the group was controlled as a single entity, as shown in Fig. 9.5d. In Ghosh and Fischer [37], Fischer and Ghosh reported an alternative fabrication approach of helical swimmers using glancing angle deposition (GLAD) of chiral structures, as shown in Fig. 9.5e. GLAD method provides porous thin films with engineered structures, such as straight pillars, zigzags structures, and helices [47]. The SEM image in Fig. 9.5f shows as-fabricated vertical array of SiO_2 helical nanostructures on self-assembled silica beads with a diameter of 200–300 nm, which had an ultra-high density of – 10^9 helices per cm^2. The helical nanostructures were released by sonication and subsequently were coated with a 30-nm thick Co thin film by evaporation. In order to generate a corkscrew motion, the helical devices were also actuated in a weak strength (6 mT) rotating magnetic field. The maximum swimming speed of approximately 40 µm/s (20 body-lengths per second) was reported. The helical swimming microrobots can also be utilized to manipulate microobjects with or without a physical contact [41, 44, 48, 49]. Once functionalized by liposomes loaded with chemical or biological substances, these helical devices have the potential to perform targeted delivery of energy and controlled drug releasing in vivo. Previously SiO_2 helical devices coated with a fluorophore were demonstrated as a proof of concept [37].

9.2.3 Scaling Effects of Micro/Nanorobots Driven by Magnetic Force and Torque

Stokes flow is the flow at very low Reynolds number (Re ≪ 1), which indicates the ratio of inertia and viscous forces. For instance, the drag force and torque of a sphere in Stokes flow are given by

$$\mathbf{F_{drag}} = 3\pi\mu d\mathbf{U} \tag{9.3}$$

and

$$\mathbf{T_{drag}} = \pi\mu d^3\Omega \tag{9.4}$$

where d is the diameter of the sphere, $\mathbf{U}$ is the translational velocity, and Ω is the angular velocity of the rotating sphere. Comparing with Eqs. 9.1 and 9.2, in which both magnetic force and torque are directly proportional to the volume of the magnetic body (for a sphere, i.e. $\pi d^3/6$), the maximum velocity and angular velocity by applying a gradient field and a uniform rotating field can be computed. The translational velocity **U** of a sphere induced by a gradient pulling is proportional to the term of d^2, whereas the angular velocity Ω by a uniform rotating field is independent of d. Therefore, when the robots are scaled-down, with the same applied field, the translational velocity will reduced rapidly, whereas the angular velocity can be maintained as a constant. This calculations also implies that if a rotational motion can be converted to a translational motion with a converting ration of less than d^2, the corkscrew locomotion based on rotational motions can be more efficient than that based on direct translational pulling using a gradient field [50].

The propelling motion of helical swimmers is described with the force f, torque τ, angular velocity ω, and translational velocity v as Purcell [33],

$$\begin{bmatrix} f \\ \tau \end{bmatrix} = \begin{bmatrix} A & B \\ B & C \end{bmatrix} \begin{bmatrix} v \\ \omega \end{bmatrix} \tag{9.5}$$

where the parameters A, B and C are given by geometrical parameters of a helix and the viscosity of environmental fluid [50, 51]. Since the parameters A, B and C are directly proportional to the terms L, L^2 and L^3 (here, L is the characteristic length of the helix) respectively, the velocity v is computed to be proportional to the characteristic length L. In comparison to the fact that the velocity induced by direct pulling is proportional to the second power of its characteristic length, the helical propulsion is more feasible for small scale locomotion due to the scaling effect.

9.3 Bacterial Actuation

Bacteria, as natural "microrobots," can perform robotic tasks with high intelligence. In nature, bacteria can respond to various environmental cues, for example, magnetic field (magnetotaxis), light (phototaxis), electric field (galvanotaxis), chemical concentration (chemotaxis), and heat (thermotaxis). Here we show two examples of micromanipulation using bacteria based on their magnetotaxis and phototaxis.

Magnetotactic bacteria (MTBs), which possess flagella and magnetosome, swim in the direction of a magnetic field [52]. In Martel et al. [53], Martel and coworkers demonstrated that the magnetic steering of MTBs, and explained that their motion can be potentially trackable by a clinical MRI system. Figure 9.6a and b show controlled swimming of a MTB in the presence of an external magnetic field with and without a cargo [54]. Martel's group also demonstrated micromanipulation using a swarm of MTBs, by which large thrust force can be generated for pick-and-place microobjects [56].

Using phototactic bacteria enables light control of micro/nanorobotic components [55, 57]. In Weibel et al. [55], Whitesides and co-workers demonstrated pick-and-place manipulation of micro cargos by phototaxis of bacteria and photoreactive chemical linkages. Figure 9.6c shows phototaxis of *Chlamydomonas reinhardtii* (*CR*) under LED lights. In order to transport microscale cargo, chemical

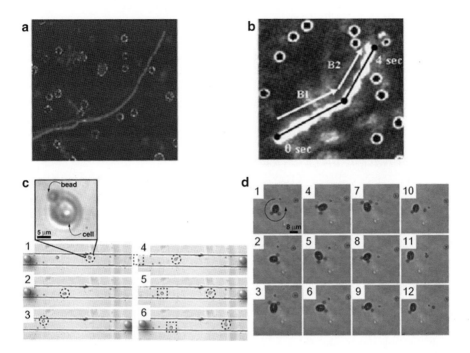

Fig. 9.6 Bacterial actuation. (**a**) Movement path of the MTB without cargos steered by an external field. (**b**) The single MTB pushed the 3-μm-diameter microbead with a control of external field. The white arrows B1 and B2 represent the direction of the field, and the black line indicates the path of microbead transported by the MTB. Images have edges of 36.0 μm ((**a, b**) Reproduced with permission from Martel et al. [54]. Copyright 2006 American Institute of Physics). (**c**) A sequence of images showing how the cell in C can be steered by using positive phototaxis. Illuminated LEDs are indicated by the presence of a cartoon of the LED. (**d**) A series of chronological images showing the photochemical release of a bead from a cell with two PS beads (3 μm diameter) attached by exposing UV light (λ=365 nm, 80 W) for 20 s. The arrow in frame 1 indicates the direction in which the cell is rotating. The time that had elapsed between the frames was 2 s. In (d9), the bead was released from the cell and slowly diffused away ((**c, d**) Adapted with permission from Weibel et al. [55]. Copyright 2005 National Academy of Science, U.S.A.)

links were formed between cargos and *CR* bodies. The attached-cargos can be photochemically released by exposing a UV light of a specific wavelength. Figure 9.6d shows the optical microscope image of photochemical releasing of the attached microbead from *CR* by illuminating the UV light.

9.4   Chemical-Fuel-Driven Actuation

Like biomotors, such as myosins, kinesins, and dyneins, utilize ATP as chemical fuels, manmade micro/nanomotors that propel themselves by employing chemical fuels have been developed for about a decade. The first bimetal nanomotors were demonstrated by two groups independently, one from Pennsylvania State University [58] and the other from University of Toronto [59] in 2004. A number of excellent reviews published recently regarding catalytic nanomotors from their fabrication, consideration of materials, and functionalization to their practical applications [60–63]. Here, we introduce the basic principle and some recently reported biomedical applications of these chemical-fuel-driven micro/nanomotors.

9.4.1 Principle and Fabrication

To date self-propelled micro/nanomotors exhibit locomotion in two types: bubble-induced propulsion and self-electrophoretic motion, as shown in Fig. 9.7a. The bubble-induced propulsion is based on the oxygen bubbles generated in decomposition of hydrogen peroxide. The tubular micro/nanomotors with asymmetric open ends, with conical channel, lead the bubbles emitted from the largely-opened ends. The conical microengines can be fabricated with self-rolling technique or template-assisted electroplating. The self-electrophoretic motion occurs with multisegment nanorods integrated with Au-Pt bimetals. The hydrogen peroxide is oxidized at the platinum segment, and the generated electrons travel to the gold segment, where reduction reaction occurs. Because of this asymmetrical redox reaction, the nanorods with Au-Pt bisegments move toward the direction of platinum segments. Ni segments are often integrated for steering by an external magnetic field. The multi-segmented nanorods are synthesized into the templates with cylindrical nanopores using template-assisted electrodeposition.

9.4.2 Applications

The chemical-fuel-driven micro/nanomotors can reach extremely high speeds, up to approximately 3 mm/s [67], thus the thrust force is sufficient to transport microobjects. Figure 9.7b shows transport and assembly of magnetic microplates using a microtubular motor [64]. Controlled manipulation of animal cells, i.e. CAD cells (cathecolaminergic cells), was also demonstrated, as shown in Fig. 9.7c [65]. To realize stable and selective pick-and-place manipulation, ferromagnetic layers can be integrated into the motor and the cargo for magnetic attraction [68, 69]. Releasing of the magnetic cargo is possible by turning the magnetic motor with a high speed, which is attributed to the different drag force applied on the motor and the cargo. Alternatively, UV lights can trigger photochemical releasing of the streptavidin-coated polystyrene cargo from the catalytic nanomotor [70], which is the same concept with the photochemical release of cargos from CR bacteria in Weibel et al. [55].

Surface-functionalized micro/nanomotors are applied for isolation of targeted biological samples, such as circulating tumor cells [71], nucleic acid [66], proteins [72], and bacteria [73], from raw complex biological media [11]. Figure 9.7d1

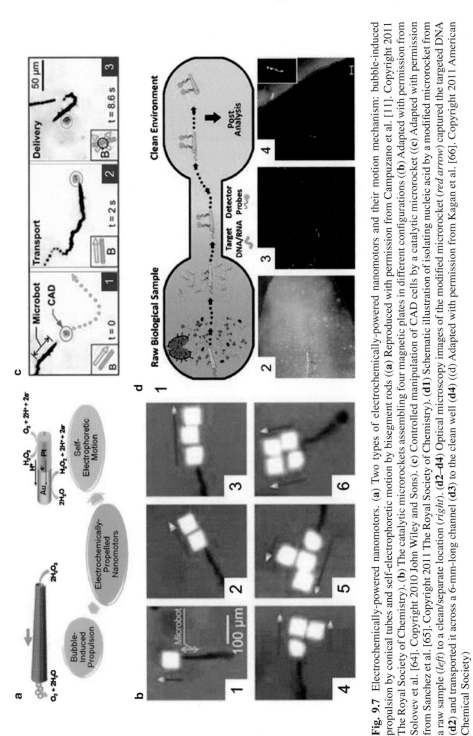

Fig. 9.7 Electrochemically-powered nanomotors. (**a**) Two types of electrochemically-powered nanomotors and their motion mechanism: bubble-induced propulsion by conical tubes and self-electrophoretic motion by bisegment rods ((**a**) Reproduced with permission from Campuzano et al. [11]. Copyright 2011 The Royal Society of Chemistry). (**b**) The catalytic microrockets assembling four magnetic plates in different configurations ((**b**) Adapted with permission from Solovev et al. [64]. Copyright 2010 John Wiley and Sons). (**c**) Controlled manipulation of CAD cells by a catalytic microrocket ((**c**) Adapted with permission from Sanchez et al. [65]. Copyright 2011 The Royal Society of Chemistry). (**d1**) Schematic illustration of isolating nucleic acid by a modified microrocket from a raw sample (*left*) to a clean/separate location (*right*). (**d2**–**d4**) Optical microscopy images of the modified microrocket (*red arrow*) captured the targeted DNA (**d2**) and transported it across a 6-mm-long channel (**d3**) to the clean well (**d4**) ((**d**) Adapted with permission from Kagan et al. [66]. Copyright 2011 American Chemical Society)

shows the schematic illustration of nucleic acid isolation by using functionalized motor from a raw biological sample to a clean microwell [66]. The outer surface of the microtubular motor was coated with a gold layer and modified with a binary self-assembled monolayer (SAM) of specific thiolated capture probe (SHCP) and a short-chain 6-mercapto-1-hexanol (MCH). The functionalized micromotors were then driven in the mixture of hydrogen peroxide and a biological sample that contains the nucleic acid (synthetic 30-mer DNA or bacterial 16S rRNA) tagged with fluorescent nanoparticles for optical visualization. Figure 9.7d2–d4 shows the optical microscopy images of capturing and transporting the targeted nucleic acids to the clean well using a micromotor.

Chemical-fuel-driven motors can also be applied for sensing based on the velocity change of their motion: the speed increases in high concentration of Ag ions in the hydrogen peroxide fuel [74]. The detection of nucleic acid was as well demonstrated by combining Ag nanoparticle-tagged detector probe [75].

To date in vivo application of chemical-fuel powered nanomotors is rare unless the current chemical fuels (mainly H_2O_2) can be replaced with biocompatible chemicals or chemicals which are inherently contained in the human body. Previously, as a proof-of-concept, bioelectrochemical motor driven by oxygen and glucose were reported [76].

9.5 Wirelessly-Actuated Robotic Tools

One of the next steps from wireless locomotion is wireless manipulation, such as holding, pinching, penetrating, etc, using nano-/microrobots. Variety of fabrication methods have been investigated and applied to create such functional wireless micro/nanorobots [77]. In this section, we explain some examples of recently reported fabrication methods and the wirelessly-actuated microrobotic tools developed with these method.

The microholders with comparable sizes enable pick-and-place manipulation of cargos without dropping off during transportation. In Tottori et al. [41], the crawl-like structures were implemented at one end of the helical micromachines. The SEM image of as-fabricated functionalized micormachine is shown in Fig. 9.8a. The structures were fabricated by means of three-dimensional lithography, in which the focused UV laser is scanned in the thin layers of photocurable polymer. In comparison to the conventional two-dimensional lithography, three-dimensional lithography provides almost arbitrary shapes with a submicron resolution. The combination of electron beam evaporation of ferromagnetic thin films allows magnetic wireless actuation of the devices. The functionalized helical micromachines were capable of transport polystyrene microbeads not only horizontally but also against gravity. The process is divided into the following four stages: approaching, loading, transporting, and releasing, as illustrated in Fig. 9.8b. A polystyrene microparticle with a diameter of 6 μm was transported in three dimensions above the Si substrate with two different heights, as shown in Fig. 9.8c. The loading and releasing of the microparticle were realized by pushing it forward and swimming backward, respectively.

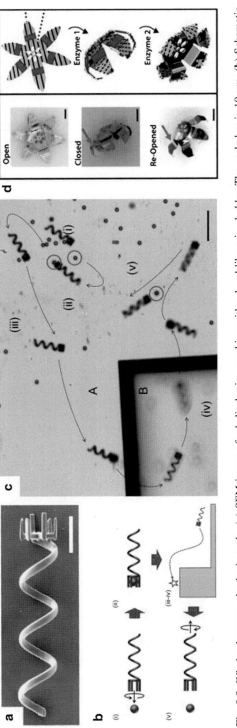

Fig. 9.8 Wirelessly-actuated robotic tools. (**a**) SEM image of a helical micromachine with a hand-like microholder. The scale bar is 10 μm. (**b**) Schematic illustration of cargo transport by the magnetic helical micromachine with a hand-like microholder. (**c**) Time lapse image of transporting 6-μm-diameter micro-bead in 3D. The scale bar is 50 μm ((**a–c**) Adapted with permission from Tottori et al. [41]. Copyright 2012 John Wiley and Sons). (**d**) Biochemically activated microgrippers. The scale bars are 200 μm ((**d**) Adapted with permission from Bassik et al. [78]. Copyright 2010 American Chemical Society)

For microsurgery applications, microtools capable of gripping may be highly useful. Gripping normally requires open-and-close motion of the devices. Based on the development of a micro-origami technique, reconfigurable microgrippers, capable of closing and re-opening wirelessly by external cues, have been fabricated with multilayer flexible joints possessing internal stress [79, 80]. By resolving or altering mechanical property of the reactive layers selectively, the microgrippers can close from a planar shape. Re-opening is also feasible by dissolve the other side of layer with different chemical etchant. A more elegant approach is to engineer the internal or surface stress of the films by biological stimuli. Such kind of autonomous microgrippers are realized recently, e.g. enzyme-responsive microgrippers [78], as shown in Fig. 9.8d. Proteases and glucosidases induce closing and re-opening, respectively. Remote motion control of the microgrippers integrated with a magnetic film can be done by using external magnets. To improve the positioning precision of miniaturized robotic tools, such as microgrippers, an electromagnetic control system is required. A good example of such an electromagnetic setup is OctoMag, which is capable of performing 5-DOF wireless micromanipulation of magnetic device in a large workspace [81].

9.6   Conclusions and Future Outlook

This chapter addressed the recent development of micro/nanorobots with focus on tetherless actuations for locomotion and/or manipulation. Precise motion control and functionalization of the micro/nanorobots by suitable wireless actuation methods enhances the quality of targeted medical treatments, i.e. high concentrating drug delivery and/or microsurgery with minimal side effects. The micro/nanorobots can also be integrated with in vitro biomedical applications, which may provide an alternative processing procedure in comparison to that of a conventional approach. Further investigations on not only wireless actuation technology but also imaging, biocompatibility, etc, are required in the future for practical medical applications.

References

1. Feynman R (1960) There's plenty of room at the bottom. Caltech's Eng Sci 23:22–36
2. Bhat A (2014) Nanobots: the future of medicine. Int J Manage Eng Sci 5(1):44–49
3. Manjunath A, Kishore V (2014) The promising future in medicine: nanorobots. Biomed Sci 2(2):42–47
4. Popa D (2014) Robust and reliable microtechnology research and education through the mobile microrobotics challenge [competitions]. Robot Autom Mag IEEE 21(1):8–12
5. Rao V (2014) Nanorobots in medicine – a new dimension in bio nanotechnology. Trans Netw Commun 2(2):46–57
6. Thangavel K, Balamurugan A, Elango M et al (2014) A survey on nano-robotics in nano-medicine. J NanoSci NanoTechnol 2(1):525–528

7. Nelson B, Kaliakatsos I, Abbott J (2010) Microrobots for minimally invasive medicine. Annu Rev Biomed Eng 12:55–85
8. Patel G, Patel G, Patel R et al (2006) Nanorobot: a versatile tool in nanomedicine. J Drug Target 14(2):63–67
9. Tibbals H (2010) Medical nanotechnology and nanomedicine. CRC Press, Boca Raton
10. Vogel V (2009) Nanotechnology. Nanomedicine, vol 5. Wiley-VCH, Weinheim
11. Campuzano S, Kagan D, Orozco J et al (2011) Motion-driven sensing and biosensing using electrochemically propelled nanomotors. Analyst 136(22):4621–4630
12. Whitesides G (2006) The origins and the future of microfluidics. Nature 442(7101):368–373
13. Dobson J (2006) Magnetic nanoparticles for drug delivery. Drug Dev Res 67(1):55–60
14. Pankhurst Q, Connolly J, Jones S et al (2003) Applications of magnetic nanoparticles in biomedicine. J Phys D Appl Phys 36(13):R167–R181
15. Dames P, Gleich B, Flemmer A et al (2007) Targeted delivery of magnetic aerosol droplets to the lung. Nat Nanotechnol 2(8):495–499
16. Torchilin V (2000) Drug targeting. Eur J Pharm Sci 11:S81–S91
17. Arbab A, Jordan E, Wilson L et al (2004) In vivo trafficking and targeted delivery of magnetically labeled stem cells. Hum Gene Ther 15(4):351–360
18. Wilson M, Kerlan R, Fidelman N et al (2004) Hepatocellular carcinoma: regional therapy with a magnetic targeted carrier bound to doxorubicin in a dual MR imaging/conventional angiography suite – initial experience with four patients. Radiology 230(1):287–293
19. Chertok B, David A, Yang V (2010) Polyethyleneimine-modified iron oxide nanoparticles for brain tumor drug delivery using magnetic targeting and intra-carotid administration. Biomaterials 31(24):6317–6324
20. Liu H, Hua M, Yang H (2010) Magnetic resonance monitoring of focused ultrasound/magnetic nanoparticle targeting delivery of therapeutic agents to the brain. Proc Natl Acad Sci U S A 107(34):15205–15210
21. Yellen B, Forbes Z, Halverson D et al (2005) Targeted drug delivery to magnetic implants for therapeutic applications. J Magn Magn Mater 293(1):647–654
22. Aviles M, Ebner A, Ritter J (2008) Implant assisted-magnetic drug targeting: comparison of in vitro experiments with theory. J Magn Magn Mater 320(21):2704–2713
23. Chorny M, Fishbein I, Yellen B et al (2010) Targeting stents with local delivery of paclitaxel-loaded magnetic nanoparticles using uniform fields. Proc Natl Acad Sci U S A 107(18):8346–8351
24. Kempe H, Kempe M, Snowball I et al (2010) The use of magnetite nanoparticles for implant-assisted magnetic drug targeting in thrombolytic therapy. Biomaterials 31(36):9499–9510
25. Pislaru S, Harbuzariu A, Gulati R et al (2006) Magnetically targeted endothelial cell localization in stented vessels. J Am Coll Cardiol 48(9):1839–1845
26. Polyak B, Fishbein I, Chorny M et al (2008) High field gradient targeting of magnetic nanoparticle-loaded endothelial cells to the surfaces of steel stents. Proc Natl Acad Sci U S A 105(2):698–703
27. Vartholomeos P, Fruchard M, Ferreira A et al (2011) MRI-guided nanorobotic systems for therapeutic and diagnostic applications. Annu Rev Biomed Eng 13:157–184
28. Mathieu J, Martel S (2010) Steering of aggregating magnetic microparticles using propulsion gradients coils in an MRI scanner. Magn Reson Med 63(5):1336–1345
29. Pouponneau P, Leroux J, Martel S (2009) Magnetic nanoparticles encapsulated into biodegradable microparticles steered with an upgraded magnetic resonance imaging system for tumor chemoembolization. Biomaterials 30(31):6327–6332
30. Pouponneau P, Leroux J, Soulez G et al (2011) Co-encapsulation of magnetic nanoparticles and doxorubicin into biodegradable microcarriers for deep tissue targeting by vascular MRI navigation. Biomaterials 32(13):3481–3486
31. Riegler J, Wells J, Kyrtatos P et al (2010) Targeted magnetic delivery and tracking of cells using a magnetic resonance imaging system. Biomaterials 31(20):5366–5371
32. Ludwig W (1930) Zur theorie der flimmerbewegung (Dynamik, Nutzeffekt, Energiebilanz). Physiol 13:397–504

33. Purcell E (1977) Life at low Reynolds-number. Am J Phys 45(1):3–11
34. Dreyfus R, Baudry J, Roper M et al (2005) Microscopic artificial swimmers. Nature 437(7060):862–865
35. Fischer P, Ghosh A (2011) Magnetically actuated propulsion at low Reynolds numbers: towards nanoscale control. Nanoscale 3(2):557–563
36. Gao W, Sattayasamitsathit S, Manesh K et al (2010) Magnetically powered flexible metal nanowire motors. J Am Chem Soc 132(41):14403–14405
37. Ghosh A, Fischer P (2009) Controlled propulsion of artificial magnetic nanostructured propellers. Nano Lett 9(6):2243–2245
38. Peyer K, Tottori S, Qiu F et al (2013) Magnetic helical micromachines. Chem Eur J 19(1):28–38
39. Qiu F, Zhang L, Tottori S et al (2012) Bio-inspired microrobots. Mater Today 15(10):463
40. Tottori S, Sugita N, Kometani R et al (2011) Selective control method for multiple magnetic helical microrobots. J Micro-Nano Mech 6(3–4):89–95
41. Tottori S, Zhang L, Qiu F et al (2012) Magnetic helical micromachines: fabrication, controlled swimming, and cargo transport. Adv Mater 24(6):811–816
42. Zhang L, Abbott J, Dong L et al (2009) Artificial bacterial flagella: fabrication and magnetic control. Appl Phys Lett 94(6):064107
43. Zhang L, Abbott J, Dong L et al (2009) Characterizing the swimming properties of artificial bcterial flagella. Nano Lett 9(10):3663–3667
44. Zhang L, Peyer KE, Nelson B (2010) Artificial bacterial flagella for micromanipulation. Lab Chip 10(17):2203–2215
45. Schmidt O, Eberl K (2001) Nanotechnology – thin solid films roll up into nanotubes. Nature 410(6825):168
46. Zhang L, Ruh E, Grützmacher D (2006) Anomalous coiling of SiGe/Si and SiGe/Si/Cr helical nanobelts. Nano Lett 6(7):1311–1317
47. Hawkeye M, Brett M (2007) Glancing angle deposition: fabrication, properties, and applications of micro- and nanostructured thin films. J Vac Sci Technol A 25(5):1317–1335
48. Peyer K, Zhang L, Nelson B (2011) Localized non-contact manipulation using artificial bacterial flagella. Appl Phys Lett 99(17):174101
49. Peyer K, Zhang L, Nelson B (2013) Bio-inspired magnetic swimming microrobots for biomedical applications. Nanoscale 5(4):1259–1272
50. Abbott J, Peyer K, Lagomarsino M et al (2009) How should microrobots swim? Int J Robot Res 28(11–12):1434–1447
51. Behkam B, Sitti M (2006) Design methodology for biomimetic propulsion of miniature swimming robots. J Dyn Syst-T ASME 128(1):36–43
52. Blakemore R (1975) Magnetotactic bacteria. Science 190(4212):377–379
53. Martel S, Mohammadi M, Felfoul O et al (2009) Flagellated magnetotactic bacteria as controlled MRI-trackable propulsion and steering systems for medical nanorobots operating in the human microvasculature. Int J Robot Res 28(4):571–582
54. Martel S, Tremblay C, Ngakeng S et al (2006) Controlled manipulation and actuation of micro-objects with magnetotactic bacteria. Appl Phys Lett 89(23)
55. Weibel D, Garstecki P, Ryan D et al (2005) Microoxen: microorganisms to move microscale loads. Proc Natl Acad Sci U S A 102(34):11963–11967
56. Martel S, Mohammadi M (2010) Using a swarm of self-propelled natural microrobots in the form of flagellated bacteria to perform complex micro-assembly tasks. Paper presented at the IEEE International Conference on Robotics and Automation (ICRA), May
57. Steager E, Kim C, Patel J et al (2007) Control of microfabricated structures powered by flagellated bacteria using phototaxis. Appl Phys Lett 90(26):263901
58. Paxton W, Kistler K, Olmeda C et al (2004) Catalytic nanomotors: autonomous movement of striped nanorods. J Am Chem Soc 126(41):13424–13431
59. Fournier-Bidoz S, Arsenault A, Manners I et al (2005) Synthetic self-propelled nanorotors. Chem Commun 4:441–443

60. Gibbs J, Zhao Y (2011) Catalytic nanomotors: fabrication, mechanism, and applications. Front Mater Sci 5(1):25–39
61. Mei Y, Solovev A, Sanchez S et al (2011) Rolled-up nanotech on polymers: from basic perception to self-propelled catalytic microengines. Chem Soc Rev 40(5):2109–2119
62. Mirkovic T, Zacharia N, Scholes G et al (2010) Fuel for thought: chemically powered nanomotors out-swim nature's flagellated bacteria. ACS Nano 4(4):1782–1789
63. Wang J (2009) Can man-made nanomachines compete with nature biomotors? ACS Nano 3(1):4–9
64. Solovev A, Sanchez S, Pumera M et al (2010) Magnetic control of tubular catalytic microbots for the transport, assembly, and delivery of micro-objects. Adv Funct Mater 20(15):2430–2435
65. Sanchez S, Solovev A, Schulze S et al (2011) Controlled manipulation of multiple cells using catalytic microbots. Chem Commun 47(2):698–700
66. Kagan D, Campuzano S, Balasubramanian S et al (2011) Functionalized micromachines for selective and rapid isolation of nucleic acid targets from complex samples. Nano Lett 11(5):2083–2087
67. Gao W, Sattayasamitsathit S, Orozco J et al (2011) Highly efficient catalytic microengines: template electrosynthesis of polyaniline/platinum microtubes. J Am Chem Soc 133(31):11862–11864
68. Burdick J, Laocharoensuk R, Wheat P et al (2008) Synthetic nanomotors in microchannel networks: directional microchip motion and controlled manipulation of cargo. J Am Chem Soc 130(26):8164–8165
69. Kagan D, Laocharoensuk R, Zimmerman M et al (2010) Rapid delivery of drug carriers propelled and navigated by catalytic nanoshuttles. Small 6(23):2741–2747
70. Sundararajan S, Sengupta S, Ibele M et al (2010) Drop-off of colloidal cargo transported by catalytic Pt-Au nanomotors via photochemical stimuli. Small 6(14):1479–1482
71. Balasubramanian S, Kagan D, Hu C et al (2011) Micromachine-enabled capture and isolation of cancer cells in complex media. Angew Chem Int Edit 50(18):4161–4164
72. Orozco J, Campuzano S, Kagan D et al (2011) Dynamic isolation and unloading of target proteins by aptamer-modified microtransporters. Anal Chem 83(20):7962–7969
73. Campuzano S, Orozco J, Kagan D et al (2012) Bacterial isolation by lectin-modified microengines. Nano Lett 12(1):396–401
74. Kagan D, Calvo-Marzal P, Balasubramanian S et al (2009) Chemical sensing based on catalytic nanomotors: motion-based detection of trace silver. J Am Chem Soc 131(34):12082–12083
75. Wu J, Balasubramanian S, Kagan D et al (2010) Motion-based DNA detection using catalytic nanomotors. Nat Commun 1
76. Ozin G, Manners I, Fournier-Bidoz S et al (2005) Dream nanomachines. Adv Mater 17(24):3011–3018
77. Leong T, Zarafshar A, Gracias D (2010) Three-dimensional fabrication at small size scales. Small 6(7):792–806
78. Bassik N, Brafman A, Zarafshar A et al (2010) Enzymatically triggered actuation of miniaturized tools. J Am Chem Soc 132(46):16314–16317
79. Leong T, Randall C, Benson B et al (2009) Tetherless thermobiochemically actuated microgrippers. Proc Natl Acad Sci U S A 106(3):703–708
80. Randhawa J, Leong T, Bassik N et al (2008) Pick-and-place using chemically actuated microgrippers. J Am Chem Soc 130(51):17238–17239
81. Kummer M, Abbott J, Kratochvil B (2006) An electromagnetic system for 5-DOF wireless micromanipulation, IEEE. Trans Robot 26(6):1006–1017

Chapter 10
Pharmaceutical Nanotechnology: Overcoming Drug Delivery Challenges in Contemporary Medicine

Srinivas Ganta, Amit Singh, Timothy P. Coleman, David Williams, and Mansoor Amiji

10.1 Challenges in Delivery of Contemporary Therapeutics

Drug discovery process has been in forefront utilizing recent advances in molecular biology, -together with medicinal chemistry, protein structure based screening, and computational analysis, as part of rational approach to discovering drug molecules that will address unmet clinical needs. For example, proteins identified from structural biology platform can serve as targets for discovering new drug molecules. The discovery of antisense oligonucleotides (ASN), plasmid DNA (pDNA), peptides and protein therapeutics has also shown a greater potential in treating several complex diseases. A recent development in drug discovery is RNA interference (RNAi) which uses small stretches of double stranded RNA with 21–23 nucleotides in length, to inhibit the expression of a gene of interest bearing its complementary sequence [1]. Small interfering RNA (siRNA) can induce RNAi in human cells. This RNAi technology has many advantages over other posttranscriptional gene silencing methods, such as gene knockouts and antisense technologies [2]. In addition, only a few molecules of siRNA need to enter a cell to inactivate a gene at almost any stage of development. MicroRNA (miRNA), advancement from siRNA, is a new class of drugs still in the investigative stage based on nucleic acid chemistry. miRNA with 19–25 nucleotides in length, interfere pathways that involve in disease process [3]. In general, all these recent drugs have shown a great potential in the clinical management of several complex diseases like cancer, metabolic diseases, auto-immune diseases, cardiovascular diseases, eye diseases, neurodegenerative disorders and other illness [1].

S. Ganta • T.P. Coleman • D. Williams
Nemucore Medical Innovations, Inc., Worcester, MA 01608, USA

A. Singh • M. Amiji (✉)
Department of Pharmaceutical Sciences, School of Pharmacy, Northeastern University, Boston, MA 02115, USA
e-mail: m.amiji@neu.edu

© Springer Science+Business Media New York 2014
Y. Ge et al. (eds.), *Nanomedicine*, Nanostructure Science and Technology,
DOI 10.1007/978-1-4614-2140-5_10

Despite the diversity and size of therapeutic libraries are continually increasing, delivering them to the disease sites has been hampered by physico-chemical attributes of drugs and physiological barriers of the body. For example, many small and macromolecular drugs (ASN, pDNA, peptides, proteins, siRNA and miRNA) often fail to reach cellular targets because of several chemical and anatomical barriers that limit their entry into the cells [4, 5]. Therefore, the outcome of therapy with that contemporary therapeutics is often unpredictable, ranging from beneficial effects to lack of efficacy to serious adverse effects. These challenges have been discussed in the following sections with an attempt to apply nanotechnology-based concepts in designing of drug delivery systems (DDS) that overcome barriers in drug delivery.

10.1.1   Chemical Challenges

Physico-chemical properties impact on both pharmacokinetics and pharmacodynamics of the drug in vivo, and must be considered when selecting a suitable delivery method. The chemical challenges faced by small and macromolecular drugs (ASN, pDNA, peptides, proteins, siRNA and miRNA) are many folds, which mainly include:

  (i) Molecular size
 (ii) Charge
(iii) Hydrophobicity
 (iv) In vivo stability
  (v) Substrate to efflux transporters

Size, Charge and Hydrophobicity The chemical properties that mainly affect drug permeability through anatomical barriers are molecular size and solubility. High molecular size and increased hydrophobicity are the predominant problems particularly associated with combinatorial synthesis and high throughput screening methods [6]. These methods allow for identification of lead molecules faster based on their best fit into receptors, but shift the molecules towards high molecular size and increased hydrophobicity, resulting in poor aqueous solubility [6]. One estimate shows that around 40 % of the newly discovered molecules are poorly aqueous soluble, thus need a suitable delivery method to achieve pharmacologically relevant concentrations in the body [7, 8]. Oral route for drug delivery remains popular due to ease of administration and patient compliance. However, oral absorption can be hindered by poor aqueous solubility of therapeutics in GI fluids. The rate of dissolution, which is a prerequisite for oral absorption, depends on the drug solubility in the GI fluids. In addition, drug molecule must possess adequate lipophilicity (logP) in order to efficiently permeate across intestinal epithelial cells [9]. This is one of the reasons for hydrophilic macromolecules such as proteins, peptides, and nucleic acid constructs do not show any oral bioavailability and resulting in limited clinical success [10].

Drug transport mechanisms involving in intestinal epithelium are transcellular and paracellular transport [11]. Transcellular mechanism involving in transport of drug molecules across the cell membrane, which occurs by (1) passive diffusion, (2) facilitated diffusion, (3) active transport, and (4) transcytosis. Lipophilic drug molecules can diffuse freely across the epithelial membrane barrier while hydrophilic and charged molecules need specialized transport carriers to facilitate cellular uptake. Transcytosis process involving endocytosis and exocytosis mechanisms is mainly for macromolecular (proteins, peptides) drugs. Recent studies show that orally given nanoparticles can pass through the epithelial membranes in GI tract through the endocytosis process [12, 13], and this can be a potential route for transport of macromolecular therapeutics.

Paracellular route, on the other hand, involves diffusion of hydrophilic drugs between the cells of epithelial or endothelial membrane through sieving mechanism. The formation of tight junction between the epithelial and endothelial cells strictly limits the paracellular drug transport. Molecular cut-off for the paracellular transport is approximately 400–500 Da [14]. Molecular mass less than the cell junction can easily pass through paracellular route regardless of polarity, for example, water and ions. It has been observed that the diffusion of drugs with molecular size <300 Da is not significantly affected by the physicochemical properties of the drug, and which will mostly pass through aqueous channels of the membrane. However, the rate of permeation is highly dependent on molecular size for compounds with MW >300 Da. The Lipinski rule of five suggest that an upper limit of 500 Da as being the limit for orally administered drugs [15]. Numerous studies are focused on identifying the nature of cellular tight junctions and the signaling molecules involved in preserving the barrier function in order to find right approach to promote oral drug absorption.

Increased hydrophobicity of a molecule also causes greater protein binding. Protein binding is both help and hindrance to the disposition of drugs in the body [16]. Elimination and metabolism may be delayed because of highly protein bind. Therefore, protein binding affects both the duration and intensity of drug action in the body.

Stability In vivo stability is also a critical chemical property of the drug that affects drug levels in the body. For example, the extent of drug ionization, stability in the acidic environment of the stomach or stability in the presence of gut enzymes, as well as presence of food and gastric emptying can reduce oral bioavailability of many small and macromolecular drugs. On the other hand, drugs are subjected to metabolism in the body by different sequential and competitive chemical mechanisms involving oxidation, reduction, hydrolysis (phase I reactions) and glucouronidation, sulfation, acetylation and methylation (phase II reactions). Cytochorme P450 enzymes which catalyze oxidation reaction are mainly responsible for first-pass biotransformation of majority of the drugs in the body, thus limiting oral absorption and systemic availability of the drugs [17]. Cytochrome P450 abundantly present in the intestinal epithelium and liver tissue, and metabolizes several chemically unrelated drugs from major therapeutic classes [17].

Besides this, macromolecular drugs such as proteins and peptides, ASN, pDNA, siRNA and miRNA have poor biological stability and a short half-life resulting in unpredictable pharmacokinetics and pharmacodynamics. Proteins and peptidal drugs are highly prone to enzymatic cleavage in the blood circulation and tissues, whereas nucleic acid therapeutics are highly susceptible to degradation by intra- and extra-cellular nucleases, leading to degradation and a short biological half-life [5, 18]. DDS have the potential to overcome the challenges of degradation and short biological half-life, and can provide safe and efficient delivery of macromolecular therapeutics.

Expression of Membrane-Bound Drug Efflux Pumps If the drug molecules are substrates to efflux pumps, their transport through cellular membranes is severely restricted [20–21]. The ATP-binding cassette (ABC) efflux pumps are trans-membrane proteins present at various organ sites within the body, and use ATP as a source of energy to actively transport drug molecules across the lipid cell membranes [23]. Among the ABC family of efflux pumps, P-glycoprotein (P-gp) is highly expressed in epithelial cells of the small intestine, which is the primary site of absorption for the majority of the orally given drugs [24]. Efflux pumps also present on the luminal side of the endothelial cells of BBB, and restrict entry of hydrophobic molecules into the brain [25, 26]. The multi-drug resistance in many cancers is linked to the ABC efflux transporters which express on cell membranes and produce intracellular drug levels below the effective concentrations necessary for cytotoxicity [19]. All these efflux transporters preset a broad overlap in substrate specificities and act as a formidable barrier to drug absorption and availability at target sites [24].

DDS can be employed to overcome most of these chemical challenges. For example, paclitaxel is a potent anticancer drug, is poorly absorbed after oral administration and its bioavailability is <6 % [27]. The obvious reason for its low bioavailability are high molecular weight, poor aqueous solubility, the affinity to drug efflux pumps, and rapid metabolism by cytochrome P450 enzymes in the gut [24]. Nanoemulsions and self-emulsifying DDS have been employed recently for the successful oral and parenteral delivery of paclitaxel [20–22, 28, 29]. Similarly, to protect RNA based therapeutics from enzymatic cleavage, several DDS have been proposed and they are at different stages of preclinical and clinical development.

10.1.2 Remote Disease Targets

Anatomical and physiological barriers involved in the body restrict the direct entry of small and macromolecular drugs into the target extracellular or intracellular tissue locations [4, 30] resulting in sub-optimal doses at target site and reduced efficacy. However, cytotoxic drugs and RNA therapeutics have their target sites inside the cells, therefore need to be delivered intracellular in sufficient doses to produce therapeutic effect. The first limiting anatomical barrier for orally administered drugs is epithelial lining of gut walls, where from drugs will permeate through either by transcellular or paracellular transport. This transport is in turn dictated by the chemical properties of drugs as alluded above. Therefore, altering the chemical properties by making the drugs in salt form, encapsulating in DDS based on cyclodextrins, lipid or polymeric carriers, or using permeability enhancers could promote bioavailability of drugs [20–22, 29]. Cytochrome P450 and efflux transporters present in the enterocytes of intestinal walls also forms as another limiting barrier to drug permeability [24]. Use of cytochrome P450 and efflux pump inhibitors can promote oral drug absorption. For example, pre-treatment with curcumin results in inhibition of P-gp and cytochrome P450 expression in the GI tract, leading to increased oral bioavailability and efficacy of drugs [20–22, 31].

For the drug molecules given intravenously, the limiting anatomical barrier is that of vascular endothelium and basement membrane. In addition, blood serum proteins, proteolytic enzymes, RNases etc. limit the effective drug delivery to the target sites [4, 30]. CNS disease are likely to rise to 14 % by 2020 mainly due to the ageing population, however, many newly discovered small and macromolecular therapeutics do not cross into the brain after systemic delivery [32]. Because brain is protected by blood-brain-barrier (BBB), which is composed of very tight endothelial cell junction and presence of several efflux transporters, resulting in formation of dynamic formidable barrier to drug transport [33]. However, once at the BBB, hydrophobic drugs with the size below <500 Da generally do transport through lipid cell membranes by passive diffusion, but if they are substrates to drug efflux pumps, they will be pumped out from the brain. Hydrophilic molecules also cannot transport efficiently as there is very limited paracellular transport present in the tight junctions of the BBB [20–22].

Cancer mass is another complex anatomical barrier in drug delivery. For example, solid tumor microenvironment is heterogeneous and structurally complex and presents a challenging barrier in drug delivery. The cytotoxic drugs which are intended to kill a large proportion of tumor cells in a solid tumor, must uniformly distribute through the vascular network, pass through capillary walls, and traverse the tumor tissue [34]. Nevertheless, the drug distribution in tumors is not uniform, and only a fraction of tumor cells is exposed to lethal doses of cytotoxic agents [34]. Tumor microenvironment is composed of tumor cells with varying proliferation rate and stromal cells (fibroblasts and inflammatory cells) that are surrounded in an extracellular matrix and nourished by a vascular network, and regions of hypoxia and acidity [34–37]. Each of these components may different from one site to another in the same tumor mass, and all of these factors effects tumor cell sensitivity to drug treatment [34]. In addition, stromal components in tumors contribute to an increase in interstitial fluid pressure, which limit the penetration of macromolecular drugs [38]. Furthermore, the three-dimensional nature of solid tumor tissue itself affects the sensitivity of constituent cells to chemo and radiation treatments [34, 39]. For instance, the tumor cells grown as spheroids in cell culture or tumors grown in animals, are more resistant to cisplatin and alkylating agents than the corresponding cell dispersions [40].

In addition, certain intracellular infections, like leishmaniasis and listeria, where macrophages are directly involved in the disease are not accessible to drug delivery, thus necessitating specific drug delivery strategies [41]. To overcome all these challenges, it is highly important to develop DDS that render protection to the drug from biodegradation in the body, while allowing their transport through the anatomic and physiological barriers to increase their bioavailability at the target tissue.

10.2   Nanotechnology Solutions

The science of nanotechnology has begun just in the last decade, but in this short time, it has been successfully applied in several fields ranging from electronics to engineering to medicine. Recent understanding of cellular barriers and molecular profile of diseases, and controlled manipulations of material at the nanometer length scale, nanotechnology offers great potential in the disease prevention, diagnosis, and treatment [30, 31]. Nanotechnology has also allowed for challenging innovations in drug delivery, which are in the process of transforming the delivery of drugs. Nanosystems fabricated using controlled manipulation of material are exploited for carrying the drug in a controlled manner from the site of administration to the target site in the body. They are colloidal carriers with dimensions <1,000 nm and can traverse through the small capillaries into a targeted organ down to target cell and intracellular compartments, which represent the most challenging barrier in drug targeting. The critical attributes of any nanoparticle DDS are to (1) protect a labile drug molecule from both in vitro and in vivo degradation, (2) maintain the effective pharmacokinetic and biodistribution pattern, (3) promote drug diffusion through the epithelium, and/or (4) enhance intracellular distribution. However, the specificity, sensitivity and simplicity are very important for any nanosystem to be clinically successful as a DDS. Several types of nanoparticle DDS have been evaluated for their potential drug delivery applications are in various stages of clinical development, these are discussed in the next sections.

10.2.1   Enhancing Solubility and Permeability

Solubility and permeability are two of the most critical biopharmaceutical characteristics impacting the successful delivery of drug molecules through anatomical membranes in the body. If the drug molecule is not a substrate to efflux transporters and metabolizing enzymes, then the solubility (hydrophilic and hydrophobic) plays a major role in determining oral intestinal permeability. Biopharmaceutical Classification System (BCS) is proposed based on the solubility and permeability properties of the drugs [42] which classifies drugs into one of four classes. Class I drugs are highly soluble and permeable in the GI tract, therefore, bioavailability is not an issue with Class I drugs. Class II drugs are poorly aqueous soluble but highly lipophilic. They are well permeable across the GI tract due to high lipophilicity, but the bioavailability is likely to dissolution rate limited due to low aqueous solubility. Class III drugs are highly soluble but have low permeability due to their low lipophilicity. In both Class II and Class III examples, DDS plays a critical role in overcoming poor solubility and permeability. On the other hand, Class IV drugs show low solubility and low permeability, and exhibit poor and variable bioavailability. Methods to enhance both solubility and permeability should be adopted for these drugs.

To improve solubility and permeability, several methods have been employed over the years. Such as preparation of prodrugs, use of chemical or physical permeability enhancers to transient openings of the tight junctions, or direct administration to the target site. However, formulation efforts can best exemplify in improving poor solubility and permeation profiles of both small and macromolecular drugs. Nanoparticle DDs like, liposomes, nanoemulsions, nanosuspensions, solid-lipid nanoparticles (SLN), micelles or polymeric nanoparticles are highly useful over the current methods to deliver the highly hydrophilic or highly lipophilic molecules across the intestines and BBB. For example, drug nanocrystal suspensions (nanosuspension) allow for increased dissolution velocity and saturation solubility of poorly aqueous soluble drugs, which is accompanied of an increase in oral bioavailability [43]. In addition, nanocrystals can be delivered intravenously for controlled drug release, and their surface can be tailored for both passive and active targeting. On the other hand, lipid-based systems like nanoemulsions and SLN could allow for the delivery of lipophilic drugs, by incorporating them in the lipid core of the formulation. These DDS can enable direct transfer of drug to the intestinal membranes and excluding the dissolution of drugs in aqueous fluids in GI tract. In once such study, we have formulated highly lipophilic paclitaxel into deoxycholic acid modified nanoemulsion, which showed increased oral bioavailability compared to paclitaxel solution [20–22, 44]. In another example, saquinavir, an anti-HIV protease inhibitor incorporated in nanoemulsion, showed enhanced oral absorption [45].

Recent studies show that nanoemulsions made using oils rich in omega-3 and omega-6 polyunsaturated fatty acids (PUFA) can promote drug delivery to the brain [45]. This is some extent attributed to the presence of PUFA transporters on the abluminal membrane side of the endothelial cells of BBB [46]. Tissue and cell permeability also altered by surface modification of the nanoparticles with targeting ligands which can facilitate the nanoparticle uptake along with its payload into the cells. These aspects have been discussed in the next sections.

10.2.2 Targeted Delivery to Disease Sites

Targeted delivery exploiting the structural changes and cellular markers of a given pathophysiology can potentially reduce the toxicity and increase the efficacy of drugs. This is highly important in case of diseases like cancer, where dose-limiting toxicities and drug resistance constitute major barriers to drug success. General targeting mechanisms consists of passive and active targeting [30].

Upon parenteral delivery, passive targeting depends on the size of the DDS and the disease vascular pathophysiology in order to preferentially accumulate the drug at the site of interest and avoid distribution to normal tissue [30]. For example, nanosystems escape from the blood circulation and accumulate in sites where the blood capillaries have open fenestrations as in the sinus endothelium of the liver [47] or when the integrity of the vascular endothelial membrane is perturbed by inflammation due to infections, rheumatoid arthritis or infarction [48] or by tumors [49]. In the liver, the size of capillary fenestrae can be as large as 150 nm [50] and liposomal nanocarriers showed extravasation to hepatic parenchyma [47]. Nanosystems in the size range of 50–200 in size can extravasate and accumulate inside the tumor tissue and inflammatory sites [51, 52]. Therefore, the nanomedicine in the size range is expected to provide therapeutic benefits for treating these diseases. In case of solid tumors, passive targeting involves in the transport of nanosystems through a newly formed leaky tumor microvasculature into the tumor interstium and cells (Fig. 10.1). This phenomenon has named as "enhanced permeability and retention" (EPR) effect, first discovered in murine tumors for macromolecules accumulation by Maeda and Matsumura [53]. EPR effect is observed in many human solid tumors with the exception of hypovascular tumors (prostate or pancreatic cancer) [54, 55]. This effect will be optimal if nanosystems can escape reticulo-endothelial system (RES) and show longer circulation half-life in the blood. Poly(ethylene glycol) (PEG) grafting on nanosystems will evade RES uptake, allow for prolonged circulation in the blood and enhance tumor accumulation through EPR. Besides, the RES uptake of non-PEG grafted nanosystems also offers an opportunity for passive targeting against intracellular infections such as leishmaniasis, candiasis, and listeria which reside in macrophages [41].

The specificity of passive targeting can be remarkably improved when the targeting ligands are used with nanosystems, which selectively bind to cellular markers overexpressed on the disease site [56] termed as active targeting (Fig. 10.1). For example, folic acid-nanoparticles can be used to target tumor cells that over express folate receptors, such particles internalize via folate receptor mediated endocytosis [57]. In another example, arginine-glycine-aspartic acid (RGD) sequence containing peptides can be conjugated to nanoparticle to target $\alpha_5\beta_5$ or $\alpha_5\beta_3$ integrin receptors over express on endothelial cells of the newly formed angiogenic blood vessels and also on tumor cells. Furthermore, the targeting ligands anchored to nanosystems will allow for carrying of many drug molecules compared to direct conjugation of targeting ligands with drug molecules.

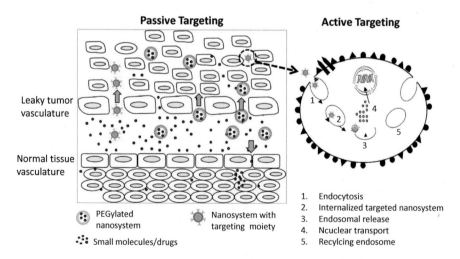

Fig. 10.1 Schematic illustration of passive and active targeting strategies in tumor drug delivery

10.2.3 Intracellular and Subcellular Delivery

The nanosystems once in the disease vicinity, they need to enter the cells and transfer the payload to sub-cellular organelles. There are two mechanisms playing a role in intracellular and subcellular delivery are non-specific or specific uptake of nanosystems by cells [30, 58]. In case of non-specific uptake, cells surround the nanosystems and forms a vesicle in the cell called an endosome. The endosomes then fuse with the highly acidic organelles called lysosome, which are rich in degrading enzymes. Endosomes usually travels in a specific direction and join at the nuclear membrane. Specific uptake on the other hand, involves receptor mediated endocytosis, where the actively targeted nanosystem binds to the cell-surface receptor, resulting in internalization of the entire nanoparticle-receptor complex and vesicular transport through the endosomes. The receptor can be re-cycled back to the cell surface following dissociation of complex. After the cellular internalization, stability of the payload in the cytosol and delivery to specific organelles, such as mitochondria, nucleus etc, is also essential for therapeutic activity. However, many drugs do not survive in the lysosomal environment. For example, 99 % of the internalized gene molecules undergoes degradation in endosomes. Thus buffering the endosomes for safe release of its contents helps in efficient gene delivery. Towards this, polycationic nanosystems have been explored, which causes endosomes to swell and burst, leading to the safe release of trapped content [59]. In another strategy, instead of trafficking drug carrier to the lysosome, the endosomal contents were released into the cytoplasm, thus bypassing the lysosomal degradation of the drug molecules [60, 61]. For example, a cyclic RGD functionalized polyplex micelles were taken into the cellular perinuclear space selectively through caveolae mediated endocytosis, thus escaping the lysosomal degradation of its active content [61].

Cellular uptake could be enhanced using of arginine rich cell penetrating peptides (CPP's) [62]. For example, HIV-1 Tat peptide was used to promote non-specific intracellular delivery of various therapeutics following systemic administration [63]. A number of cationic CPP's like penetratin also have been identified to promote intracellular drug delivery. In addition to intracellular delivery, use of delocalized cationic amphiphiles or mitochondriotropic nanosystems can promote mitochondrial drug delivery [64, 65].

10.2.4 Enabling Non-invasive Delivery

Non-invasive delivery is an alternate to systemic delivery of drugs, and mainly includes drug delivery via intranasal, pulmonary, transdermal, buccal/sublingual, oral and trans-ocular routes [66, 67]. Patient compliance has been found to be much higher when drugs given by non-invasive routes and therefore they are considered to be a preferred route of drug delivery. However, the preferred route of administration for a given drug selected based on several factors, such as biopharmaceutical properties (solubility, permeability and stability) of a drug molecule, disease state, onset of action, dose frequency and adverse effects. For example, sumatriptan and zolmitriptan administered via intranasal route provide rapid-onset of relief from migraine related pain in minutes compared to oral tablet in hours. Similarly, potent peptidal drugs like calcitonin, desmopressin allows therapeutic blood levels that are not achieved with oral route of administration. In another example, selegiline and fentanyl transdermal products eliminate GI related adverse effects. In addition, non-invasive insulin products for inhalation and buccal administration improve patient compliance by reducing multiple daily injections.

In general, oral route is much convenient for high doses of administration. However, macromolecular drugs are not stable in the GI fluids, where intranasal, buccal/sublingual or pulmonary offers a non-invasive route of choice. These routes also favor treatments that need faster absorption of drug and where a rapid systemic exposure is well tolerated. Transdermal delivery is useful in chronically administered treatments (chronic pain, depression, Parkinson's, dementia, attention deficit-hyperactivity disorder and hormonal therapies), where sustained plasma profiles and low C_{max} to C_{min} ratio are required.

10.3   Illustrative Examples of Nanotechnology Products

Nanotechnology based concepts have been extensively applied in engineering of nanosystems for delivery of contemporary therapeutics in a controlled manner from the site of administration to the target disease in the body. The history of nanosystems reaches back to 1950s when the first polymer-drug conjugate was reported with N-vinyl pyrrolidine conjugated to glycyl-L-leucine-mescaline [68]. However, the most relevant nanosystems were conceptualized only after the first report of liposomal preparations in 1964 [69] and their subsequent use as vehicle for drug delivery application [70]. Soon after, synthesis of albumin nanoparticle was reported in early 1970s [71] with a subsequent early attempt of exploiting them as the first protein based DDS [72]. The pharmacological effects of polymer-based nanoparticles were studied [73] and their application as DDS envisioned around the same time [74]. As alluded earlier, ground breaking discovery of EPR effect in tumors by Matsumura and Maeda further emphasized on relevance of the size of delivery vehicle [53]. These seminal works drew tremendous attention on nanosystems for a sustained and controlled delivery of drugs. It was realized that for an optimized delivery system, the size of the payload vehicle should be between 10 and 100 nm. Kidneys easily clear off particles smaller than 10 nm while the particles larger than 100 nm are removed by the RES [73]. Since then, several different types of nanosystems have been researched and much focus has specifically been on tailoring the size, physical properties and surface functionality of the delivery systems for varying therapeutic applications. The collective research input on the nanotechnology based improvement of DDS has enabled several products in to the market in the past two decades (Table 10.1).

Sandimmune® and Taxol® are US Food and Drug Administration (FDA) approved dosage forms of cyclosporine and paclitaxel respectively, formulated

Table 10.1 Nanotechnology-based products in clinical application

Nanotechnology platform	Trade name	Active agent	Indication(s)	Approval year
Liposomes	Abelcet	Amphotericin B	Fungal infection	1995
	AmBisome	Amphotericin B	Fungal infection	1997
	Amphotec	Amphotericin B	Fungal infection	1996
	Daunoxome	Daunorubicin	Antineoplastic	1996
	DepoCyt	Cytarabin	Lymphomatous meningitis	1999
	Doxil/Caelyx	Doxorubicin	Antineoplastic	1995
	Myocet	Doxorubicin	Antineoplastic	2000
	OncoTCS	Vincristine	Non-Hodgkin's lymphoma	2004
Micelles	Estrasorb	Estradiol	Vasomotor symptoms	2003
Nanocrystal	Emend	Aprepitant	Antiemetic	2003
	Tricor	Fenofibrate	Hypercholesterolemia and hypertriglyceridemia	2004
	Triglide	Fenofibrate	Hypercholesterolemia and hypertriglyceridemia	2005
	Megace ES	Magesterol acetate	Anorexia, cachexia or an unexplained significant weight loss in AIDS patients	2005
	Rapamune	Sirolimus	Immunosuppressant	2000
Nanoemulsion	Tocosol	Paclitaxel	Nonsuperficial urothelial cancer	2003
Nanoparticle	Abraxane	Paclitaxel	Metastatic breast cancer	2005
Nanotube	Somatuline depot	Lanreotide	Acromegaly	2007
Superparamagnetic iron oxide	Feraheme injection	Ferumoxytol	Treatment of iron deficiency anemia in patients with chronic kidney disease	2009
	Feridex	Ferumoxide	MRI contrast agent	1996
	GastroMARK	Ferumoxsil	Imaging of abdominal structures	1996

using Cremophor®EL as solubilizing nonionic surfactant. However, due to hypersensitivity reactions associated with these products, Cremophor®-free formulations based on nanosystems have been developed and commercialized. Genexol^{PM} is one such example of Cremophor-free polymeric micelles formulated paclitaxel where poly-(ethylene glycol) is used as a nonimmunogenic carrier while biodegradable poly-(D,L-Lactic acid) forms the drug solubilizing hydrophobic core [75, 76]. Several such DDS including liposomes, nanoemulsions, polymeric nanoparticles, micelles and nanocrystals (Fig. 10.2) have been developed, granted regulatory approval and have been marketed since then. The following section will focus on each of such DDS with illustrative examples of commercialized products.

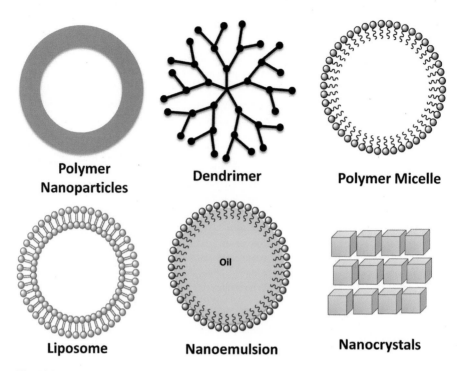

Fig. 10.2 Different types of pharmaceutical nanosystems used in drug and gene delivery

10.3.1 Lipid-Based Nanosystems

Lipid based carriers are extremely popular since they facilitate a controlled administration of both small and macromolecular drugs at therapeutically relevant doses. Liposomes and nanoemulsions are two most commonly used lipid based nanosystems for drug delivery application.

Liposomes Liposomes are vesicles formed of a lipid bi-layer, first developed by Alec Bangham in 1961, and their lipid bi-layer membrane is similar to that of cellular membranes. The lipid bi-layer of liposomes is composed of phospholipids with a hydrophilic head and a hydrophobic long-chain tail [77]. The hydrophilic core of the liposomes facilitates in compartmentalizing water-soluble drugs into the aqueous core while the hydrophobic bi-lipid membrane has been exploited to load water-insoluble drugs. Initial attempts using liposomes as nanosystems focused largely on improving their circulation time in the blood and targeting efficiency. PEG-modification of liposomes, first reported in 1990 [78] has by far been the most promising approach to achieve longer circulation of the liposomes in the blood. There has been a plethora of literature since then on the application of PEG-modified liposomes to achieve a selective delivery of drugs post-administration [79–81]. However, several other surface modifications of liposomes such as poly[N-(2-hydroxypropyl) methacrylamide] [81] poly-N-vinylpyrrolidones [82] polyvinyl alcohol [83] and amino acid-based polymer–lipid conjugates [84] have been explored. Many studies showed that the opsonization of the liposomes might be dependent on the hydrophobicity of the surface, charge of the lipid and the molecular weight of the modifying polymer [85]. Antibody [86], folate [87] and peptide [88] mediated surface receptor targeting has been primarily enabled directing the liposome based drug delivery to the target organ.

The first liposome based formulation, PEG-liposome encapsulated doxorubicin was approved in 1995 (Doxil™, Orthobiotech) initially for the treatment of HIV-related Kaposi Sarcoma [89, 90] and later for ovarian cancer and myeloma. Doxil has remarkably reduced the cardiotoxicity by lowering cardiac exposure to free doxorubicin [77, 91]. Besides, it also increased half-life and tumor accumulation compared to free doxorubicin [92]. Furthermore, antibody modification of Doxil has shown a much higher tumor accumulation and enhances the cytoxicity of the doxorubicin [93]. In a study conducted on 53 patients suffering from advanced Kaposi's sarcoma, 19 patients showed partial and 1 patient showed complete response on administration of Doxil™ once every 3 weeks [94]. The success of liposomal doxorubicin has led to several liposomal-based drug formulations that are either approved for clinical application or are undergoing different phases of clinical trial. Some of the key drugs that have been exploited for liposomal formulation are shown in Table 10.1.

Nanoemulsions Nanoemulsions are heterogeneous system of two immiscible liquids; typically oil-in-water (o/w) or water-in-oil (w/o) with a droplet size in the range of 50–200 nm. These kinetically stabilized nano-sized droplets have several advantages over macroemulsions such as higher surface area and hence more free energy, higher stability with lower creaming effects, coalescence, flocculation and sedimentation [95]. The formation of nanoemulsions however requires an external shear force to decrease the droplet size to desired range and their productions methods are broadly classified as high-energy and low-energy methods. The high-energy methods could include laboratory or industrial scale high-pressure homogenization, microfluidization or laboratory scale ultrasonication [96]. However, these methods may not be conducive for applications involving thermolabile drugs, nucleic acids and proteins. Low-energy methods such as spontaneous emulsification, the solvent-diffusion method and the phase-inversion temperature (PIT) method are used for such payloads [95, 97]. The nanoemulsions serve as an excellent vehicle for solubilizing lipophilic drugs into the oil phase or hydrophilic drugs in the aqueous phase. The application of nanoemulsions as DDS has been envisaged only in the past decade and several attempts have been realized to increase their stability, circulation time and achieve a targeting efficiency [20–22, 98, 99]).

For example, propofol was first formulated in Cremophor® EL by Imperial Chemical Industries as ICI35868, and went into clinical use. However, due to the toxicity of Cremophor®, it was withdrawn from the market, reformulated in oil-in-water emulsion and launched by the trade name Diprivan® (ICI, now AstraZeneca). Apart from propofol as active pharmaceutical ingredient, the formulation contains generally regarded as safe grade excipients (GRAS) such as soyabean oil, glycerol, egg lecithin and disodium edetate [100]. Diprivan® is used as a short acting, intravenous sedative used in intensive care medicine. It is known to have low toxicity, controlled sedation effect, rapid onset, a short duration of action and quick recovery despite prolonged usage [100, 101] TOCOSOL is another Cremophor® EL-free nanoemulsion formulation of paclitaxel that was approved by FDA in 2003 for the treatment of nonsuperficial urothelial cancer. Dexamethasone (Limethason®, Mitsubishi Pharmaceuticals), alprostadil palmitate (Liple®, Mitsubishi Pharmaceuticals), flurbiprofen axetil (Ropion®, Kaken Pharmaceuticals) and Vit A, D, E, K (Vitalipid®, Fresenius Kabi) are some other examples of therapeutically relevant compounds that have been formulated in nanoemulsions for clinical applications. Recently, NanoBio Corporation has formulated an emulsion-based antiviral drug NB 001 that shows potent activity against HSV-1 virus and antifungal drug NB 002 for the treatment of distal subungual onychomycosis (DSO). Both these formulations are currently in phase II/III trails.

10.3.2 Polymer-Based Nanosystems

Polymeric nanoparticles clearly are the most studied system for drug delivery applications. Different polymeric materials, natural, semi-synthetic and synthetic, have been exploited as polymer-drug conjugate or polymer-based nanoparticle for drug encapsulation to facilitate therapeutic applications. It is important to realize that while polymer-drug conjugate is a system which involves a single polymer chain conjugate to the drug, polymer-based nanoparticles are actually made up of several polymer chains which encapsulate the drug of interest.

Polymer-drug Conjugate Polymer-drug conjugates preparation date back to early 1950 [68] and the field has rapidly evolved since then [102]. Most drug molecules suffer from permeability through biological membranes, short half-life, non-specific distribution and dose dependent toxicities. Polymer conjugates on the contrary not only tremendously improves the in vivo circulation time of the drug but also facilitates passive delivery of these conjugates through leaky vasculature in diseases like cancer and inflammation [103]. They however also suffer from certain drawbacks such as polymer dependent toxicity, immunogenicity, rapid drug release, conjugate instability and poor drug loading. Several endeavors have been taken to overcome some of these shortcomings with much success. Besides, many bio-inspired polymers such as proteins (albumin, antibodies etc.) have also been looked upon as promising candidates for drug delivery applications.

The first polymer conjugate to undergo clinical trial was SMANCS where anti-tumor protein neocarzinostatin was (NCS) was covalently conjugated to two styrene maleic anhydride (SMA) [53]. SMANCS was approved subsequently in Japan in 1994 to treat advanced and recurrent hepatocellular carcinoma [104]. PEG-conjugate were the first candidate to get US FDA approval when PEG-L-asparaginase conjugate (Oncaspar) was accepted to treat acute lymphoblastic leukaemia [105]. Several other PEG -onjugates of drugs such as Neulasta (PEG-G-CSF; neutropaenia associated with cancer chemotherapy), PEG-asys (PEG-IFNα2a; Hepatitis B and C),

PEG-Intron (PEG-IFNα2b; Hepatitis C) have been approved to clinical treatment while several others are under various preclinical development. Besides, several other polymers (or their derivatives) conjugates (products names) such as polyglutamate (CT-2103, CT-2106), dextran (DOX-OXD, DE-310), N-(2-hydroxypropyl) methacrylamide (PK1, PK2, MAG-CPT, AP-5280, AP-4346) are being looked upon as promising candidates in their preclinical trial stages.

Though first protein nanoparticle based drug conjugation was reported in as early as 1974 [72] the first approved conjugate was realized only in 2005 when paclitaxel bound to albumin (Abraxane, AstraZeneca) was approved by FDA for treatment of metastatic breast cancer [106]. It is a non-targeted formulation with particle size around 130 nm, which is localized into the tumor partly through EPR effect and partly through albumin-binding protein. Clinical studies have demonstrated that Abraxane increases the therapeutic response, reduces the rate of disease progression and improves the survival rate among the patients. Antibodies have also been explored for drug conjugation and some examples of products from this class of nanovector includes Gemtuzumab (Mylotarg), Tositumomab and ibritumomab tiuxetan (Zevalin) [107, 108].

Micellar Delivery Systems Micelles are submicroscopic structures formed in an aqueous phase by amphiphilic surfactants or polymers that have a polar and a non-polar group. The typical size of these structures for delivery application ranges from 10 to 100 nm. These structures have a hydrophobic core, which facilitates the solubility of a lipophilic therapeutic agent and a hydrophilic corona that is exploited for surface functionalization to improve their tumor accumulation. These properties render them an attractive choice as carriers for drug delivery applications. Conventional surfactants however have a very high critical micellar concentration, and therefore are prone to disintegration on dilution in the blood stream [109]. Alternatively, polymeric micelles are usually prepared by self-assembly of a copolymer having hydrophobic moiety forming the biodegradable core while hydrophilic component for the surface. These polymers form micelles in aqueous media but at a much lower concentration compared to conventional surfactants [110]. Such polymeric micelles have been extensively researched for drug encapsulation, enhanced tumor targeting and longer in vivo circulation to aid an improved delivery system. Various approaches have been utilized to prepare polymeric micelles of desired properties using block copolymer, their lipid [111] or cyclodextrin [112] derivatives, diblock copolymers [113], triblock copolymers [114], pluronic polymer [115] and graft polymers [116].

Genexol-PM, a cremophor-free PLA-PEG copolymer-based micellar formulation completed its preclinical Phase I trial in 2004 [75]. Currently, the formulation is in its Phase II trial for the treatment of the patients suffering from taxane-pre-treated recurrent breast cancer. SP1049C is another doxorubicin encapsulated pluronic polymer micelle based formulation that is under Phase II preclinical trial for the treatment of advanced level inoperable adenocarcinoma of esophagus [117]. NK911 is yet another example of a micellar formulation of PEG and doxorubicin conjugated poly (aspartic acid) which is under preclinical development [118].

Dendrimer Delivery System Dendrimers are roughly spherical nanoparticles made of several monomers, which branch out radially from the center. The advantages associated with dendrimers such as their controlled size, multiple valency, water solubility, modifiable functionality and an internal core render them a promising choice as drug carriers. They are therefore applied as delivery vehicles in several administration routes such as intra-venous, ocular, dermal and oral [119]. Their biocompatibility and immunogenicity has been studied in vitro as well as in vivo and similar to cationic macromolecules like liposomes and micelles, cationic surface groups render dendrimers cytotoxic to cells [120, 121]. Surface functionalization of dendrimers with PEG [122] or fetal calf serum [123] however has shown to reduce the cytotoxicity effects. The drug could be loaded on the dendrimers mainly by physical interaction or by covalent attachment. Physical adsorption of drug could suffer from poor drug loading and less control on drug release kinetics. Alternatively, the pro-drug approach is far more viable where the drug is chemically attached to the dendrimer directly or using a linker giving a much better pharmacokinetic and pharmacodynamic profile [124].

The field of dendrimer-based DDS has evolved greatly in the last decade and several dendrimer-drug conjugates are in their preclinical testing. One of the key examples is conjugation of PEO modified 2,2-bis (hydroxymethyl) propionic acid based biodegradable dendrimer to doxorubicin, which shows 9-fold higher tumor accumulation and 10-fold less cytotoxicity than free drug. The intra-venous administration of prodrug to doxorubicin-nonresponsive tumor showed a rapid tumor regression in a single dose [125]. Poly(glycerol-succinic acid) dendrimer (PGLSA)-camptothecin prodrug similarly has shown an enhanced solubility, cellular uptake and retention [126]. Since these initial success reports, several drugs such as artemether, cisplatin, diclofenac, mefenamic acid, dimethoxycurcumin, diflunisal, etoposide, ibuprofen, 5-florouracil, indomethacin and many more have been conjugated to dendrimer and are undergoing preclinical/clinical trials [127].

10.3.3 Nano-sized Drug Crystals

Poor aqueous solubility is one of the key problems with many small drug molecules, which affects their delivery and therapeutic applications. It is a well-established fact that with size reduction to nanometer scale, the properties of a material is governed by quantum laws and entirely different from its macro/micro size counterpart. A drug nanocrystal is therefore drug particle with its size in the nano-range i.e. 10–100 nm, and a suspension of such nanocrystals is popularly known as nanosuspension [219]. The suspension of these nanocrystals can be achieved in aqueous solutions as well as non-aqueous medium (liquid PEG, oil) with help of stabilizers like amphiphilic surfactants (poloxamers, PVP, phospholipids, polysorbate 80) or polymeric (hydroxypropyl methyl cellulose) materials. The hallmark of drug nanocrystals is that these crystals are pure drug particles with no carrier system. Similar to typical nanoparticle preparation, drug nanocrystals could be prepared by a "bottom-up approach" (molecular level to nanocrystals) such as precipitation method or "top-down approach" (macro/micro level to nanocrystals) such as pearl milling (technology owned by Elan Nanosystems), high-pressure homogenization in water (technology owned by Skyepharma as well as Baxter) and in non-aqueous medium (technology owned by Pharmasol). Sometimes, a combination of the two approaches is used for nanocrystal production e.g. Nanoedge® (Baxter) that uses precipitation followed by homogenization. The major advantages of nanocrystallized drug are increased rate of drug dissolution and saturation solubility, improved oral bioavailability, reduced dose variations and general applicability to all routes of administration.

Rapamune® was the first nanocrystalline drug to obtain FDA approval in 2000 and was licensed to Wyeth Pharmaceuticals. It was produced by pearl milling method developed by Elan Nanosystems and contains rapamycin as the active drug. The formulation is marketed in two forms as tablets and oral suspensions. Soon after, Emend® was approved in 2003, which contains Aprepitant and is marketed by Merck. The production process has been developed by Elan Nanosystems and it is used for the treatment of emesis. Tricor® (drug Fenofibrate), Megace ES® (drug Megestrol acetate) and Theralux® (drug Thymectacin) are three other drugs which have been developed by Elan Nanosystems and have been licensed to Abbott, Par Pharmaceuticals and Celmed respectively. Several other products have however been introduced by other companies which include Semapimod® (Guanylhydrazone, Cytokine Pharmasciences), Paxceed® (Paclitaxel, Angiotech) and Nucryst® (Silver, Nucryst Pharmaceuticals).

10.4 Multifunctional Nanotechnology

As detailed in previous sections, biological system presents several barriers to effective drug delivery. It is therefore germane to develop drug delivery strategies to circumvent these barriers. This could be achieved by making the right choice of material as delivery vehicle, surface modification to increase targeting and intracellular availability of the drug and improving the functionality of the delivery system to achieve the diagnostic applications [128]. The nanosystem with these multifunctional abilities (Fig. 10.3) offer new possibilities in diagnosis, treatment and disease monitoring. The following sections will provide an in depth discussion on these aspects of drug delivery systems.

10.4.1 Choice of Materials for Nanotechnology

The material property of the delivery system is essentially the most important factor that governs the biocompatibility of formulation, stability and bioavailability of the drug and its clearance from the body. It is also equally important to

Fig. 10.3 A conceptual
model of multifunctional
nanomedicine with targeting
ability, imaging capability,
and drug/gene delivery in a
single platform

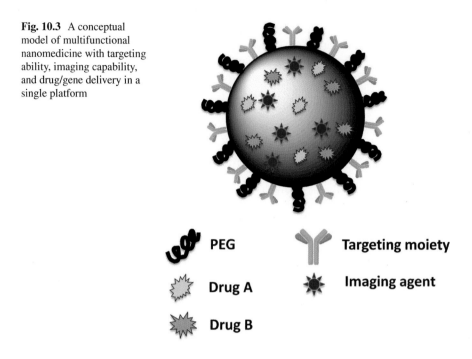

PEG

Targeting moiety

Drug A

Imaging agent

Drug B

understand the microenvironment of the target where the drug has to be delivered
to achieve an effective therapeutic concentration. Design of nanosystems governed
by microenvironment of the disease site results into a class of delivery systems that
are popularly known as stimuli-responsive DDS. The delivery payload, route of
administration and material safety profile, would also govern the components of
such delivery vehicles.

pH Responsive Delivery Systems The physiological profile of an infected, cancerous or inflammatory body tissues differ drastically compared to the normal tissue. It has also been noted that various cellular compartments maintain their own
characteristic pH levels; such as a lysosomal pH is around 4.5 where as in a mitochondria, the pH is around 8. These physiological differences result into a trans-
membrane pH gradient within the cellular compartments in a cell as well as among
the cells. Such subtle differences in physiological environment could be actively
exploited to design a pH responsive delivery system, which would be stable at physiological pH of 7.4 but actively degrade to release the drug under other conditions
[129]. For example, a tumor is composed of rapidly dividing and metabolizing cells
that are always short of the desired food and oxygen supply and thus rely on glyco-
lytic pathways for harvesting energy to sustain [130]. The lack of oxygen in the
tissue results in development of acidic condition within the tumor cells that could be
exploited to achieve the delivery of a desired payload. The physicochemical properties of the delivery vehicles in response to difference in pH therefore are important
characteristic, which has been actively focused in the past two decades.

Poly(β-amino ester) (PbAE) is a biodegradable cationic polymer which has been used for pH stimuli responsive delivery of drug. The polymer rapidly degrades under acidic environment with pH levels below 6.5 to release its payload into the cells. Significantly enhanced accumulation of drugs in the tumors has been demonstrated using PbAE polymer, leveraging pH stimuli-responsive delivery compared to a non-responsive polymer based delivery [131–133]. It has also been shown that the pH sensitivity of the polymeric delivery systems can be tailored by altering the length of the hydrophobic carbon chain length [134]. The pH responsiveness of poly (alkyl acrylic acid) polymer can be controlled by the choice of the monomer as well as the ratio of carboxylated to non-carboxylated alkylacrylate monomer. This polymeric system has been used for enhanced and effective in vitro transfection of lipoplex formulations. In yet another study, pullulan acetate-sulfadimethoxine polymer conjugate has been utilized to develop pH responsive, self-assembled hydrogels for an enhanced delivery of doxorubicin [135].

Polymers have also been directly conjugated to the target drug using pH responsive spacers, which would degrade under the low pH environment inside the tumors or lysosomes/endosomes to release the drug. In one such attempt, poly (vinylpyrrolidone-co-dimethyl maleicanhydride) (PVD) was conjugated to doxorubicin and its pH responsive controlled release increased the accumulation of the drug in to the tumor site [136]. Similarly, copolymer N-(2-hydroxypropyl) methacrylamide (HPMA) [137] and linear PEG based nanosystems are other candidates which have shown promise in delivery of drug to the tumor targets [138, 139]. Hydrolytically labile hydrazone linkage has been used for the drug release by enzymatic action in the lysosomes/endosomes from the polymeric or protein-based conjugate [138]. Serum albumin conjugates of anticancer drugs such as chlorambucil and anthracyclines have shown an enhanced antiproliferative activity compared to free drug [140]. Polyacetals are other pH labile candidates, which have been exploited for developing polymer based pH-responsive DDS [141].

Liposomes have similarly been suitably modified to achieve pH stimuli and controlled drug delivery. The intact pH-sensitive liposomes are internalized into the cells by endocytosis and fuse to the endosomes to deliver its contents inside the cytoplasm [77]. The desired modification of the liposomes is mainly achieved by using new lipid candidates, which provides acid sensitivity to liposomes or by conjugation of pH sensitive polymers on liposome surface to render them prone to pH sensitive degradation. Mildly acidic amphiphiles have been used to design such phosphotidylethanolamine based liposomes where at physiological pH, these amphiphiles act as stabilizers [142] but get protonated under acidic conditions causing a destabilization of the liposome and facilitating the delivery of the payload [143]. These delivery systems have been successfully researched to show in vitro delivery of antitumor drugs, toxins, DNA, antisense oligonucleotides and antigens [144]. Other lipids such as cholesteryl hemisuccinate (CHEMS), poly(organophosphazenes) and dioleoyl phosphatidyl ethanolamine (DOPE) have also been used for pH-sensitive liposomal formulations [145–147].

Micelles are yet another class of nanocarriers which have been extensively investigated to develop pH-responsive delivery. One approach to realize this aim has been the employment of titratable amines or carboxylic groups on the copolymer surface such that the micelle formation relies on the protonation of these groups [148, 149]. In certain cases, water-soluble block copolymers exist in different forms depending on the pH of their aqueous solution and thus have been manipulated for drug delivery applications [150]. Besides, several other water-soluble copolymers have been extensively used to develop long circulating, pH responsive micelles. Some of the common examples include block copolymers based on poly[4- vinyl-benzoic acid (VBA) and 2-N-(morpholino)ethyl methacrylate (MEMA), poly (acrylic acid)-b-polystyrene-b-poly(4-vinyl pyridine) (PAA-b-PS-b-P4VP), Poly[2 (dimethylamino)ethylmethacrylate]-block-poly[2-(N-morpholino)ethyl methacry-late] (DEA-MEMA), poly(L-lactide)-b poly(2-ethyl-2-oxazoline)-b-poly(L-lactide) (PLLA-PEOz-PLLA) ABA triblock copolymers and diblock copolymers (PEOz-PLLA) etc., have been used for such applications [30].

Dendrimers are relatively new class of materials that are being investigated to develop pH-responsive delivery systems. One promising report has been the use of dendrimer composed of 2,2-bis(hydroxymethyl)propanoic acid monomer which has been conjugated to doxorubicin to produce a pH responsive delivery [151]. In another recent attempt, the terminal ends of core-forming PEO dendrimers have been modified with hydrophobic groups using acid-sensitive acetal groups. The hydrophobic groups are cleaved off the dendrimer in acidic environment resulting in the release of the drug [152].

Thermo-responsive Delivery Systems The cancerous cells are known to be highly fragile and sensitive to heat-induced damage (compared to normal cells) largely due to their rapid dividing nature. Incorporation of components that facilitate heat induction in presence of external stimuli such as magnetic field has therefore been looked upon as attractive choices to pursue. These facts have led to the development of hyperthermia as an adjunct to the radiation and chemotherapy for treatment of cancer cells. Several recent research efforts have shown that loading of superparamagnetic iron oxide particles to a delivery system leads to hyperthermia induced cell death at tumor sites [153, 154]. Use of drug delivery vehicle to localize these magnetic particles in the tumor sites ensure that only cancerous cells are subjected to elevated temperatures without affecting the normal cells. The tumor ablation by hyperthermia coupled with incorporation of an antitumor drug in the formulation leads to enhanced efficacy and accumulation of the drug [155, 156].

The thermo-sensitive polymers display a low critical solution temperature (LCST) in aqueous solution, below which they are water-soluble but become insoluble above it. This interesting property makes them an exciting choice as thermo-responsive DDS. One such example has been the accumulation of rhoda-mine– poly(N-isopropyl acrylamide-co-acrylamide) conjugate at the tumor site using targeted hyperthermia [156]. Certain amphiphilic polymers exhibit thermo-sensitivity where they have a temperature sensitive hydrophilic component and a hydrophobic component. Poly (N-isopropylacrylamide) (NIPAAm) and its other copolymers have been the most researched thermo-sensitive amphiphilic polymers [157]. In an interesting report, gold nanoparticles coated cross-linked Pluronic®

(poloxamer) micelles that showed a thermo-sensitive reversible swelling-shrinking behavior caused by hydrophobic interactions of copolymer chains in the micells [158]. Several other illustrations of such polymer based thermo-responsive nanocarriers have been accounted in details in literature for further reading [27].

Fabrication of temperature-sensitive liposomes has been an area of tremendous interest to the researchers due to the simple known fact that the membranes of different phospholipids are known to undergo phase-transition from gel-to-liquid crystalline and lamellar-to-hexagonal transition and are release small water-soluble components during such transitions. One popular example is use of dipalmitoylphosphatidylcholine as primary lipid for liposome formation. It shows a leaky behavior at gel-to-liquid transition at 41 °C and this transition can be tailored by adding distearoylphosphatidylcholine as a co-lipid [159]. Polymers have also been employed to design thermo-sensitive liposomes that also show LCST. These polymer chains exhibit a coil-to-globule transition with a change in temperature and thus impart temperature-regulated functionality to the liposomes [160]. Such polymers stabilize the liposomes in their hydrated form below the LCST but their dehydrated form destabilizes the liposomal structural integrity resulting in delivery of the drug [161]. Several reports exploit the modification of liposomes with NIPAAm copolymers for the fabrication of thermo-responsive substitutes [160, 162].

Redox-Responsive Delivery Systems Nucleic acid based therapeutics has acquired considerable interest lately and numerous attempts have been made to deliver ASN, pDNA, siRNA and miRNA, peptides and proteins for treatment of many genetic diseases. However, successful delivery of these biomolecules to the target cells is an important challenge considering the fact that these agents are highly prone to degradation. A stimuli-responsive system will be of tremendous application as DDS for these biomolecules to ascertain their structural integrity and therefore the therapeutic functionality. It has been established that there is a redox potential difference between the reducing extracellular space and the oxidizing intracellular compartment, which can be potentially exploited to guide the DDS into the cells [163]. Redox-sensitive delivery systems largely rely on components containing disulfide linkage that are taken up in the cell by endocytosis and the disulfide linkage is disrupted in the lysosomes to facilitate payload delivery [164]. The glutathione pathway plays a key role in reduction of the disulfide linkage in the reducing intracellular environment by maintaining an elevated level of reducing glutathione. Besides, the disulfide crosslinking also ensures more stable and robust structural integrity of the nanosystem that decreases the chances of early release of the payload.

One of the strategies to exploit the redox stimuli has been the use of polyaspartamide that uses positively charged groups in the polymer to electrostatically entrap DNA while the thiol groups on the polymer chain form the disulfide linkage resulting in formation of thiopolyplexes [165]. Thiolated gelatin particles have also been shown to form gelatin thiopolyplexes and have been used as potential redox-responsive nanosystem for pDNA delivery [166, 167]. Thiolated polyethylene imine has been directly conjugate to DNA to form polyplexes [168, 169] or have been used with a crosslinking agent [170] to successfully delivery DNA into the cells with high transfection efficiency. In yet another report, glutathione sensitive polymer coated chitosan particles were used for designing of nanosystems stabilized by disulfide bond to provide gene delivery [171]. FDA has recently approved redox-responsive anti-DC33 antibody conjugate (Mylotarg®) for the treatment of acute myeloid leukemia [172].

Disulfide bond based redox-responsive liposomes have also been explored to enhance liposomal stability and delivery efficiency. Such liposomes are formed by a standard phospholipid along with a small chain lipid of which the hydrophobic and hydrophilic ends are linked by disulfide bond. These liposomes show tremendous structural stability until they reach the reducing environment inside the cells where the disulfide bond cleavage results in destabilization and delivery of the gene [173]. Thiocholesterol lipid based liposomes have been shown to successfully delivery gene into the cell in the reducing environment of the cells [174]. Mitomycin C conjugate with a cleavable disulfide bond incorporated into liposomes has shown lesser toxicity and better therapeutic potential than the free drug [175].

10.4.2   Surface Modification to Increase Availability at Tissue and Cell Levels

A careful designing of the nanosystems will enable them to deliver the drugs successfully to the target disease through active or passive targeting. However, to do so successfully, the DDS should be available in the blood stream for longer period of time by avoiding recognition by the components of immune system, circumventing the process of opsonization and preventing subsequent clearance by the RES. The longevity of nanosystem in the circulation not only allows their deposition at the target site through EPR effect but also improves targeting ligand to interact to its receptor. Suitable surface modifications of the nanocarriers for a prolonged and sustained presence in the body have therefore garnered tremendous interest.

Water-soluble polymers have been most commonly used to improve the retention time of the nanosystem in the blood and PEG is found to be most efficient in this regard. The PEG coating on the nanosystem surface provides a steric hindrance that prevents the interaction and binding of blood proteins to nanoparticle surface. The fact that RES recognition of a foreign object in the body largely depends on the binding on these plasma proteins to the surface, the sterically stabilized nanocarriers successfully escape body clearance [176]. This property to evade the immune system is popularly known as the "stealth" effect of the polymer. PEG is an excellent choice as surface protection moiety due to its high solubility in aqueous medium, flexibility of chain length, low immunogenicity and low toxicity. Besides, it does not interfere with the biological performance of the drug loaded in the delivery vehicle. PEG therefore by far is the most studied surface modifying agent to improve the residence time of the pharmaceutically relevant nanosystems. It has also been observed that while the particles modified with brush-like PEG effectively escape the immune response, surfaces modified with mushroom-like PEG molecules seem to activate the immune system against the particles [177, 178]. Literature serves several derivatives of PEG that have actively been used to enable the surface functionalization of the delivery vehicles [179].

Besides PEG alone, copolymers of PEG have also been explored for surface modification of drug delivery constructs. Block copolymer of PEG-polylactide glycolide (PLGA) forms a hydrophobic core of PLGA and a hydrophilic shell of PEG that shows a longer residence time in the blood circulation [180]. Such polymeric preformed particles of PLGA could also be functionalized by PEG derivatives to prevent recognition by the immune system and therefore an enhanced retention time in the body. For example, the PLGA particles functionalized with polylysine-PEG copolymers shows a considerably reduced opsonization [181] while PEG modified poly (cyanoacrylate) particles provided longer-circulation as well as permeation into the brain tissue [182]. In a similar attempt, surface modification of polystyrene nanosystems by hydrophobically-modified dextran and PEG-dextran was studied to show that the stability of construct could be tailored by the density and also the nature of the surface modifying polymer [183]. Lipid derivatives of PEG have similarly been used to prepare PEG modified liposomes for enhanced circulation and improved performance of the delivery system [184].

Even though use of PEG has largely dominated the surface modification of DDS to increase retention time, several other alternatives have also been explored. The pre-requisite for a substitute of PEG has to be a water-soluble, biocompatible and non-immunogenic material. Polyoxomers, polyoxamines, polysorbate 80 and many more polymers have been used to modify the surface of nanoparticles to improve the bioavailability inside body. Lipid derivatives of poly (acryl amide) and poly (vinyl pyrrolidone) as well as other amphiphilic polymers such as poly (acryloyl morpholine) (PAcM), phospholipid (PE)-modified poly(2-methyl-2-oxazoline) or poly(2-ethyl-2-oxazoline), phosphatidyl poly glycerols, and polyvinyl alcohol have been successfully employed for surface modification of the liposomes.

10.4.3 Image-Guided Therapy

Imaging is an indispensable component of therapy and has been routinely used in hospitals and clinics for diagnosis of diseases and defects in the body. Conventional methods such as computerized axial tomography (CAT), magnetic resonance imaging (MRI), X-Ray imaging etc. have been employed in medical science for past several decades. Therefore, it was only fitting that with the advent of nanotechnology and more specifically nano-pharmaceutics, the concept of "molecular imaging" has been envisioned. Ability to image a DDS has therefore been an integral aspect of drug delivery application since it provides a visual feature to locate the site and extend of a disease in the body. Besides, it also enables a real-time assessment of the site of localization of a delivery vehicle in the body, its extent of sequestration in a particular organ and more specifically within a cell in question. For instance, presence of an imaging modality in a delivery vehicle customized to target a metastatic tumor could be essentially tracked to the end site of its localization providing a direct visual evidence of the efficiency of a targeted or non-targeted system as well as the location of the tumor in the patient. Owing to the versatility of such a delivery system, extensive endeavors have been exercised to develop multifunctional nano-system (Fig. 10.3) comprising of targeting ligands, therapeutic agent(s) as well as imaging agents. To this date, several organic and inorganic imaging agents have been explored including liposomes [185], dye-conjugated silica [186], quantum dots [187], gold nanoparticle and nano-shells [188] magnetic nanoparticles [189] and many other contrast enhancing agents. Along with advances in conventional techniques like CAT and MRI scan, many new molecular imaging approaches such as radioactivity-based imaging (gamma scintigraphy, positron emission tomography (PET), single-photon emission computed tomography (SPECT)), surface enhanced raman scattering (SERS), optical coherence tomography (OCT), near-infrared fluorescence imaging etc., are been actively researched.

Radiolabelled probes are the most commonly used imaging agents in the drug delivery systems. Gamma scintigraphy provides a 2-dimension imaging ability while SPECT and PET enable a 3-D scanning. These techniques have their own advantages and disadvantages [190]. However, radioactivity based imaging systems are plagued by difficulties such as handling radioactive material, regulations concerned with their administration, their residence and clearance time from the body. Alternatively, improvement in MRI by the use of magnetic nanoparticles [191] or contrast enhancing agents [192] in the delivery system has been explored with vigor because of the non-invasive nature of the technique. Complexes of gadolinium, manganese, ferrofluids as well as superparamagnetic iron oxide are some of the most commonly applied contrast enhancing agents in MRI scans. Other popular imaging modalities include application of fluorescent dyes and quantum dots [193], SERS agents such as gold and silver nanoparticles [194].

10.4.4 Combination Therapeutics

Reports of multiples drug resistance (MDR) against antibacterial, antiviral, antifungal and anticancer drugs have become regularity in the previous decade. Numerous research endeavors have been applied to understand the origin of MDR and design therapeutic agents against them. However, the more we strive to overcome the medical enigmas by new drug discovery, the more complex the problem of MDR becomes. The gravity of the situation can be envisaged by a fact that the probability of MDR tuberculosis infection in acquired immunodeficiency syndrome (AIDS) patient is many folds more than a normal person. The inception of drug resistance has triggered the use of combination of drugs targeting a disease causing organism/process. The components of combination therapy may impact different independent targets, complement each other effect on the same target or bind independent of each other to give a combined effect for containment of the disease. Such combination therapy has successfully been realized in the treatment of cancer, diabetes, bacterial and viral infections and asthma.

Co-administration of paclitaxel and ceramide using nanosystems has been proven to be extremely effective against MDR ovarian cancer [131, 132] as well as brain tumor cells [195] compared to the effect of individual drugs. Similarly, the use of a combination of paclitaxel and curcumin [28, 196] as well as doxorubicin and curcumin [197] enables to overcome the MDR in cancer cells. Several commercialized drugs such as Vytorin®, Caduet®, Lotrel®, Glucovance®, Avandamet®, Truvada®, Kaletra®, Rebetron®, Bactrim® and Advair® are actually a combination of two drugs [198]. Celetor Pharmaceuticals have developed CombiPlex® technology to launch combination chemotherapies for treatment of cancer. The technology uses high throughput screening, mathematical algorithm for synergy analysis and advanced nanosystems to predict right drug combination for therapy. This platform is meant to design chemotherapies so as to maintain an optimized ratio of the drugs in the body for enhanced efficacy. Their formulation CPX-1 is a fixed ratio combination of irinotecan and floxuridine that has shown positive results in its Phase-1 trial and is currently under Phase-2 trial for treatment against colorectal cancer [199]. CPX-351 similarly is a combination of cytarabine and daunorubicin and is under Phase-1 trial for the treatment of acute myeloid leukemia [200].

10.5 Regulatory Issues in Nano-pharmaceuticals

10.5.1 Approval of Pharmaceutical Products in the US

Despite the advances in nanomaterial application in disease diagnosis and drug delivery, significant amount of work still to be done in terms of characterizing nanomedicine safety and long term effects on biological system. Currently, all nanomedicine go through the FDA's traditional regulatory pathway within the Center for Drug Evaluation and Research (CDER) or Center for Devices and Radiological Health (CDRH). This pathway includes the following general requirements prior to approval.

  (i) CDER reviews applications for new drugs.
 (ii) Prior to clinical testing, laboratory and animal testing is performed to determine pharmacokinetic and pharmacodynamic attributes of the drug to determine a likely safety and toxicology profile in humans.
(iii) Clinical trials are performed in stages to determine if the drug is safe in healthy, then sick patients, and whether it provides a significant health benefit.
 (iv) A team of FDA physicians, chemists, toxicologists, pharmacologists, and other pertinent scientists evaluates clinical data, and if safety and efficacy are established, the drug is approved for marketing.

Prior to the initiation of clinical trials, pre-clinical testing and manufacturing are regulated by several levels of regulation or guidance. These are FDA internally generated guidance documents, codified regulations listed in Title 21 Code of Federal Regulations (CFR) and International Conference on Harmonization (ICH) guidelines. Guidance documents are not codified law, but represent the Agency's current thinking on a particular subject. They do not create or confer any rights for, or on any person and do not operate to bind FDA or the public. An alternative approach may be used if such approach satisfies the requirements of the applicable statute, regulations, or both [201].

Title 21 is the portion of the CFR that governs food and drugs within the United States for the FDA. It is divided into three chapters: Chapter I – Food and Drug Administration, Chapter II – Drug Enforcement Administration, and Chapter III – Office of National Drug Control Policy.

Most of the Chapter I regulations are based on the Federal Food, Drug, and Cosmetic Act. Notable sections in Chapter I are:

(a) 11 Electronic records and electronic signature related
(b) 50 Protection of human subjects in clinical trials
(c) 54 Financial Disclosure by Clinical Investigators [33]
(d) 56 Institutional Review Boards that oversee clinical trials
(e) 58 Good Laboratory Practices (GLP) for nonclinical studies

The 200 and 300 series sections are regulations pertaining to pharmaceuticals:

(a) 202–203 Drug advertising and marketing
(b) 210 cGMP's for pharmaceuticals
(c) 310 Requirements for new drugs
(d) 328 Specific requirements for over-the-counter (OTC) drugs

The 600 series covers biological products (e.g. vaccines, blood):

(a) 601 Licensing under section 351 of the Public Health Service Act
(b) 606 cGMP's for human blood and blood products

The 700 series includes the limited regulations on cosmetics:

(a) 701 Labeling requirements

The 800 series are for medical devices:

(a) 803 Medical Device Reporting
(b) 814 Premarket Approval of Medical Devices [104]
(c) 820 Quality system regulations (analogous to cGMP, but structured like ISO) [128]
(d) 860 Listing of specific approved devices and how they are classified

ICH guidelines are the result of The International Conference on Harmonization of Technical Requirements for Registration of Pharmaceuticals for Human Use and are unique in bringing together the regulatory authorities and pharmaceutical industry of Europe, Japan and the US to discuss scientific and technical aspects of drug registration. Since its inception in 1990, ICH has evolved, through its ICH Global Cooperation Group, to respond to the global face of drug development, so that the benefits of international harmonization for better global health can be realized worldwide [202]. The FDA has adopted ICH guidance within four main categories as described below.

1. *ICH – Efficacy*

   (a) Clinical Safety E1–E2F
   (b) Clinical Study Reports E3
   (c) Dose-response Studies E4
   (d) Ethic factors E5
   (e) Good Clinical Practice E6
   (f) Clinical Trials E7–E11
   (g) Clinical Evaluation by therapeutic Category E12
   (h) Clinical Evaluation E14
   (i) Pharmacogenomics E15–E16

2. *ICH – Joint Safety/Efficacy (Multidisciplinary)*

   (a) MedDRA Terminology M1
   (b) Electronic Standards M2
   (c) Nonclinical Safety Studies M3
   (d) Common Technical Document M4
   (e) Data Elements and Standards for Drug Dictionaries M5
   (f) Gene Therapy M6
   (g) Genotoxic Impurities M7
   (h) Electronic Common Technical Document (eCTD) M8

3. *ICH – Quality*

   (a) Stability Q1A–Q1F
   (b) Analytical Validation Q2
   (c) Impurities Q3A–Q3D
   (d) Pharmacopoeias Q4–Q4B
   (e) Quality of Biotechnological Products Q5A–Q5E
   (f) Specifications Q6A–Q6B
   (g) Good Manufacturing Practice Q7
   (h) Pharmaceutical Development Q8
   (i) Quality Risk Management Q9
   (j) Pharmaceutical Quality System Q10
   (k) Development and Manufacture of Drug substance Q11

4. *ICH – Safety*

   (a) Carcinogenicity Studies S1A–S1C
   (b) Genotoxicity Studies S2
   (c) Toxicokinetics and Pharmacokinetics S3A–S3B
   (d) Toxicity Testing S4

(e) Reproductive Toxicology S5
(f) Biotechnology Products S6
(g) Pharmacology Studies S7A–S7B
(h) Immunotoxicology Studies S8
(i) Nonclinical Evaluation for Anticancer Pharmaceuticals S9
(j) Photo-safety Evaluations S10

The most relevant FDA regulatory document associate with nanomedicine manufacturing is the 'Liposome Drug Products' guidance document proposed in August of 2002 [203] This document currently guides development of liposomal based drugs, which generally fall into the definition of nanomedicine based on particle size. The guidance provides recommendations for drug development applicants on chemistry, manufacturing and controls (CMC), human pharmacokinetics and bioavailability; and labeling documentation for liposome drug products submitted in new drug applications (NDAs). The guidance recommendations are segmented as follows.

1. *Chemistry, Manufacturing, and Controls*

(a) Description and composition
(b) Physiochemical Properties
(c) Description of Manufacturing Processes and Controls
(d) Control of excipients: Lipid Components
(e) Control of Drug Product Specifications
(f) Stability
(g) Changes in Manufacturing

2. *Human Pharmacokinetics and Bioavailability*

(a) Bioanalytical Methods
(b) In Vivo Integrity (Stability) Considerations
(c) Protein Binding
(d) In Vitro Stability
(e) Pharmacokinetics and Bioavailability

3. *Labeling*

(a) Product Name
(b) Cautionary Notes and Warnings
(c) Dosage Administration

Nanomedicine platforms have a number of common issues that are related to regulatory oversight. Some of these include functional qualities such as significantly different chemical properties than corresponding small or large molecules, different PK/PD/ADMET properties, delivery, targeting, release, stabilization, and bioavailability. Characterization, in terms of physiochemical attributes and general CMC issues (stability, sterility, etc.), are also common to many of the nanomedicine platforms, but differ greatly from the traditional small/large molecule drug [204].

While nanomedicine are becoming more prevalent in the areas of cancer, AIDS, and brain disorders, there are concerns that the unique properties of nanoparticles, such as size, shape, affinity, and surface chemistry may not fit the traditional safety and quality evaluation protocol proposed under current regulations.

The FDA and European Medicines Agency (EMA) have begun to address the lack of a more comprehensive regulatory framework for nanomedicine through the establishment of international scientific workshops such as the EMA 1st International Workshop on Nanomedicine in September of 2010 [205]. The FDA has also recognized the need for specific nanomedicine guidance, and is working toward that goal. In August 2006, the FDA established a Nanotechnology Task Force to determine the regulatory framework needed to develop safe and effective FDA-regulated products that use nanotechnology materials. The resulting Nanotechnology Task Force Report recommended that the FDA pursue the development of nanotechnology guidance for manufacturers and researchers, and that because of the emerging and uncertain nature of nanotechnology and the potential for multiple medical applications, there was a requirement for transparent, consistent and predictable regulatory pathways.

Current FDA recommendations, until specific guidance documents are developed, are to follow current FDA guidance including all normal testing procedures, normal drug stability testing, and those associated with CMC, in vivo, and in vitro analysis. Though understanding specific technical and scientific aspects of the drug product, tests should be designed accordingly. All parts of the drug product should be tested for stability, both individually and formulated. It will be critical for nanomedicine drug companies to communicate and develop acceptable procedures in concert with the FDA as early in the product development process as possible [204].

10.5.2 Preclinical and Clinical Development

There are more than twenty FDA approved products that contain nanomaterials (Table 10.1). To date, all of these products have been approved through the traditional regulatory pathway. As previously described nanomedicines are becoming more prevalent in the areas of cancer, AIDS, and brain disorders. There are currently hundreds of nanotechnology companies and research facilities trying to benefit from the emerging nanomedicine marketplace. Within the life sciences industry sector, funding has been primarily focused on those companies that apply nanotechnologies to 'conventional' therapeutics (i.e. drugs as either chemicals or biologics) to increase or extend their application; for example, targeted drug delivery systems (Nemucore Medical Innovations, BioDelivery Sciences International, CytImmune Sciences Inc., NanoBioMagnetics Inc., Nanobiotix, Nanotherapeutics Inc.), diagnostics (Nanosphere Inc., Oxonica Ltd) and medical imaging systems (Life Technologies Inc.- Qdots). These products and applications have a relatively well-defined route to commercialization (subject to the regulatory hurdles facing nanotechnologies in general) [206].

Most notable of the nanomedicine products are the combinatorial drugs that combine targeting, drug delivery, stability, protection, and imaging. Figure 10.3 illustrates a typical combinatorial nanomedicine unit. The multifunctional nanoparticle is by nature a complex mixture of hydrophobic/hydrophilic molecules, inorganic components, peptides, and/or small molecule organic drug molecules. Many issues, regarding in vivo and in vitro assays need to be developed to segregate different properties of a multifunctional drug product. Some of these are:

(a) Synergies or interactions between the nanoparticle components
(b) Biocompatibility
(c) Long-term/chronic exposure assays/data
(d) General toxicology assays and analytics
(e) Animal models
(f) Molecular weight
(g) Particle size
(h) Charge distribution
(i) Purity
(j) Contaminants
(k) Stability – individual components and formulated
(l) Consistency in manufacturing
(m) PK/PD/ADMET assays/profiles
(n) Aseptic processing/sterilization
(o) Immunogenicity

10.5.3   Knowledge Management, Manufacturing and Scale-Up

Process development and manufacturing of nanomedicine is at its early stages of development and thus is also in its seminal stages of preparing to respond to the guidance of the FDA. With FDA's push to move from quality by testing to quality-by-design (QbD) (Fig. 10.4) for nanomedicine community to succeed in this new environment it is imperative to develop robust documented, process knowledge for the fabrication of nanomedicine. Acquisition and development of process knowledge will enable practitioners to bring novel therapies to the clinic with unique multifunctional capabilities. Articulation of the key variables (equipment, materials, idiosyncratic protocols etc.) at an early stage (i.e. the discovery lab) involved in production process will lead to a better understanding of how to translate good lab scale synthesis into scale processes for future clinical translation and assist manufacturing partners to produce material according to FDA's QbD principles.

QbD first implemented in pilot capacity by the FDA in 2005 has been formally adopted as a way to harmonize the development lifecycle of biopharmaceuticals and

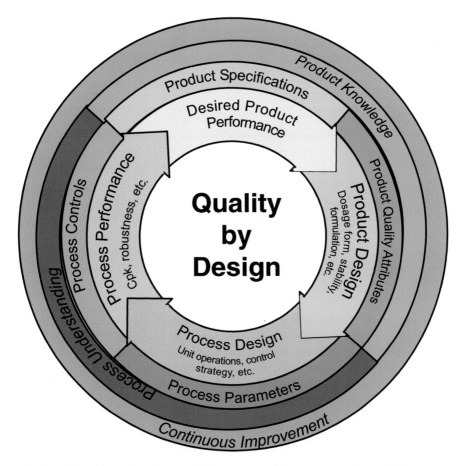

Fig. 10.4 The United States Food and Drug Administration recommended quality-by-design (QdB) approach to link product knowledge with process knowledge and create a continuous improvement product development environment

move away from sampling to find product defects to an environment were "in control" validated processes drive a data rich environment where variations within specification are acceptable. QbD is based on the underlying principle that quality, safety and efficacy must be designed into a product and that quality specifically cannot be tested or inspected into a product. Officially defined as "a systematic approach to development that begins with predefined objectives and emphasizes product and process understanding and process control, based on sound science and quality risk management" [207]. We consider QbD to be essential for the development of manufacturing processes for nanomedicine. QbD creates a continuous knowledge cycle, an important concept for advancing beyond the seminal steps for identifying innovative means to scale production of complex nanomedicine products [208]. The FDA has come out with guidance that covers Pharmaceutical Development, Quality Risk Management, and Quality System with a predisposition that the future state of bio-pharmaceutical manufacturing, of which nanomedicine will be a part, will be an environment governed by QbD [207, 209]. Table 10.2 illustrates the differences in approach and the information requirements of QbD over traditional biopharmaceutical manufacturing which is dependent upon inspection, testing, locked processes and reproducibility [207].

It is important that nanomedicine manufacturers understand that QbD is knowledge rich environment dependent upon user definition of critical quality attributes (CQA), such that the physical, chemical, or biological property or characteristic of the intended nanomedicine should be within a proper range or distribution to ensure product quality. Linking CQA to process inputs (raw materials, chemicals, biologics etc), and process parameters (temperature, pressure, pH, etc) is performed in the early stage experimentation defined as the "design space" which is defined as the range of input variables or parameters for a single operation or it can span multiple operations. Early articulation of the design space, CQA and process inputs can provide a very flexible operational environment with the desired attributes for scale-up manufacturing.

Importance of Knowledge Management in Nanomedicine Nanomedicine holds the promise to cure complex diseases like cancer and save lives [213]. Today, academic scientists lead the development of the complex multifunctional nanomedicine, but for all their promise, there is a striking lag in clinical translation. This lag rests on the fact that nanomedicine investigators under appreciate the value of target product profiles (TPP), a key component of QbD, for ensuring that processes used in the laboratory are compatible with commercial scale-up processes and regulatory guidance [210]. A solution to this problem is at very early stage, put information into the hands of investigators to guide efforts towards nanomedicine that will have a chance to make it to the clinic. Innovation in informatics is another essential area and is complementary to the NIH's proposed investment to create National Center for Advancing Translational Sciences [211].

Nanomedicine translation faces substantial challenges related to managing the complex data streams emerging from the work at the bench, from process development work, and from preclinical studies all with important attributes required to drafting a TPP. The critical information developed during these activities is required to navigate a complex regulatory environment. Without effective data capture solutions and subsequent translation of large quantities of data into shared information, it will be "challenging" to coordinate the bench level process with scale-up process development, risk management and regulatory compliance. We are currently developing a software package, Fig. 10.5, designed to assist academics in overcoming this translational bottleneck for nanomedicine by consolidating existing drug development best practices into a single package for use as a guide to further advance nanomedicine development.

"Nanolytics", developed by Nemucore Medical Innovations, Inc. (NMI) is a knowledge management system for information pertinent to development of TPP, processes development plans, validation plans and risk management assessment

Table 10.2 Quality-by-design (QdB) approach in manufacturing

Aspect	Traditional approach	Enhanced QbD approach	Informatics requirements
Overall pharmaceutical development	a. Mainly empirical b. Developmental research often conducted one variable at a time	a. Systematic, relating mechanistic understanding of input material attributes and process parameters to drug product CQAs b. Multivariate experiments to understand product and process c. Establishment of design space d. Process Analytical Technology (PAT) tools utilized	a. Knowledge management across entire life cycle b. Process traceability and change management from development through manufacturing c. One-point access to all phases and all levels of data
Manufacturing process	a. Fixed b. Validation primarily based on initial full-scale batches c. Focus on optimization and reproducibility	a. Adjustable within design space b. Lifecycle approach to validation and, ideally, continuous process verification c. Focus on control strategy and robustness d. Use of statistical process control methods	a. Full documentation of process analysis and verification decisions b. Integration of these decisions with PAT tool configuration setups
Process controls	a. In-process tests primarily for go/no go decisions b. Off-line analysis	a. PAT tools utilized with appropriate feed forward and feedback controls b. Process operations tracked and trended to support continual improvement efforts post-approval	a. Web-based "Digital Dashboard" providing remote process monitoring b. Record of all significant parameter variances and trends
Product specifications	a. Primary means of control b. Based on batch data available at time of registration	a. Part of the overall quality control strategy b. Based on desired product performance with relevant supportive data	a. All lifecycle documents (from URS though PBR) interlinked, with traceability of changes
Control strategy	a. Drug product quality controlled primarily by intermediate and end product testing	a. Drug product quality ensured by risk-based control strategy for well understood product and process b. Quality controls shifted upstream, with the possibility of real-time release or reduced end-product testing	a. Inventories and risk assessments of all systems, equipment and processes b. Decisions on parameter prioritization and acceptable variances c. Integration with PAT records, for refinement analysis
Lifecycle management	a. Reactive (i.e., problem solving and corrective action)	a. Preventive action b. Continual improvement facilitated	a. Change management integrated across all lifecycle phases

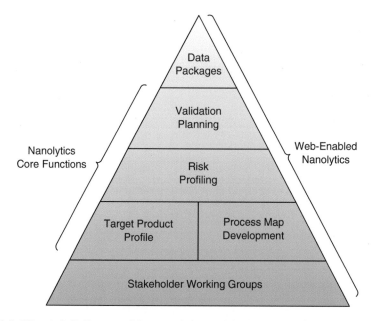

Fig. 10.5 *"Nanolytics"*: Conceptual framework for combination of informatics with processing technology for optimization of nano-pharmaceutical formulations

needed to support effective nanomedicine translation. Nanolytics allows academic investigators early in research to contextualize how a nanomedicine could move to the clinic. Unlike either small molecule or biologic development the creation of nanomedicine, which are complex molecular entities, is very process and design intensive. A manual process already demonstrated value of an informatics approach to identify barriers (use of equipment not compatible with scale-up) and risks (regulatory, material, etc.) to translating these nanomedicine to the clinic. Nanolytics software consists of three suites: a TPP Suite, a Process Suite and Validation Suite. These suites and the knowledge they will manage should mitigate cost and reduce time of development of scale-up processes, lower barriers to clinical development for nanomedicine and leverage research costs more effectively [211, 212]. Nanolytics allows for the input of key information based on initial research and outputs documentation on how to achieve for the pilot scale production of the target nanomedicine. As always is the case, better information, begets a more realistic product development plans. This development of information "outside" of the typical areas of focus of a nanomedicine researcher will reduce risk and clarify efforts in translating nanomedicine from bench to bedside.

Significance to Nanomanufacturing Practices Developing manufacturing capability in the past has been capital intensive and typically relegated to a commercial responsibility. But with many of the advances happening in nanomedicine there is a discreet need to lower the barrier to access manufacturing capabilities on a molecule agnostic platform. In an effort to create such an environment we have begun the process to establish the first in the nation FlexFactory™ nanomedicine manufacturing facility compliant with QbD principles. FlexFactory™ was developed by Xcellerex, Inc, (Marlborough, MA) to transition from single molecule manufacturing footprint to a modular, single use backbone which is agnostic to molecule. FlexFactory™ provides the ideal manufacturing environment for nanomedicine as the controlled environmental units (CEMs) are able to maintain a single unit operation of a manufacturing process with the ability to grow with the progress of the molecule from preclinical thru commercial launch. The innovation of the FlexFactory™, briefly, is if a unit operation needs to change for the development of a new nanomedicine manufacturing process the modular CEMs can be opened a new unit operation installed, the new step and the new process validated allowing for the production of a nanomedicine that conforms to different CQA. While there are other modular platforms that can be used in a similar manner they often have to be pieced together. The FlexFactory™ system has withstood numerous FDA audits, inspections, and license applications for a variety of biologics. The sophistication required for biologic therapeutic manufacturing is suspected to be similar to the complexity required for multifunctional nanomedicines. This level of complexity and novelty of scaling nanomedicine production is why we have taken a two-step approach to aggregate knowledge using Nanolytics and the molecule agnostic manufacturing platform FlexFactory™, Fig. 10.6.

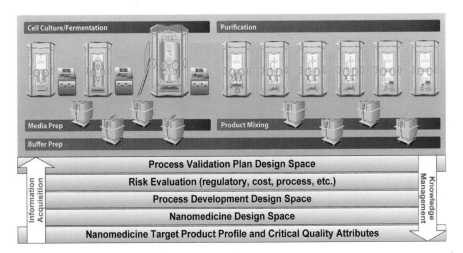

Fig. 10.6 NMI FlexFactory™ footprint shown to illustrate that data captured in Nanolytics serves as foundation for manufacturing Information and knowledge about product characteristics, process, and systems drive manufacturing design to optimize manufacturing of nanomedicine

10.6   Conclusions and Future Outlook

With greater understanding of chemical and physiological barriers associated in drug delivery and advances in nanomedicine design, there is an opportunity to efficient delivery of small and macromolecular drugs to complex diseases. Along these lines, the nanosystems have been engineered with specific attributes such as biocompatibility, suitable size and charge, longevity in blood circulation, targeting ability and image guided therapeutics, which can deliver the drug/imaging agent to the specific site of interest, based on active and passive targeting mechanisms. These systems cannot only improve the drug delivery to the target disease, but also the resolution of detection at cellular and sub-cellular levels.

To fully realize the potential of nanosystems for delivery of contemporary therapeutics in clinical setting, it is imperative that researchers also address the material safety, scale-up and quality control issues. Scale-up and quality control becomes extremely challenging especially when dealing with nanosystem designed to carry multiple drugs, imaging agents and targeting moieties. Furthermore, in vivo fate of nanomedicine engineered using novel nanomaterials are need to be fully assessed before being used in clinical application.

References

 1. Pushparaj PN, Aarthi JJ, Manikandan J, Kumar SD (2008) siRNA, miRNA, and shRNA: in vivo applications. J Dent Res 87:992–1003
 2. Baumann K (2014) Gene expression: RNAi as a global transcriptional activator. Nat Rev Mol Cell Biol 15(5):298–299
 3. Ha M, Kim VN (2014) Regulation of microRNA biogenesis. Nat Rev Mol Cell Biol 15:509–524
 4. Alonso MJ (2004) Nanomedicines for overcoming biological barriers. Biomed Pharmacother 58:168–172
 5. Pecot CV, Calin GA, Coleman RL, Lopez-Berestein G, Sood AK (2011) RNA interference in the clinic: challenges and future directions. Nat Rev Cancer 11:59–67
 6. Stegemann S, Leveiller F, Franchi D, de Jong H, Linden H (2007) When poor solubility becomes an issue: from early stage to proof of concept. Eur J Pharm Sci 31:249–261
 7. Lipinski CA (2000) Drug-like properties and the causes of poor solubility and poor permeability. J Pharmacol Toxicol Methods 44:235–249
 8. Merisko-Liversidge EM, Liversidge GG (2008) Drug nanoparticles: formulating poorly water-soluble compounds. Toxicol Pathol 36:43–48
 9. Aungst BJ (1999) P-glycoprotein, secretory transport, and other barriers to the oral delivery of anti-HIV drugs. Adv Drug Deliv Rev 39:105–116
10. Goldberg M, Gomez-Orellana I (2003) Challenges for the oral delivery of macromolecules. Nat Rev Drug Discov 2:289–295
11. Salama N, Eddington N, Fasano A (2006) Tight junction modulation and its relationship to drug delivery. Adv Drug Deliv Rev 58:15–28
12. Florence AT (2005) Nanoparticle uptake by the oral route: fulfilling its potential? Drug Discov Today 2:75–81
13. Yang SC, Benita S (2000) Enhanced absorption and drug targeting by positively charged submicron emulsions. Drug Dev Res 50:476–486

14. Artursson P, Ungell AL, Lofroth JE (1993) Selective paracellular permeability in two models of intestinal absorption: cultured monolayers of human intestinal epithelial cells and rat intestinal segments. Pharm Res 10:1123–1129

15. Lipinski CA, Lombardo F, Dominy BW, Feeney PJ (2001) Experimental and computational approaches to estimate solubility and permeability in drug discovery and development settings. Adv Drug Deliv Rev 46:3–26

16. Lindup WE, Orme MC (1981) Clinical pharmacology: plasma protein binding of drugs. Br Med J (Clin Res Ed) 282:212–214

17. Shen DD, Kunze KL, Thummel KE (1997) Enzyme-catalyzed processes of first-pass hepatic and intestinal drug extraction. Adv Drug Deliv Rev 27:99–127

18. Patil SD, Rhodes DG, Burgess DJ (2005) DNA-based therapeutics and DNA delivery systems: a comprehensive review. AAPS J 7:E61–E77

19. Ejendal KF, Hrycyna CA (2002) Multidrug resistance and cancer: the role of the human ABC transporter ABCG2. Curr Protein Pept Sci 3:503–511

20. Ganta S, Deshpande D, Korde A, Amiji M (2010) A review of multifunctional nanoemulsion systems to overcome oral and CNS drug delivery barriers. Mol Membr Biol 27:260–273

21. Ganta S, Devalapally H, Amiji M (2010) Curcumin enhances oral bioavailability and anti-tumor therapeutic efficacy of paclitaxel upon administration in nanoemulsion formulation. J Pharm Sci 99:4630–4641

22. Ganta S, Sharma P, Paxton JW, Baguley BC, Garg S (2010) Pharmacokinetics and pharmacodynamics of chlorambucil delivered in long-circulating nanoemulsion. J Drug Target 18:125–133

23. Jones PM, George AM (2004) The ABC transporter structure and mechanism: perspectives on recent research. Cell Mol Life Sci 61:682–699

24. Zhang Y, Benet LZ (2001) The gut as a barrier to drug absorption: combined role of cyto-chrome P450 3A and P-glycoprotein. Clin Pharmacokinet 40:159–168

25. Demeule M, Regina A, Jodoin J, Laplante A, Dagenais C, Berthelet F, Moghrabi A, Beliveau R (2002) Drug transport to the brain: key roles for the efflux pump P-glycoprotein in the blood–brain barrier. Vascul Pharmacol 38:339–348

26. Loscher W, Potschka H (2005) Blood–brain barrier active efflux transporters: ATP-binding cassette gene family. NeuroRx 2:86–98

27. Malingre MM, Beijnen JH, Schellens JH (2001) Oral delivery of taxanes. Invest New Drugs 19:155–162

28. Ganta S, Amiji M (2009) Coadministration of paclitaxel and curcumin in nanoemulsion for-mulations to overcome multidrug resistance in tumor cells. Mol Pharm 6:928–939

29. Yang S, Gursoy RN, Lambert G, Benita S (2004) Enhanced oral absorption of paclitaxel in a novel self-microemulsifying drug delivery system with or without concomitant use of P-glycoprotein inhibitors. Pharm Res 21:261–270

30. Ganta S, Devalapally H, Shahiwala A, Amiji M (2008) A review of stimuli-responsive nano-carriers for drug and gene delivery. J Control Release 126:187–204

31. Zhang W, Tan TM, Lim LY (2007) Impact of curcumin-induced changes in P-glycoprotein and CYP3A expression on the pharmacokinetics of peroral celiprolol and midazolam in rats. Drug Metab Dispos 35:110–115

32. Pardridge WM (2007) Blood–brain barrier delivery. Drug Discov Today 12:54–61

33. Abbott NJ, Patabendige AA, Dolman DE, Yusof SR, Begley DJ (2010) Structure and function of the blood–brain barrier. Neurobiol Dis 37:13–25

34. Tredan O, Galmarini CM, Patel K, Tannock IF (2007) Drug resistance and the solid tumor microenvironment. J Natl Cancer Inst 99:1441–1454

35. Berns A, Pandolfi PP (2014) Tumor microenvironment revisited. EMBO Rep 15(5): 458–459

36. Mittal K, Ebos J, Rini B (2014) Angiogenesis and the tumor microenvironment: vascular endothelial growth factor and beyond. Semin Oncol 41(2):235–251

37. Vaupel P (2004) Tumor microenvironmental physiology and its implications for radiation oncology. Semin Radiat Oncol 14:198–206

38. Netti PA, Berk DA, Swartz MA, Grodzinsky AJ, Jain RK (2000) Role of extracellular matrix assembly in interstitial transport in solid tumors. Cancer Res 60:2497–2503
39. Olive PL, Durand RE (1994) Drug and radiation resistance in spheroids: cell contact and kinetics. Cancer Metastasis Rev 13:121–138
40. Teicher BA, Herman TS, Holden SA, Wang YY, Pfeffer MR, Crawford JW, Frei E 3rd (1990) Tumor resistance to alkylating agents conferred by mechanisms operative only in vivo. Science 247:1457–1461
41. Davis SS (1997) Biomedical applications of nanotechnology–implications for drug targeting and gene therapy. Trends Biotechnol 15:217–224
42. Amidon GL, Lennernas H, Shah VP, Crison JR (1995) A theoretical basis for a biopharmaceutic drug classification: the correlation of in vitro drug product dissolution and in vivo bioavailability. Pharm Res 12:413–420
43. Muller RH, Keck CM (2004) Challenges and solutions for the delivery of biotech drugs–a review of drug nanocrystal technology and lipid nanoparticles. J Biotechnol 113:151–170
44. Tiwari SB, Amiji MM (2006) Improved oral delivery of paclitaxel following administration in nanoemulsion formulations. J Nanosci Nanotechnol 6:3215–3221
45. Vyas TK, Shahiwala A, Amiji MM (2008) Improved oral bioavailability and brain transport of Saquinavir upon administration in novel nanoemulsion formulations. Int J Pharm 347:93–101
46. Edmond J (2001) Essential polyunsaturated fatty acids and the barrier to the brain: the components of a model for transport. J Mol Neurosci 16:181–193, discussion 215–121
47. Roerdink F, Regts J, Van Leeuwen B, Scherphof G (1984) Intrahepatic uptake and processing of intravenously injected small unilamellar phospholipid vesicles in rats. Biochim Biophys Acta 770:195–202
48. Turner N, Wright N (1992) The response to injury. Oxf Textb Pathol 351–390
49. Jain RK (1989) Delivery of novel therapeutic agents in tumors: physiological barriers and strategies. J Natl Cancer Inst 81:570–576
50. Braet F, De Zanger R, Baekeland M, Crabbe E, Van Der Smissen P, Wisse E (1995) Structure and dynamics of the fenestrae-associated cytoskeleton of rat liver sinusoidal endothelial cells. Hepatology 21:180–189
51. Dams ET, Oyen WJ, Boerman OC, Storm G, Laverman P, Kok PJ, Buijs WC, Bakker H, van der Meer JW, Corstens FH (2000) 99mTc-PEG liposomes for the scintigraphic detection of infection and inflammation: clinical evaluation. J Nucl Med 41:622–630
52. Danhier F, Feron O, Préat V (2010) To exploit the tumor microenvironment: passive and active tumor targeting of nanocarriers for anti-cancer drug delivery. J Control Release 148:135–146
53. Matsumura Y, Maeda H (1986) A new concept for macromolecular therapeutics in cancer chemotherapy: mechanism of tumoritropic accumulation of proteins and the antitumor agent smancs. Cancer Res 46:6387–6392
54. Jain RK (1987) Transport of molecules in the tumor interstitium: a review. Cancer Res 47:3039–3051
55. Maeda H, Sawa T, Konno T (2001) Mechanism of tumor-targeted delivery of macromolecular drugs, including the EPR effect in solid tumor and clinical overview of the prototype polymeric drug SMANCS. J Control Release 74:47–61
56. Marcucci F, Lefoulon F (2004) Active targeting with particulate drug carriers in tumor therapy: fundamentals and recent progress. Drug Discov Today 9:219–228
57. Rihova B (1998) Receptor-mediated targeted drug or toxin delivery. Adv Drug Deliv Rev 29:273–289
58. Torchilin VP (2006) Recent approaches to intracellular delivery of drugs and DNA and organelle targeting. Annu Rev Biomed Eng 8:343–375
59. Kichler A, Leborgne C, Coeytaux E, Danos O (2001) Polyethylenimine-mediated gene delivery: a mechanistic study. J Gene Med 3:135–144
60. Low PS, Antony AC (2004) Folate receptor-targeted drugs for cancer and inflammatory diseases. Adv Drug Deliv Rev 56:1055–1058

61. Oba M, Aoyagi K, Miyata K, Matsumoto Y, Itaka K, Nishiyama N, Yamasaki Y, Koyama H, Kataoka K (2008) Polyplex micelles with cyclic RGD peptide ligands and disulfide cross-links directing to the enhanced transfection via controlled intracellular trafficking. Mol Pharm 5:1080–1092

62. Gupta B, Torchilin VP (2006) Transactivating transcriptional activator-mediated drug delivery. Expert Opin Drug Deliv 3:177–190

63. Snyder EL, Dowdy SF (2001) Protein/peptide transduction domains: potential to deliver large DNA molecules into cells. Curr Opin Mol Ther 3:147–152

64. Weissig V, Torchilin VP (2001) Cationic bolasomes with delocalized charge centers as mitochondria-specific DNA delivery systems. Adv Drug Deliv Rev 49:127–149

65. Weissig V, Torchilin VP (2001) Drug and DNA delivery to mitochondria. Adv Drug Deliv Rev 49:1–2

66. Kushwaha SKS, Keshari RK, Rai A (2011) Advances in nasal trans-mucosal drug delivery. J Appl Pharm Sci 1:21–28

67. Mathias NR, Hussain MA (2010) Non-invasive systemic drug delivery: developability considerations for alternate routes of administration. J Pharm Sci 99:1–20

68. Jatzkewitz H (1955) An ein kolloidales blutplasmaersatzmittel (polyvinylpyrrolidon) gebundenes peptamin (glycyl l-leucyl-mezcalin) als neuartige depotform fur biologisch aktive primare amine (mezcalin). Z Naturforsch B 10:27–31

69. Bangham AD, Horne RW (1964) Negative staining of phospholipids and their structural modification by surface-active agents as observed in the electron microscope. J Mol Biol 8:660–668

70. Gregoriadis G (1973) Drug entrapment in liposomes. FEBS Lett 36:292–296

71. Scheffel U, Rhodes BA, Natarajan TK, Wagner HN Jr (1972) Albumin microspheres for study of the reticuloendothelial system. J Nucl Med 13:498–503

72. Kramer PA (1974) Letter: Albumin microspheres as vehicles for achieving specificity in drug delivery. J Pharm Sci 63:1646–1647

73. Ringsdorf H (1975) Structure and properties of pharmacologically active polymers. J Polym Sci Polym Sym 51:135–153

74. Kreuter J (2007) Nanoparticles–a historical perspective. Int J Pharm 331:1–10

75. Kim TY, Kim DW, Chung JY, Shin SG, Kim SC, Heo DS, Kim NK, Bang YJ (2004) Phase I and pharmacokinetic study of Genexol-PM, a cremophor-free, polymeric micelle-formulated paclitaxel, in patients with advanced malignancies. Clin Cancer Res 10:3708–3716

76. Lee KS, Chung HC, Im SA, Park YH, Kim CS, Kim SB, Rha SY, Lee MY, Ro J (2008) Multicenter phase II trial of Genexol-PM, a Cremophor-free, polymeric micelle formulation of paclitaxel, in patients with metastatic breast cancer. Breast Cancer Res Treat 108:241–250

77. Torchilin VP (2005) Recent advances with liposomes as pharmaceutical carriers. Nat Rev Drug Discov 4:145–160

78. Klibanov AL, Maruyama K, Torchilin VP, Huang L (1990) Amphipathic polyethyleneglycols effectively prolong the circulation time of liposomes. FEBS Lett 268:235–237

79. Blume G, Cevc G (1993) Molecular mechanism of the lipid vesicle longevity in vivo. Biochim Biophys Acta 1146:157–168

80. Torchilin VPT (1995) Which polymers can make nanoparticulate drug carriers long-circulating? Adv Drug Deliv Rev 16:141–155

81. Whiteman KR, Subr V, Ulbrich K, Torchilin VP (2001) Poly(Hpma)-coated liposomes demonstrate prolonged circulation in mice. J Liposome Res 11:153–164

82. Torchilin VP, Levchenko TS, Whiteman KR, Yaroslavov AA, Tsatsakis AM, Rizos AK, Michailova EV, Shtilman MI (2001) Amphiphilic poly-N-vinylpyrrolidones: synthesis, properties and liposome surface modification. Biomaterials 22:3035–3044

83. Takeuchi H, Kojima H, Yamamoto H, Kawashima Y (2001) Evaluation of circulation profiles of liposomes coated with hydrophilic polymers having different molecular weights in rats. J Control Release 75:83–91

84. Metselaar JM, Bruin P, de Boer LW, de Vringer T, Snel C, Oussoren C, Wauben MH, Crommelin DJ, Storm G, Hennink WE (2003) A novel family of L-amino acid-based biodegradable polymer-lipid conjugates for the development of long-circulating liposomes with effective drug-targeting capacity. Bioconjug Chem 14:1156–1164

85. Levchenko TS, Rammohan R, Lukyanov AN, Whiteman KR, Torchilin VP (2002) Liposome clearance in mice: the effect of a separate and combined presence of surface charge and polymer coating. Int J Pharm 240:95–102

86. Allen TM, Sapra P, Moase E, Moreira J, Iden D (2002) Adventures in targeting. J Liposome Res 12:5–12

87. Gabizon A, Shmeeda H, Horowitz AT, Zalipsky S (2004) Tumor cell targeting of liposome-entrapped drugs with phospholipid-anchored folic acid-PEG conjugates. Adv Drug Deliv Rev 56:1177–1192

88. Gupta B, Levchenko TS, Torchilin VP (2005) Intracellular delivery of large molecules and small particles by cell-penetrating proteins and peptides. Adv Drug Deliv Rev 57:637–651

89. Berry G, Billingham M, Alderman E, Richardson P, Torti F, Lum B, Patek A, Martin FJ (1998) The use of cardiac biopsy to demonstrate reduced cardiotoxicity in AIDS Kaposi's sarcoma patients treated with pegylated liposomal doxorubicin. Ann Oncol 9:711–716

90. Northfelt DW, Dezube BJ, Thommes JA, Miller BJ, Fischl MA, Friedman-Kien A, Kaplan LD, Du Mond C, Mamelok RD, Henry DH (1998) Pegylated-liposomal doxorubicin versus doxorubicin, bleomycin, and vincristine in the treatment of AIDS-related Kaposi's sarcoma: results of a randomized phase III clinical trial. J Clin Oncol 16:2445–2451

91. Davis ME, Chen ZG, Shin DM (2008) Nanoparticle therapeutics: an emerging treatment modality for cancer. Nat Rev Drug Discov 7:771–782

92. Hamilton A, Biganzoli L, Coleman R, Mauriac L, Hennebert P, Awada A, Nooij M, Beex L, Piccart M, Van Hoorebeeck I, Bruning P, de Valeriola D (2002) EORTC 10968: a phase I clinical and pharmacokinetic study of polyethylene glycol liposomal doxorubicin (Caelyx, Doxil) at a 6-week interval in patients with metastatic breast cancer. European Organization for Research and Treatment of Cancer. Ann Oncol 13:910–918

93. Lukyanov AN, Elbayoumi TA, Chakilam AR, Torchilin VP (2004) Tumor-targeted liposomes: doxorubicin-loaded long-circulating liposomes modified with anti-cancer antibody. J Control Release 100:135–144

94. Northfelt DW, Dezube BJ, Thommes JA, Levine R, Von Roenn JH, Dosik GM, Rios A, Krown SE, DuMond C, Mamelok RD (1997) Efficacy of pegylated-liposomal doxorubicin in the treatment of AIDS-related Kaposi's sarcoma after failure of standard chemotherapy. J Clin Oncol 15:653–659

95. Tadros T, Izquierdo P, Esquena J, Solans C (2004) Formation and stability of nano-emulsions. Adv Colloid Interface Sci 108–109:303–318

96. Shafiq-un-Nabi S, Shakeel F, Talegaonkar S, Ali J, Baboota S, Ahuja A, Khar RK, Ali M (2007) Formulation development and optimization using nanoemulsion technique: a technical note, AAPS PharmSciTech 8, Article 28

97. Anton N, Saulnier P, Beduneau A, Benoit JP (2007) Salting-out effect induced by temperature cycling on a water/nonionic surfactant/oil system. J Phys Chem B 111:3651–3657

98. Talekar M, Ganta S, Singh A, Amiji M, Kendall J, Denny WA, Garg S (2012) Phosphatidylinositol 3-kinase inhibitor (PIK75) containing surface functionalized nanoemulsion for enhanced drug delivery, cytotoxicity and pro-apoptotic activity in ovarian cancer cells. Pharm Res 29:2874–2886

99. Talekar M, Kendall J, Denny W, Jamieson S, Garg S (2012) Development and evaluation of PIK75 nanosuspension, a phosphatidylinositol-3-kinase inhibitor. Eur J Pharm Sci 47:824–833

100. Cockshott ID (1985) Propofol ('Diprivan') pharmacokinetics and metabolism–an overview. Postgrad Med J 61(Suppl 3):45–50

101. Langley MS, Heel RC (1988) Propofol. A review of its pharmacodynamic and pharmacokinetic properties and use as an intravenous anaesthetic. Drugs 35:334–372

102. Duncan R (2006) Polymer conjugates as anticancer nanomedicines. Nat Rev Cancer 6:688–701

103. Tanaka T, Shiramoto S, Miyashita M, Fujishima Y, Kaneo Y (2004) Tumor targeting based on the effect of enhanced permeability and retention (EPR) and the mechanism of receptor-mediated endocytosis (RME). Int J Pharm 277:39–61

104. Abe S, Otsuki M (2002) Styrene maleic acid neocarzinostatin treatment for hepatocellular carcinoma. Curr Med Chem Anticancer Agents 2:715–726

105. Graham ML (2003) Pegaspargase: a review of clinical studies. Adv Drug Deliv Rev 55:1293–1302

106. Gradishar WJ, Tjulandin S, Davidson N, Shaw H, Desai N, Bhar P, Hawkins M, O'Shaughnessy J (2005) Phase III trial of nanoparticle albumin-bound paclitaxel compared with polyethylated castor oil-based paclitaxel in women with breast cancer. J Clin Oncol 23:7794–7803

107. Allen TM (2002) Ligand-targeted therapeutics in anticancer therapy. Nat Rev Cancer 2:750–763

108. Milenic DE, Brady ED, Brechbiel MW (2004) Antibody-targeted radiation cancer therapy. Nat Rev Drug Discov 3:488–499

109. Torchilin VP (2001) Structure and design of polymeric surfactant-based drug delivery systems. J Control Release 73:137–172

110. Nishiyama N, Kataoka K (2006) Current state, achievements, and future prospects of polymeric micelles as nanocarriers for drug and gene delivery. Pharmacol Ther 112:630–648

111. Gaber NN, Darwis Y, Peh KK, Tan YT (2006) Characterization of polymeric micelles for pulmonary delivery of beclomethasone dipropionate. J Nanosci Nanotechnol 6:3095–3101

112. Dong H, Li Y, Cai S, Zhuo R, Zhang X, Liu L (2008) A facile one-pot construction of supramolecular polymer micelles from alpha-cyclodextrin and poly(epsilon-caprolactone). Angew Chem Int Ed Engl 47:5573–5576

113. Satoh T, Higuchi Y, Kawakami S, Hashida M, Kagechika H, Shudo K, Yokoyama M (2009) Encapsulation of the synthetic retinoids Am80 and LE540 into polymeric micelles and the retinoids' release control. J Control Release 136:187–195

114. Wei X, Gong C, Shi S, Fu S, Men K, Zeng S, Zheng X, Gou M, Chen L, Qiu L, Qian Z (2009) Self-assembled honokiol-loaded micelles based on poly(epsilon-caprolactone)-poly(ethylene glycol)-poly(epsilon-caprolactone) copolymer. Int J Pharm 369:170–175

115. Wang Y, Li Y, Wang Q, Fang X (2008) Pharmacokinetics and biodistribution of polymeric micelles of paclitaxel with pluronic P105/poly(caprolactone) copolymers. Pharmazie 63:446–452

116. Opanasopit P, Ngawhirunpat T, Rojanarata T, Choochottiros C, Chirachanchai S (2007) Camptothecin-incorporating N-phthaloylchitosan-g-mPEG self-assembly micellar system: effect of degree of deacetylation. Colloids Surf B Biointerfaces 60:117–124

117. Valle JW, Armstrong A, Newman C, Alakhov V, Pietrzynski G, Brewer J, Campbell S, Corrie P, Rowinsky EK, Ranson M (2011) A phase 2 study of SP1049C, doxorubicin in P-glycoprotein-targeting pluronics, in patients with advanced adenocarcinoma of the esophagus and gastroesophageal junction. Invest New Drugs 29:1029–1037

118. Matsumura Y, Hamaguchi T, Ura T, Muro K, Yamada Y, Shimada Y, Shirao K, Okusaka T, Ueno H, Ikeda M, Watanabe N (2004) Phase I clinical trial and pharmacokinetic evaluation of NK911, a micelle-encapsulated doxorubicin. Br J Cancer 91:1775–1781

119. Cheng Y, Xu Z, Ma M, Xu T (2008) Dendrimers as drug carriers: applications in different routes of drug administration. J Pharm Sci 97:123–143

120. Fischer D, Li Y, Ahlemeyer B, Krieglstein J, Kissel T (2003) In vitro cytotoxicity testing of polycations: influence of polymer structure on cell viability and hemolysis. Biomaterials 24:1121–1131

121. Jevprasesphant R, Penny J, Jalal R, Attwood D, McKeown NB, D'Emanuele A (2003) The influence of surface modification on the cytotoxicity of PAMAM dendrimers. Int J Pharm 252:263–266

122. El-Sayed M, Ginski M, Rhodes C, Ghandehari H (2002) Transepithelial transport of poly(amidoamine) dendrimers across Caco-2 cell monolayers. J Control Release 81:355–365

123. Yoo H, Juliano RL (2000) Enhanced delivery of antisense oligonucleotides with fluorophore-conjugated PAMAM dendrimers. Nucleic Acids Res 28:4225–4231

124. Gurdag S, Khandare J, Stapels S, Matherly LH, Kannan RM (2006) Activity of dendrimer-methotrexate conjugates on methotrexate-sensitive and -resistant cell lines. Bioconjug Chem 17:275–283

125. Lee CC, Gillies ER, Fox ME, Guillaudeu SJ, Frechet JM, Dy EE, Szoka FC (2006) A single dose of doxorubicin-functionalized bow-tie dendrimer cures mice bearing C-26 colon carcinomas. Proc Natl Acad Sci U S A 103:16649–16654

126. Morgan MT, Nakanishi Y, Kroll DJ, Griset AP, Carnahan MA, Wathier M, Oberlies NH, Manikumar G, Wani MC, Grinstaff MW (2006) Dendrimer-encapsulated camptothecins: increased solubility, cellular uptake, and cellular retention affords enhanced anticancer activity in vitro. Cancer Res 66:11913–11921

127. Svenson S (2009) Dendrimers as versatile platform in drug delivery applications. Eur J Pharm Biopharm 71:445–462

128. Abeylath SC, Ganta S, Iyer AK, Amiji M (2011) Combinatorial-designed multifunctional polymeric nanosystems for tumor-targeted therapeutic delivery. Acc Chem Res 44:1009–1017

129. Yatvin MB, Kreutz W, Horwitz BA, Shinitzky M (1980) pH-sensitive liposomes: possible clinical implications. Science 210:1253–1255

130. Pelicano H, Martin DS, Xu RH, Huang P (2006) Glycolysis inhibition for anticancer treatment. Oncogene 25:4633–4646

131. Devalapally H, Duan Z, Seiden MV, Amiji MM (2007) Paclitaxel and ceramide co-administration in biodegradable polymeric nanoparticulate delivery system to overcome drug resistance in ovarian cancer. Int J Cancer 121:1830–1838

132. Devalapally H, Shenoy D, Little S, Langer R, Amiji M (2007) Poly(ethylene oxide)-modified poly(beta-amino ester) nanoparticles as a pH-sensitive system for tumor-targeted delivery of hydrophobic drugs: part 3. Therapeutic efficacy and safety studies in ovarian cancer xenograft model. Cancer Chemother Pharmacol 59:477–484

133. Shenoy D, Little S, Langer R, Amiji M (2005) Poly(ethylene oxide)-modified poly(beta-amino ester) nanoparticles as a pH-sensitive system for tumor-targeted delivery of hydrophobic drugs. 1. In vitro evaluations. Mol Pharm 2:357–366

134. Stayton PS, El-Sayed ME, Murthy N, Bulmus V, Lackey C, Cheung C, Hoffman AS (2005) 'Smart' delivery systems for biomolecular therapeutics. Orthod Craniofac Res 8:219–225

135. Na K, Lee ES, Bae YH (2003) Adriamycin loaded pullulan acetate/sulfonamide conjugate nanoparticles responding to tumor pH: pH-dependent cell interaction, internalization and cytotoxicity in vitro. J Control Release 87:3–13

136. Kamada H, Tsutsumi Y, Yoshioka Y, Yamamoto Y, Kodaira H, Tsunoda S, Okamoto T, Mukai Y, Shibata H, Nakagawa S, Mayumi T (2004) Design of a pH-sensitive polymeric carrier for drug release and its application in cancer therapy. Clin Cancer Res 10:2545–2550

137. Shigeta K, Kawakami S, Higuchi Y, Okuda T, Yagi H, Yamashita F, Hashida M (2007) Novel histidine-conjugated galactosylated cationic liposomes for efficient hepatocyte-selective gene transfer in human hepatoma HepG2 cells. J Control Release 118:262–270

138. Ulbrich K, Etrych T, Chytil P, Jelinkova M, Rihova B (2004) Antibody-targeted polymer-doxorubicin conjugates with pH-controlled activation. J Drug Target 12:477–489

139. Ulbrich K, Subr V, Strohalm J, Plocova D, Jelinkova M, Rihova B (2000) Polymeric drugs based on conjugates of synthetic and natural macromolecules. I. Synthesis and physico-chemical characterisation. J Control Release 64:63–79

140. Beyer U, Roth T, Schumacher P, Maier G, Unold A, Frahm AW, Fiebig HH, Unger C, Kratz F (1998) Synthesis and in vitro efficacy of transferrin conjugates of the anticancer drug chlorambucil. J Med Chem 41:2701–2708

141. Tomlinson R, Heller J, Brocchini S, Duncan R (2003) Polyacetal-doxorubicin conjugates designed for pH-dependent degradation. Bioconjug Chem 14:1096–1106

142. Wang CY, Huang L (1989) Highly efficient DNA delivery mediated by pH-sensitive immuno-liposomes. Biochemistry 28:9508–9514

143. Litzinger DC, Huang L (1992) Phosphatidylethanolamine liposomes: drug delivery, gene transfer and immunodiagnostic applications. Biochim Biophys Acta 1113:201–227

144. Connor J, Huang L (1986) pH-sensitive immunoliposomes as an efficient and target-specific carrier for antitumor drugs. Cancer Res 46:3431–3435

145. Couffin-Hoarau AC, Leroux JC (2004) Report on the use of poly(organophosphazenes) for the design of stimuli-responsive vesicles. Biomacromolecules 5:2082–2087

146. Ellens H, Bentz J, Szoka FC (1984) pH-induced destabilization of phosphatidylethanolamine-containing liposomes: role of bilayer contact. Biochemistry 23:1532–1538

147. Simoes S, Moreira JN, Fonseca C, Duzgunes N, de Lima MC (2004) On the formulation of pH-sensitive liposomes with long circulation times. Adv Drug Deliv Rev 56:947–965

148. Lee ES, Na K, Bae YH (2003) Polymeric micelle for tumor pH and folate-mediated targeting. J Control Release 91:103–113

149. Leroux J, Roux E, Le Garrec D, Hong K, Drummond DC (2001) N-isopropylacrylamide copolymers for the preparation of pH-sensitive liposomes and polymeric micelles. J Control Release 72:71–84

150. Shen H, Eisenberg A (2000) Control of architecture in block-copolymer vesicles we thank the petroleum research fund, administered by the American Chemical Society, for the support of this work. Angew Chem Int Ed Engl 39:3310–3312

151. Ihre HR, Padilla De Jesus OL, Szoka FC Jr, Frechet JM (2002) Polyester dendritic systems for drug delivery applications: design, synthesis, and characterization. Bioconjug Chem 13:443–452

152. Gillies ER, Jonsson TB, Frechet JM (2004) Stimuli-responsive supramolecular assemblies of linear-dendritic copolymers. J Am Chem Soc 126:11936–11943

153. Gupta AK, Naregalkar RR, Vaidya VD, Gupta M (2007) Recent advances on surface engineering of magnetic iron oxide nanoparticles and their biomedical applications. Nanomedicine 2:23–39

154. Jin H, Kang KA (2007) Application of novel metal nanoparticles as optical/thermal agents in optical mammography and hyperthermic treatment for breast cancer. Adv Exp Med Biol 599:45–52

155. Ahmed M, Lukyanov AN, Torchilin V, Tournier H, Schneider AN, Goldberg SN (2005) Combined radiofrequency ablation and adjuvant liposomal chemotherapy: effect of chemotherapeutic agent, nanoparticle size, and circulation time. J Vasc Interv Radiol 16:1365–1371

156. Meyer DE, Shin BC, Kong GA, Dewhirst MW, Chilkoti A (2001) Drug targeting using thermally responsive polymers and local hyperthermia. J Control Release 74:213–224

157. Chung JE, Yokoyama M, Okano T (2000) Inner core segment design for drug delivery control of thermo-responsive polymeric micelles. J Control Release 65:93–103

158. Bae KH, Choi SH, Park SY, Lee Y, Park TG (2006) Thermosensitive pluronic micelles stabilized by shell cross-linking with gold nanoparticles. Langmuir 22:6380–6384

159. Yatvin MB, Weinstein JN, Dennis WH, Blumenthal R (1978) Design of liposomes for enhanced local release of drugs by hyperthermia. Science 202:1290–1293

160. Kono K (2001) Thermosensitive polymer-modified liposomes. Adv Drug Deliv Rev 53:307–319

161. Kono K, Nakai R, Morimoto K, Takagishi T (1999) Thermosensitive polymer-modified liposomes that release contents around physiological temperature. Biochim Biophys Acta 1416:239–250

162. Kono K, Yoshino K, Takagishi T (2002) Effect of poly(ethylene glycol) grafts on temperature-sensitivity of thermosensitive polymer-modified liposomes. J Control Release 80:321–332

163. Saito G, Swanson JA, Lee KD (2003) Drug delivery strategy utilizing conjugation via reversible disulfide linkages: role and site of cellular reducing activities. Adv Drug Deliv Rev 55:199–215

164. Collins DS, Unanue ER, Harding CV (1991) Reduction of disulfide bonds within lysosomes is a key step in antigen processing. J Immunol 147:4054–4059

165. Cavallaro G, Campisi M, Licciardi M, Ogris M, Giammona G (2006) Reversibly stable thiopolyplexes for intracellular delivery of genes. J Control Release 115:322–334

166. Kommareddy S, Amiji M (2005) Preparation and evaluation of thiol-modified gelatin nanoparticles for intracellular DNA delivery in response to glutathione. Bioconjug Chem 16:1423–1432
167. Kommareddy S, Amiji M (2007) Poly(ethylene glycol)-modified thiolated gelatin nanoparticles for glutathione-responsive intracellular DNA delivery. Nanomedicine 3:32–42
168. Carlisle RC, Etrych T, Briggs SS, Preece JA, Ulbrich K, Seymour LW (2004) Polymer-coated polyethylenimine/DNA complexes designed for triggered activation by intracellular reduction. J Gene Med 6:337–344
169. Neu M, Germershaus O, Mao S, Voigt KH, Behe M, Kissel T (2007) Crosslinked nanocarriers based upon poly(ethylene imine) for systemic plasmid delivery: in vitro characterization and in vivo studies in mice. J Control Release 118:370–380
170. Wang Y, Chen P, Shen J (2006) The development and characterization of a glutathione-sensitive cross-linked polyethylenimine gene vector. Biomaterials 27:5292–5298
171. Schmitz T, Bravo-Osuna I, Vauthier C, Ponchel G, Loretz B, Bernkop-Schnurch A (2007) Development and in vitro evaluation of a thiomer-based nanoparticulate gene delivery system. Biomaterials 28:524–531
172. Niculescu-Duvaz I (2000) Technology evaluation: gemtuzumab ozogamicin, Celltech group. Curr Opin Mol Ther 2:691–696
173. West KR, Otto S (2005) Reversible covalent chemistry in drug delivery. Curr Drug Discov Technol 2:123–160
174. Huang Z, Li W, MacKay JA, Szoka FC Jr (2005) Thiocholesterol-based lipids for ordered assembly of bioresponsive gene carriers. Mol Ther 11:409–417
175. Gabizon AA, Tzemach D, Horowitz AT, Shmeeda H, Yeh J, Zalipsky S (2006) Reduced toxicity and superior therapeutic activity of a mitomycin C lipid-based prodrug incorporated in pegylated liposomes. Clin Cancer Res 12:1913–1920
176. Allen TM (1994) Long-circulating (sterically stabilized) liposomes for targeted drug delivery. Trends Pharmacol Sci 15:215–220
177. Bhadra D, Bhadra S, Jain P, Jain NK (2002) Pegnology: a review of PEG-ylated systems. Pharmazie 57:5–29
178. Olivier JC (2005) Drug transport to brain with targeted nanoparticles. NeuroRx 2:108–119
179. Zalipsky S (1995) Functionalized poly(ethylene glycol) for preparation of biologically relevant conjugates. Bioconjug Chem 6:150–165
180. Gref R, Minamitake Y, Peracchia MT, Trubetskoy V, Torchilin V, Langer R (1994) Biodegradable long-circulating polymeric nanospheres. Science 263:1600–1603
181. Muller M, Voros J, Csucs G, Walter E, Danuser G, Merkle HP, Spencer ND, Textor M (2003) Surface modification of PLGA microspheres. J Biomed Mater Res A 66:55–61
182. Calvo P, Gouritin B, Chacun H, Desmaele D, D'Angelo J, Noel JP, Georgin D, Fattal E, Andreux JP, Couvreur P (2001) Long-circulating PEGylated polycyanoacrylate nanoparticles as new drug carrier for brain delivery. Pharm Res 18:1157–1166
183. de Sousa Delgado A, Leonard M, Dellacherie E (2000) Surface modification of polystyrene nanoparticles using dextrans and dextran-POE copolymers: polymer adsorption and colloidal characterization. J Biomater Sci Polym Ed 11:1395–1410
184. Kenworthy AK, Hristova K, Needham D, McIntosh TJ (1995) Range and magnitude of the steric pressure between bilayers containing phospholipids with covalently attached poly(ethylene glycol). Biophys J 68:1921–1936
185. Proffitt RT, Williams LE, Presant CA, Tin GW, Uliana JA, Gamble RC, Baldeschwieler JD (1983) Liposomal blockade of the reticuloendothelial system: improved tumor imaging with small unilamellar vesicles. Science 220:502–505
186. Santra S, Zhang P, Wang K, Tapec R, Tan W (2001) Conjugation of biomolecules with luminophore-doped silica nanoparticles for photostable biomarkers. Anal Chem 73:4988–4993
187. Medintz IL, Uyeda HT, Goldman ER, Mattoussi H (2005) Quantum dot bioconjugates for imaging, labelling and sensing. Nat Mater 4:435–446
188. Loo C, Lowery A, Halas N, West J, Drezek R (2005) Immunotargeted nanoshells for integrated cancer imaging and therapy. Nano Lett 5:709–711

189. Sipkins DA, Cheresh DA, Kazemi MR, Nevin LM, Bednarski MD, Li KC (1998) Detection of tumor angiogenesis in vivo by alphaVbeta3-targeted magnetic resonance imaging. Nat Med 4:623–626

190. Newman SP, Wilding IR (1999) Imaging techniques for assessing drug delivery in man. Pharm Sci Technol Today 2:181–189

191. Veiseh O, Gunn JW, Zhang M (2010) Design and fabrication of magnetic nanoparticles for targeted drug delivery and imaging. Adv Drug Deliv Rev 62:284–304

192. Frullano L, Meade TJ (2007) Multimodal MRI contrast agents. J Biol Inorg Chem 12:939–949

193. Michalet X, Pinaud FF, Bentolila LA, Tsay JM, Doose S, Li JJ, Sundaresan G, Wu AM, Gambhir SS, Weiss S (2005) Quantum dots for live cells, in vivo imaging, and diagnostics. Science 307:538–544

194. Keren S, Zavaleta C, Cheng Z, de la Zerda A, Gheysens O, Gambhir SS (2008) Noninvasive molecular imaging of small living subjects using Raman spectroscopy. Proc Natl Acad Sci U S A 105:5844–5849

195. Desai A, Vyas T, Amiji M (2008) Cytotoxicity and apoptosis enhancement in brain tumor cells upon coadministration of paclitaxel and ceramide in nanoemulsion formulations. J Pharm Sci 97:2745–2756

196. Ganta S, Paxton JW, Baguley BC, Garg S (2009) Formulation and pharmacokinetic evaluation of an asulacrine nanocrystalline suspension for intravenous delivery. Int J Pharm 367:179–186

197. Misra R, Sahoo SK (2011) Coformulation of doxorubicin and curcumin in poly(D, L-lactide-co-glycolide) nanoparticles suppresses the development of multidrug resistance in K562 cells. Mol Pharm 8:852–866

198. Zimmermann GR, Lehar J, Keith CT (2007) Multi-target therapeutics: when the whole is greater than the sum of the parts. Drug Discov Today 12:34–42

199. Batist G, Gelmon KA, Chi KN, Miller WH Jr, Chia SK, Mayer LD, Swenson CE, Janoff AS, Louie AC (2009) Safety, pharmacokinetics, and efficacy of CPX-1 liposome injection in patients with advanced solid tumors. Clin Cancer Res 15:692–700

200. Feldman EJ, Lancet JE, Kolitz JE, Ritchie EK, Roboz GJ, List AF, Allen SL, Asatiani E, Mayer LD, Swenson C, Louie AC (2011) First-in-man study of CPX-351: a liposomal carrier containing cytarabine and daunorubicin in a fixed 5:1 molar ratio for the treatment of relapsed and refractory acute myeloid leukemia. J Clin Oncol 29:979–985

201. FDA (2013) Guidance compliance & regulatory information. http://www.fda.gov/Drugs/

202. ICH (2013) The International Conference on Harmonisation of technical requirements for registration of pharmaceuticals for human use. http://www.ich.org/

203. FDA (2002) Guidance for industry, liposome drug products, chemistry, manufacturing, and controls; human pharmacokinetics and bioavailability; and labeling documentation, draft guidance

204. Tyner K (2011) Nanomedicines and the regulatory path, roundtable presentation. Center of Innovation for Nanobiotechnology, Research Triangle Park

205. European Medicines Agency (2010) European Medicines Agency holds first scientific workshop on nanomedicines. EMA/559074/2010

206. Prescott C (2010) Regenerative nanomedicines: an emerging investment prospective? J R Soc Interface 7(Suppl 6):S783–S787

207. FDA (2008) Q8(R1) pharmaceutical development revision 1. International Conference on Harmonisation (ICH) Q8 guideline

208. Nasr M (2006) Nasr M. FDA's view on QbD. Industry guidance

209. FDA (2009) Guidance for Industry Q9 Quality Risk management. International Conference on Harmonisation (ICH) Q9 guideline; Q10 Pharmaceutical Quality System. International Conference on Harmonisation (ICH) Q10 guideline

210. Tebbey PW, Rink C (2009) Target product profile: a renaissance for its definition and use. J Med Mark 9:301

211. Vastag B (2011) Panel backs new NIH center devoted to translational medicine. Nat Med 17:5

212. DiMasi JA, Grabowski HG (2007) The cost of biopharmaceutical R&D: is biotech different? Managerial Decis Econ 28:469–479
213. Couvreur P, Vauthier C (2006) Nanotechnology: intelligent design to treat complex disease. Pharm Res 23:1417–1450
214. Rabinow BE (2004) Nanosuspensions in drug delivery. Nat Rev Drug Discov 3:785–796

Chapter 11
Harmful or Helpful, the Toxicity and Safety of Nano-sized Medicine

Sophie A. Rocks, Huijun Zhu, Robert Dorey, and Philip Holmes

11.1 Introduction

11.1.1 Chapter Overview

Engineered nanomaterials bring well known benefits to manufacturing and material properties, whether it is lower cost, novel functionality or improved dispersal within a system. However, there is a wide body of research that also suggests that there are potential issues of increased toxicity with at least some nano-scale materials. This chapter summarises the recent research into the toxicity of engineered nanomaterials, the current understanding of the different regulatory approaches that are applicable to the healthcare industry in particular, and highlights areas of concern that require addressing within the near future.

S.A. Rocks (✉) • H. Zhu
School of Energy, Environment and Agrifoods, Cranfield University,
Cranfield, Bedfordshire MK43 0AL, UK
e-mail: s.rocks@cranfield.ac.uk

R. Dorey
Department of Mechanical Engineering Sciences, Faculty of Engineering and
Physical Sciences, University of Surrey, Guildford, Surrey, GU2 7XH, UK

P. Holmes
Risk & Policy Analysts, Loddon, Norfolk NR14 6LT, UK

© Springer Science+Business Media New York 2014   237
Y. Ge et al. (eds.), *Nanomedicine*, Nanostructure Science and Technology,
DOI 10.1007/978-1-4614-2140-5_11

11.1.2   Paradox of Nanomaterials as Toxins and Therapeutic Agents

Engineered nanomaterials have been suggested by the European Commission as comprising materials with external dimensions or having internal structures or surface structures that measure between 1 and 100 nm [1]; although this recommended definition does not concur with previous recommendations from a European Scientific Committee [2, 3]. The recommended definition encompasses all kinds of nanomaterials, irrespective of their origin, and determines that materials that contain particles (in an unbound state or as an aggregate or agglomerate) where 50 % or more of the number size distribution of the particles has one or more dimensions in the stated size range [1]. The EC definition does not identify nanomaterials as hazardous (i.e. intrinsic adverse properties of an agent [4]); however materials that do not conform to the definition may still exhibit nano-related effects. Conversely the European Medicines Agency (EMA) describes nanotechnology as *"the use of tiny structures – less than 1,000 nm across – that are designed to have specific properties"* [5] suggesting that there is further disagreement between different sectors.

Nanomaterials are considered to represent a major advance in engineering, enabling functional coatings with unique properties and reducing costs for manufacturers amongst other benefits [6]. Particularly in the field of medicine, nano-scale drugs have enabled advances in cancer diagnostics, imaging and therapy [7–11]. However, as is common for all emerging technologies, there are concerns that the associated benefits from the development and commercial use of a technology will overtake the consideration of the potential inherent adverse effects of a substance (hazard) and exposure to susceptible organisms (receptors [4, 12]), and legislation must follow development.

As the use of engineered nanomaterials increases in manufacturing and industrial contexts so too does the investigation into their beneficial use within medicine; there are strong comparisons to be drawn between the study of risks associated with engineered nanomaterials and those of nanomedicines. In the published literature to date, however, there has been little discussion as to the toxicity of nanomedicines. Many researchers have reported that toxicity issues related to nanomaterials used in nanomedicine are often ignored (e.g. Kagan et al. [13], Moghimi et al. [14] and Linkov et al. [15]) but there is a widely held belief that, with the development of novel materials, there is also a moral requirement for the consideration of the toxicological and environmental effects of those materials. Whilst the majority of this chapter will draw upon the published literature on the toxicity and risk of engineered nanomaterials, the concerns and issues raised should be remembered to be highly comparable with the situation facing nanomedicines though differences will be indicated where possible.

11.2 Toxicology of Engineered Nanomaterials: Size Really Does Matter

11.2.1 Engineering Benefits of Nanomaterials

The global market for nanotechnology is expected to reach US\$ 1 trillion by 2015 and to employ approximately two million workers [16, 17]. The high level of industrial and commercial interest in nanotechnology can be considered to be a result of the opportunities presented by the changes in properties and functionality that occurs at the nano-scale [18]. These altered properties may result in, for example, improved performance, providing large changes in composite properties as a result of relatively small additions of materials, and can even allow entirely new functional properties to be incorporated into products, thereby impacting on manufacturing capabilities, methods and costs [19–21]. However, with increased commercial use, there is also an increased possibility of unknown and/or underappreciated adverse human health and environmental impacts occurring, particularly when the mechanisms and interactions with the biological environments (both human and the wider environmental receptors) are under-explored [12].

For many, the benefits represented by nano-scale manufacturing are that nano-scale or nano-structured materials have novel properties when compared to bulk forms of a material [20]. In many cases these properties do not correspond to those of the bulk material or even those of the constituent atoms or molecules. Instead, the altered properties can be considered as unique to the nano-scale and are related to the different proportions or states of surface atoms, absence of grain boundaries or even distortion of the atomic or electronic structure of the material [22].

Bulk materials are typically composed of collections of grains, each grain atomically bonded to its neighbour but possessing a unique crystallographic orientation. Only rarely are these bulk materials composed of a single grain. In these cases they are referred to as single crystals and, due to their rarity and difficulty in fabrication, tend to attract a considerable price premium. As the atomic structure and bonding are disrupted across the grain boundaries, the chemical and mechanical properties of the material become altered from that of a single crystal material. As the size of the polycrystalline body as a whole approaches that of the individual grains, changes in material properties arise due to the changes in the proportion of material near to the grain boundaries. At this stage the crystal structure is still identical to that of the bulk material and, for a given mass of material, there will be a higher proportion of surface atoms resulting in an increase in specific reactivity. Decreasing the size of the material still further into the nano-region results not only in a continued change in the ratio of surface atoms to bulk atoms and associated changes in specific reactivity, but also in a deformation of the crystal lattice as the atomic bonds are distorted as a consequence of trying to minimise the surface area relative to the bulk volume of the crystal. This has two effects: firstly the reactivity of the surface atoms is altered as atomic bonds become easier to break; and secondly the inter-atomic separations of individual atoms are altered, changing the crystal structure and giving rise to changes in physical, thermal, electronic and optical behaviour.

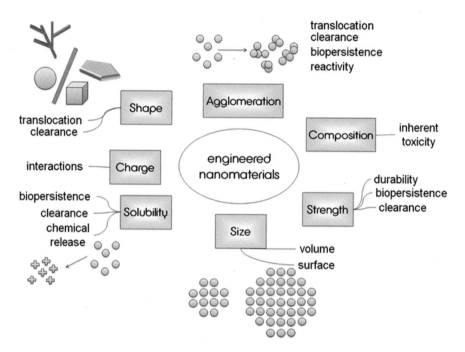

Fig. 11.1 Properties of engineered nanomaterials (ENM) and their environmental effects [48]

These changes will result in altered physiochemical properties of a nanomaterial when compared to the bulk material (Fig. 11.1) and will vary depending on the size of the nanomaterial. While such changes may be relatively innocuous and have no immediate effect for many nanomedicines, there may be far reaching consequences that arise at later stages within the life cycle of the material. Such changes may be simply an increase in strength that makes it harder to process or break down nanomedicines, changes in reactivity such that dissolution or reaction rates are altered, but may potentially extend to changes in fundamental properties that interfere with the fundamental diagnosis. Such variations will also impact the mechanical properties of the materials; whilst this may not be a clinical issue for many nanomedicines, it may cause changes in material strength, reducing degradation within the environment. Once a nanomaterial is dissolved the inherent properties of the chemicals within the particle are considered to be the toxic agent and the particle itself is no longer a concern [23].

11.2.2 Nanotoxicity

There are examples where toxicity is dependent on the size of the particle, and where quantum effects and functional properties (i.e. the optical, magnetic or electrical properties) can dominate material behaviour [16]. Studies have shown that nanoparticle toxicity can be altered by the size of the nanoparticle [24], although this may not be a linear relationship [25]. Additionally, the surface chemistry or addition of a functionalised coating of the nanoparticle has been shown to relate to changes in toxicity [26–28]. The solubility of nanoparticles influences their toxicity, with insoluble nanoparticles showing an increased measure of toxicity as the particle size decreases [29] and highly soluble nanoparticles also showing, in some cases, more toxicity to cells than less soluble nanoparticles [27]. However, given the number of variables and lack of standardised test procedures or "standard" nanoparticles to benchmark experiments, it is not currently possible to predict with any accuracy how particular physical or chemical properties of nanoparticles will influence toxicity. There are many toxicity tests that use established cell lines rather than whole body models, which may not provide relevant exposure conditions to enable conclusions to be drawn. Whilst in vitro tests may give an indication of the mechanism of toxicity and can be considered as more ethically justifiable (considering the drive towards refinement, replacement and reduction of animal models within toxicological assessment), it is possible that nanoparticles may result in unpredictably higher in vivo toxicity by causing greater damage than bulk forms due to interactions with biological fluids (e.g. blood, extracellular fluid) or may be altered within a biological environment either through the formation of agglomerates or coating with protein shells.

It has been suggested that the increased surface area to volume ratio shown by nanoparticles may be the overriding attribute that influences the biological reactivity of such compounds. Indeed this change does result in an increased proportion of atoms at the surface of the particle and thereby increases the potential number of reactive groups for a given mass or dose of material [22]. However, surface modification may also result in the presence of specific surface groups that alter the reactivity or change the lipophilicity of the particle.

A number of common toxicological pathways have been suggested for nanomaterials and nanoparticles (see Table 11.1). These mechanisms are a result of: surface properties such as electron donor/acceptor active groups or redox cycling via metal ion coatings (e.g. Fe^{3+}) or organic compounds (e.g. quinones) resulting in direct or indirect free radical generation; dissolution of the particle by the media which may be retarded or passivated by a coating (either intentional or formed as a result of interaction with the biological media); or UV radiation leading to free radical generation. Whilst such observations seem to imply a common underlying process, relevant to a number of different toxicological mechanisms, involving free radical generation

Table 11.1 Generic toxicological mechanisms identified with engineered nanomaterials (ENM) and examples of ENMs that have been shown to result in such end points

Toxicological mechanisms	Examples (from literature)
Oxidative stress and free radical production	Metal oxides, carbon nanotubes
Inflammatory reactions	Metal oxides, carbon nanotubes
Cell membrane damage (lysis/death)	Metal oxides, carbon nanomaterials
Fibrogenesis (asbestos-like damage)	High aspect ratio particles (e.g. rods), carbon nanotubes
Blood platelet and clotting interference	Silicon oxide

that leads to an increase in reactive oxygen species (ROS) [30, 31], so as to result in increased oxidative stress, this has yet to be demonstrated as the primary mechanism across the wide variety of nanomedicines or nanomaterials that are now available.

11.2.3 Specific Issues Related to Nanomedicines

The recommended definition of nanomaterials [1] identifies that special circumstances may occur for pharmaceuticals and that the definition 'should not prejudice the use of the term 'nano' when defining certain pharmaceuticals and medical devices'. The current EMA definition considers that the structures are less than 1,000 nm across and designed to have specific properties [5] – a concept that has not yet been considered by the other recommendations. The EMA specifically recognises that the Committee for Medicinal Products for Human Use (CHMP) has recommended approval of medicines based on nanotechnology, including medicines composed of liposomes (containing active substances such as doxorubicin) and nano-scale particles of an active substance (such as paclitaxel and aprepitant) [5].

Size is a major factor when considering the translocation, reactivity and fate of nano-scale materials within a body and this will directly influence their toxicity [14]. Many "non-nano" medicines are given as small molecular drugs within a carrier membrane or matrix. Once released from their carrier (i.e. the surrounding membrane is dissolved within the stomach), the molecules will stay in the blood stream until the majority are removed via first-pass clearance in the kidneys. This will reduce the active therapeutic dose but also increase the possibility of nephrotoxicity or parenteral exposure.

Conversely nanomedicine forms are larger than normal molecular-based drugs, which suggests that this comparison would run counter to the issues previously discussed. The larger particle size may mean that the nanomedicines do not undergo first-pass clearance from the bloodstream to the same extent, as size influences elimination from the body. For instance, if a nanoparticle is greater than 20 nm in diameter, it will avoid first pass elimination, thereby resulting in a prolongation of systemic exposure. Particles smaller than 150 nm will also avoid sequestration by sinusoidal fenestrae in the spleen and the liver (which are approximately 100–150 nm diameter [32]), again prolonging systemic exposure. Therefore, if the objective is to ensure maximum exposure and reduce elimination from the body, the "ideal" diameter for a nanomedicine particle is between 20 and 150 nm in diameter. The increased exposure, increasing the time in the blood stream, will enable a greater proportion of the dose to reach the target organ when compared to comparable molecular drug forms. The resultant increased availability of the nanomedicine would mean that a reduced dose would be required to achieve the same therapeutic outcome [33]. However, the reduced elimination will also expose a greater number of potential target tissues to the nanomedicine and increase the possibility of translocation within the body. This also means that, whilst current nanotoxicological studies on chemical exposure concentrates on the difficulties related to reduced particle size, nanomedicine requires consideration of the increased particle size (or presence of a particle) compared to a molecule.

The size of the nanomedicine may also determine the therapeutic targeting. Passive targeting is possible with nanomaterials where the diameter is less than 300 nm. The preferential targeting of tumours by nanomedicines is increased by enhanced permeability of the angiogenic tumour vessels and increased retention of the circulating nanomedicine within the body [34]. Active targeting requires surface modifications for specific ligands (e.g. peptides, antibodies, aptamers or small molecules) – enabling a further reduction in the dose whist ensuring that the necessary therapeutic concentration is maintained at the target site. The improved specificity would be anticipated to lead to an improved therapeutic index and reduce the potential for adverse side effects (i.e. effectively reducing the potential for toxicity of the nanomaterial [34]).

The benefits of nanomaterials in nanomedicine also directly reflect their toxicity. Whilst nanomaterials are developed for their surface properties (versus bulk materials), and the surface is in direct contact with the body tissue, it is rational that this is an indicator of the potential severity of toxicity [16]. Therefore the shape and size of nanomaterials which determined their unique beneficial effects will also influence their toxicological effects. These characteristics will also influence the translocation of nanomaterials throughout the body – a mechanism dictated by shape, size and surface coating. As toxicity tests are conducted on healthy animals and preclinical tests on nanomedicines are conducted on "healthy" subjects (volunteers) rather than on subjects suffering from disease states, the actual effects elicited in diseased tissues cannot be truly characterised before clinical application in humans suffering from the particular conditions for which the medicine is intended. This suggests that additional investigations may be required, not only to consider how physical and chemical characteristics dictate the translocation and distribution of nanomaterials around the body but also to consider how this and the effects of the nanomedicine may alter when administered to a diseased subject.

11.2.4   Toxicity Testing of Nano-scale Particles

Alongside the development of suitable approaches to characterise the toxicity of nanomaterials in general, there needs to be separate consideration as to whether, when nanovesicles are used as drug carriers, the toxicity of the incorporated drug is altered (either reduced or increased). In medicinal testing, the toxicity of the whole formulation is appropriate rather than considering the separate components (i.e. even at phase 1 testing) and this does not (at least initially) need to distinguish between the specific toxicities of the individual components.

Nanomedicine products may encompass a wide variety of structures, which are not directly considered by existing hazard characterisation approaches. In general the majority of nanomedicines currently under trial are either nano-sized carriers where the external surface (the carrier) is exposed to the body until exposure to a stimulus causes the drug to be released – i.e. delivered to the target tissue (e.g. liposomes), nanosuspensions used to improve drug solubility, or metal nanoparticles used for bioimaging of, for example, tumours [35].

Currently, uncertainties relating to the risks posed by nanoparticles (including nanomedicines) are difficult to address because of the absence of knowledge on the range of potential interactions of these at a physiological and/or molecular level, and due to the possible consequences for human health [36]. In such circumstances, use of expert judgment is essential as is common in a range of risk assessment approaches [36, 37]. Whilst drawing on such expert knowledge may provide some insights when these are not otherwise objectively measurable and, may be crucial to identifying important variables, estimating uncertainties in parameters and weighing the strengths of competing models and mechanisms [4], it does require a degree of understanding of the specific issues associated with the particular scenario under consideration. In terms of assessing the potential hazards and risks associated with nanomedicines, the associated unknowns and significant knowledge gaps mean that application of expert judgement may be unsubstantiated and potentially difficult to justify.

11.3   EC/UK Regulation

In Europe and its Member States, the regulatory assessment of medicines is covered under specific legislation, however the generic chemicals either imported or manufactured for use within this process are considered separately [12, 38]. To date, in chemical regulations, the hazard assessment of nanomaterials and nanoparticles has not been well distinguished from that of larger particles or bulk materials, and the risk assessment process has followed suit, despite indications that the behaviour of nanomaterials may be different [12]. Indeed, whilst there is now a suggested definition and guidance on testing needs from the European Commission [1], this has yet to be adopted in Member State legislative approaches and the significance with regard to nanomedicines remains uncertain. Additionally, there is the further complication that the prime current distinction between chemicals for regulatory purposes is based by their chemical formulae (CAS number), which does not allow for particle size or shape to be adequately considered. This is further complicated as some materials that are declared or marketed as "nano" are not, whilst some that are not explicitly labelled as "nano" are [6].

In Europe, regulations distinguish between medicinal products and medical devices. A medicinal product is defined as substance or combination of substances presented as having properties for treating or preventing disease in human beings (Directive 2001/83/EC) and can be considered to cover the majority of nanomedicines; these are subject to more stringent regulation. The regulation of medical devices (used for medical purposes in patients, in diagnosis, therapy or surgery) is less developed and – in large part – relies on self-regulation [39]. Nanomedicines are considered under these same regulations [39].

Pharmaceutical products are regulated by country-specific organisations. The Medicines and Healthcare products Regulatory Agency (MHRA; UK) adopts legislation from the European Community, where the legislation for medicinal products for veterinary and human use is established under the EU medicines regulatory regime (Regulation 726/2004 and Directives 2004/27/EC [human] and 2004/28/EC [veterinary]) and controlled by the EMA.

Within the EU, any new medicine (including nanomedicines) requires a national or EU Marketing Authorisation to be granted before marketing, with failure to obtain this authorisation penalised as set out in Regulation (EC) No. 658/2007 which considers the safety, quality and efficacy of the medicines. The MHRA and expert advisory bodies within the UK control new medicine production to ensure that they meet the required standards on safety and effectiveness throughout the lifetime of the product (known as pharmacovigilance). Within this process, international standards set by the World Health Organisation (WHO; resolution WHA22.50; WHA28.65) provide quality assurance for the global market and reduce the need for repetition of toxicity testing. However, for nanomedicines where the standard toxicity testing regimen may not be able to deal with the specific issues, there remains an outstanding need for additional research and guidance. It is noted that the current regulatory framework has no specific provisions for nanomaterials [37], but that the EMA is actively considering potential changes to legislation [40].

In any event, pharmaceuticals must pass through extensive preclinical assessment before entering into clinical trials (Phases 1–3; Harman 2004) (Fig. 11.2).

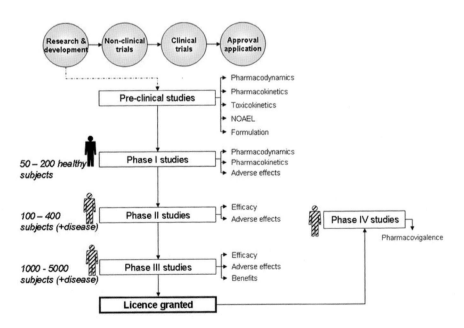

Fig. 11.2 Risk assessment in pharmaceutical development. Phase IV studies (pharmacovigilance) ensures the continuous monitoring of unexpected events over a set period of time

Within this process, guidelines must be followed that control the conduct of the supporting studies, including Good Laboratory Practice (GLP; OECD ENV/MC/CHEM(98)/17, 1998), Good Manufacturing Practice (GMP; EU Directive 2004/27/EC) and Good Clinical Practice (GCP; EU Directive 2005/28/EC).

Preclinical safety studies are intended to define both pharmacological and toxicological effects, both prior to and throughout clinical development, using both in vitro and in vivo studies to aid characterisation [39]. However, pharmaceuticals that are similar, both structurally and pharmacologically, to an available product may need less extensive toxicity testing and, since nanomaterials are not explicitly recognised within EC legislation, it may be difficult to adequately distinguish novel nanoscale medicines (e.g. where there are changes in size or shape) from previously tested medicines. The existing preclinical tests require use of high multiple doses on two animal species, extensive histopathology (on most organs), functional tests (including cardiac, neurologic, respiratory, reproductive, immune systems) and, potentially, extended treatment periods (up to 2 years or more for carcinogenicity). However existing designs do not take into account the particular issues surrounding nanoscale particles [12, 19, 38, 41, 42], which include:

- Dosimetry – many tests use concentration to determine dose levels, however this is not appropriate for use with nanoscale particles. Instead surface area:volume ratio, particle number and particle size distribution are suggested to enable a measurement of potential reactive surface available; and
- Detection of nanomaterials – nanomaterials, individually or in small aggregates, will not be detectable by light microscopy. Whilst chemical labelling will help to show the location of nanoparticles within histology samples, this will also change the properties of the material. Electron microscopy techniques may be used to image accumulation of particles, but this will be difficult to isolate. Additionally, it is necessary to confirm the chemical structure of the nanomaterials using techniques such as energy dispersive X-ray (EDX) and X-ray photoelectron spectroscopy (XPS).

In vitro assays of various types can be used to evaluate and characterise biological activity related to clinical activity, i.e. pharmacodynamic behaviour. Cell lines or primary cell cultures may also be used to examine, for example, the direct effects on cellular phenotype and look for proliferative responses [43]. Mammalian cell lines or tissue cultures can – at least to some extent – be used to predict specific aspects of in vivo activity and to assess quantitatively the relative sensitivity of various species (including humans). Studies can be used to determine receptor occupancy, receptor affinity and pharmacological effects, to assist in the selection of appropriate animal species for further in vivo pharmacological and toxicological tests. In vivo studies assessing the pharmacological activity (including defining mechanisms) are often used to support the rationale of the proposed use of the product in clinical studies.

Data on the pharmacokinetics and bioavailability of the product in the animal species will help to inform decisions as to the appropriate amount and volume that should be administered in studies to test animals. These latter investigations will provide further information on both pharmacological and toxicological dose–

response relationships, and include the definition of what constitutes a toxic dose and what represents the no observed adverse effect level (NOAEL) [4].

Subsequent clinical trials – informed by the preclinical data – allow safety and pharmacokinetic data on humans to be collected as well as – in later phases – its clinical efficacy. Thus it is possible to compare the biological properties of the product as predicted from studies in animal models with the data gained from humans. Unlike the situation in many other fields, risk assessment of medicines therefore involves not only the extrapolation of data across species from studies in animals in relation to the potential toxic effects in humans but also the direct evaluation of human toxicity and pharmacological effect data. In such circumstances, safety assessments are not based on the application of a standard (default) uncertainty factor to the NOAEL from animal studies but the findings from such studies remain important in assessing the adequacy of the safety assessment arising from the clinical trials.

The use of nanoparticles and nanotechnology in medicines are numerous [34, 35]. Commercial applications have been identified as focusing on drug delivery and targeting, as well as the bioavailability of medicines but novel applications are likely to make use of the increased functionality that the nanoscale offers. Of particular interest is the increased surface area to volume ratio of nanoparticles, which in turn increases the particle surface energy [22] and which may make such particles more biologically reactive [44]. This increased biological activity may be beneficial or harmful. However, the same properties that make a nanoparticle of medical interest also mean that it is harder to predict, based on current understanding, the behaviour of the nanoparticle and, therefore, its toxicity profile. There are currently no specific regulatory requirements to test nanoparticles for health, safety and environmental impacts separate to those for bulk materials [5, 23]. Yet, it may be argued that, in any event the regulatory requirements on the testing of pharmaceuticals are already extremely robust and designed to support cautionary developments (as opposed to precautionary [45]) and these are therefore applicable to the manufacture of nanoscale medicines.

11.4 Conclusions

Despite strenuous research efforts to date, the development of methods that permit the detailed characterisation of the potential hazards of nanomaterials (including nanomedicines) is still in its infancy [46]. The assessment of risks associated with the use of such materials – in terms of human or environmental health – remains fraught with uncertainties and knowledge gaps – some of which have yet to be even fully delineated [47]. Specifically these knowledge gaps prevent a full understanding of the potential hazards that may be associated with nanomedicines and are further complicated by contradictory guidance on labelling or for distinguishing between different morphologies of particles. Whilst nanomedicine involves the direct intentional exposure, as well as potentially indirect exposure, of humans to engineered nanomaterials, there are still a number of distinct essential knowledge gaps to be addressed before this novel field is understood, including:

- What are the specific nanomaterial physicochemical properties that need to be characterised for a nanomedicine to inform the potential fate and transport of nanomaterials within the body?
- How can the limited data that are currently available on nanomedicine's toxicity, exposure and environmental fate and transport, be used to inform experimental procedures?
- How do the specific physical and chemical characteristics of nanomaterials contribute to toxicity specifically in relation to nanomedicine applications?
- How may specific delivery mechanisms influence nanomedicine toxicity?

Additionally, there is also the possibility that unexpected interactions between nanomedicines and co-administered traditional pharmaceuticals (or indeed other exogenous substances present in the body) will result in particular concerns given the wide range of chemicals to which we are exposed in everyday life and anticipating the possible interactions, changes in chemical behaviour or in tissue behaviour that could arise.

It is clear that nanomedicines are one of many novel applications of nanomaterials in today's society and offer many potential clinical benefits. However, the implications and uncertainties that are now emerging as to the potential toxicity of non-medical nanomaterials should be considered in terms of their significance to the development of nanomedical applications, in particular to inform the development of appropriate hazard and risk assessment methodologies and risk management techniques that are suited to medical applications and to inform approaches to the communication of evidence and understanding of the balance between benefit and risk that may associate with nanomedical materials.

References

1. European Commission (2011) Commission recommendation of 18 October 2011 on the definition of nanomaterial (Text with EEA relevance) (2011/696/EU)
2. Scientific Committee on Emerging and Newly-Identified Health Risks (SCENIHR) (2010) Opinion on the scientific basis for the definition of the term "nanomaterial", 8 Dec 2010
3. Nowack B, Mueller N, Krug H et al (2014) How to consider engineered nanomaterials in major accident regulations? Environ Sci Europe 26(1):2
4. Department of Environment, Food and Rural Affairs and Collaborative Centre of Excellence in Understanding and Managing Natural and Environmental Risks, Cranfield University (2011) Guidelines for environmental risk assessment and risk management, Greenleaves III, 3rd edn. Department of Environment, Food and Rural Affairs, London
5. European Medicines Agency (2013) European Medicines Agency. http://www.ema.europa.eu/ema/index.jsp?curl=pages/special_topics/general/general_content_000345.jsp&mid=WC0b01ac05800baed9. Accessed 17 Jan 2014
6. Aitken R, Chaudhry M, Boxall A et al (2006) Manufacture and use of nanomaterials: current status in the UK and global trends. Occup Med 56:300–306

7. Wang A, Lange R, Farokhzad O (2012) Nanoparticle delivery of cancer drugs. Annu Rev Med 63:185–198
8. Zhang G, Zeng X, Li P (2013) Nanomaterials in cancer-therapy drug delivery system. J Biomed Nanotechnol 9(5):741–750
9. Bertrand N, Wu J, Xu X et al (2014) Cancer nanotechnology: the impact of passive and active targeting in the era of modern cancer biology. Adv Drug Deliv Rev 66:2–25
10. Qureshi S, Sahni Y, Singh S et al (2014) Application of nanotechnology in cancer treatment. J Pharm Res Opin 1(2):1–7
11. Sanna V, Pala N, Sechi M (2014) Targeted therapy using nanotechnology: focus on cancer. Int J Nanomedicine 9:467–483
12. Rocks S, Owen R, Pollard S et al. (2009) Risk assessment of manufactured nanomaterials. In: Lead J, Smith E (eds) Environmental and human health effects of nanoparticles. Wiley-Blackwell, Chichester
13. Kagan V, Bayir H, Shvedova A (2005) Nanomedicine and nanotoxicology: two sides of the same coin. Nanomed Nanotechnol Biol Med 1:313–316
14. Moghimi S, Hunter A, Murray J (2005) Nanomedicine: current status and future prospects. FASEB J 19:311–330
15. Linkov I, Loney D, Cormier S et al (2009) Weight of evidence evaluation in environmental assessment: review of qualitative and quantitative approaches. Sci Total Environ 407: 5199–5205
16. Wilson M, Kannangara K, Smith G et al (2002) Nanotechnology: basic science and emerging technologies. Chapman & Hall/CRC, Boca Raton
17. Department of Business, Innovation and Skills (2010) UK nanotechnologies strategy. Crown Copyright, London
18. Royal Commission on Environmental Pollution, (2008) Novel materials in the environment: the case of nanotechnology. The Stationery Office, London
19. Royal Society and Royal Academy of Engineering (2004) Nanoscience and nanotechnologies: opportunities and uncertainties. Royal Society and Royal Academy of Engineering, London
20. Council for Science and Technology (2007) Nanosciences and nanotechnologies: a review of government's progress on its policy commitments. Council for Science and Technology, London
21. Saad R, Thiboutot S, Ampleman G et al (2010) Degradation of trinitroglycerin (TNG) using zero-valent iron nanoparticles/nanosilica SBA-15 composite. Chemosphere 81:853–858
22. Gogotsi Y (ed) (2006) Nanomaterials handbook. Taylor & Francis Group, Boca Raton
23. Bleeker E, De Jong W, Geertsma R et al (2013) Considerations on the EU definition of a nano-material: science to support policy making. Regul Toxicol Pharmacol 65(1):119–125
24. Choi O, Hu Z (2008) Size dependent and reactive oxygen species related nanosilver toxicity to nitrifying bacteria. Environ Sci Tech 42:4583–4588
25. Chen J, Dong X, Zhao J (2009) In vivo acute toxicity of titanium dioxide nanoparticles to mice after intraperitioneal injection. J Appl Toxicol 29:330–337
26. Churg A (2003) Interactions of exogenous or evoked agents and particles: the role of reactive oxygen species. Biol Med 34:1230–1235
27. Colvin V (2003) The potential environmental impact of engineered nanomaterials. Nat Biotechnol 21:1166–1170
28. Zhang L, Bai R, Ge C et al (2011) Rutile TiO_2 particles exert size and surface coating dependent retention and lesions on the murine brain. Toxicol Lett 1:73–81
29. Monteiller C, Tran L, MacNee W et al (2007) The pro-inflammatory effects of low-toxicity low-solubility particles, nanoparticles and fine particles, on epithelial cells in vitro: the role of surface area. Occup Environ Med 64(9):609–615
30. Towner R (2000) Chemistry of spin trapping. In: Rhodes C (ed) Toxicology of the human environment. Taylor and Francis, London, pp 7–24
31. Rocks S (2012) What are the risks of nanoscale particles? Mater World 20:17–19
32. Wisse E, Jacobs F, Topal B et al (2008) The size of endothelial fenestrae in human liver sinusoids: implications for hepatocyte-directed gene transfer. Gene Ther 15:1193–1199
33. Rapoport N, Gao Z, Kennedy A (2007) Multifunctional nanoparticles for combining ultrasonic tumor imaging and targeted chemotherapy. J Natl Cancer Inst 99(14):1095–1106

34. Cho K, Wang X, Nie S et al (2008) Therapeutic nanoparticles for drug delivery in cancer. Clin Cancer Res 14:1310–1316
35. Roszek B, De Jong W, Geertsma R (2005) Nanotechnology in medical applications: state-of-the-art in materials and devices. RIVM report 265001001
36. Flari V, Chaudhry Q, Neslo R et al (2011) Expert judgment based multi-criteria decision model to address uncertainties in risk assessment of nanotechnology-enabled food products. J Nanopart Res 13:1813–1831
37. Bleeker E, Cassee F, Geertsma R et al (2012) Interpretation and implications of the European Commission Recommendation on the definition of nanomaterial. National Institute for Human Health and the Environment (RIVM), Bilthoven
38. Rocks S, Pollard S, Dorey R et al (2007) Comparison of risk assessment approaches for manufactured nanomaterials CB403 report for Department of Environment, Food and Rural Affairs, London
39. Chowdhury N (2010) Regulation of nanomedicines in the EU: distilling lessons from the paediatric and the advanced therapy medicinal products approaches. Nanomedicine 5:135–142
40. Committee for Medicinal Products for Human Use (2006) Reflection paper on nanotechnology-based medicinal products for human use, EMEA/CHMP/79769/2006, London
41. Scientific Committee on Emerging and Newly-Identified Health Risks (SCENIHR) (2007) Opinion on the appropriateness of the risk assessment methodology in accordance with the technical guidance documents for new and existing substances for assessing the risks of nanomaterials, European Commission
42. Organisation of Economic Cooperation and Development (2010) Series on the safety of manufactured nanomaterials no. 27. List of manufactured nanomaterials and list of endpoints for phase one of the sponsorship programme for the testing of manufactured nanomaterials: Revision env/jm/mono(2010)46
43. Greaves P (2011) Histopathology of preclinical toxicity studies: interpretation and relevance in drug safety evaluation, 4th edn. Academic Press, Amsterdam
44. Oberdörster G, Oberdörster E, Oberdörster J (2005) Nanotoxicology: an emerging discipline evolving from studies of ultrafine particles. Environ Health Perspect 113:823–839
45. Throne-Holst H, Sto E (2008) Who should be precautionary? Governance of nanotechnology in the risk society. Technol Anal Strateg Manage 20:99–112
46. Harris S (2009) The regulation of nanomedicine: will the existing regulatory scheme of the FDA suffice? XVI Richmond J Law Technol 4(2):1–25
47. Beaudrie C, Kandlikar M (2011) Horses for courses: risk information and decision making in the regulation of nanomaterials. J Nanopart Res Spec Focus Gov Nanobiotechnol 13:1477–1488
48. Stone V, Johnston H, Clift M (2008) Air pollution, ultrafine and nanoparticle toxicology: cellular and molecular interactions. IEEE Trans Nanobioscience 6:331–340

Chapter 12
Ethical Implications of Nanomedicine: Implications of Imagining New Futures for Medicine

Donald Bruce

12.1 A Historical Introduction

Nanomedicine is seen as one of the most exciting prospects amongst all the potential application of emerging nanosciences and technologies. Sophisticated and exquisitely finely focused instrumentation is providing new understandings of processes of the body and mechanisms of disease. These in turn should open up increasing possibilities for more precise diagnosis and monitoring, and for therapeutic or prophylactic interventions, at the scale of genes, proteins and cells of our bodies. Many of the anticipated benefits of nanomedicine remain as future prospects, and at times there here has been a regrettable tendency to exaggerate. Yet nanotechnologies are beginning to emerge from their initial discovery and exploratory phase. In 6 years since 2008, successive annual *Clinam* European conferences on clinical nanomedicine have reflected a growth in nanomedical techniques, products and clinical trials [1]. This is evidence that this field of nanotechnology is beginning to show at least the first fruits of its promise.

As the scope and influence of nanotechnological applications in medicine increases, there is a corresponding responsibility to consider their ethical and social implications. Ten years ago concerns were expressed at the gap between the rapid development and ethical assessment [2]. But since then numerous studies have explored these questions. A series of European Commission funded projects included a scoping review in 2005 from the Nano-2-Life European Network of Excellence [3, 4], and reports from the NanoBioRaise [5] and NanoMed Round Table [6], Observatory Nano [7] and Deepen [8] projects, and an opinion on nanomedicine of the official European Commission's European Group on Ethics [9].

D. Bruce (✉)
Edinethics Ltd., 11/6 Dundonald Street, Edinburgh EH3 6RZ, UK
e-mail: info@edinethics.co.uk

© Springer Science+Business Media New York 2014
Y. Ge et al. (eds.), *Nanomedicine*, Nanostructure Science and Technology,
DOI 10.1007/978-1-4614-2140-5_12

Early concerns over potential risks had been raised in a report from the ETC Group in 2003 [10]. A seminal scientific study by the Royal Society and Royal Academy of Engineering in 2004 highlighted how little was known scientifically about the behaviour of nano-sized particles regarding health and the environment. While this was not deemed sufficient to justify what was seen as an over-precautionary call by ETC for a general moratorium, the academies' report was notable for pointing to the need for substantial research into the risk aspects, and also for including ethics in its considerations [11].

The Nano-2-Life and NanoBio-Raise reports noted a gap of legitimation and accountability between researchers and developers and a European general public which was largely unaware of what nanotechnology was [4, 5, 12]. This gap of engagement has begun to be addressed a number of national studies, for example in the UK, Germany and Switzerland and Netherlands, as well as the above EC research projects [13–15].

These activities were energised in part by concerns of governments and the EU, that nanotechnologies might arouse public suspicions and NGO opposition comparable to that experienced with genetically modified food. It was widely acknowledged that nanotechnologies should be opened to public debate as a matter for so-called upstream engagement. For example, stakeholder groups were invited members of a UK Government forum on nanotechnologies which ran for several years [16].

But in general, nanotechnologies have not so far aroused major controversies in Europe. There have been critical reports from NGO groups in particular sectors like food and cosmetics, and the use of nanosilver as an anti-bacterial agent in hospitals [17, 18]. More worrying were a few organised protests of militant groups in France which disrupted a national programme of public debates, and a series of deeply disturbing letter bomb attacks in Mexico, which seriously injured scientists [19]. Such militant opposition does not seem to represent general public attitudes. From the various studies mentioned above, the prospects for nanomedicine so far seem to have met with a broad approval among the wider European population, but with some concerns and at times some useful insights.

For example, a public engagement study for the UK Engineering and Physical Sciences Research Council (EPSRC) [20] asked people about specific nanomedical research priorities. General support was expressed for most applications of nanomedicine, but tempered by concerns about long terms risks and what may happen when the medical research goals enter the domain of large commercial corporations. But its participants also brought important human insights that nanotechnological solutions might sometimes be too sophisticated. Nanosilver surface coatings in hospitals might be less useful than simply doing more ordinary cleaning. Implanted theranostic devices could become too 'smart' if they did not allow control by either the patient or a doctor.

Because the technologies are themselves evolving, ethical reflection is inevitably work in progress. This chapter surveys some of the main ethical and social issues which have emerged to date. Eight themes will be considered: cross-cutting issues, diagnostics, remote monitoring, targeted drug delivery, theranostics, regenerative medicine, risk and uncertainty, and the relationship of nanomedicine to notions of human enhancement. Three types of questions may arise, running as threads through these themes. In some cases there may be ethical issues with the techniques themselves. There will be ethical and social implications from their uptake by individuals and societies, including possible unintended consequences. Thirdly some more fundamental questions are asked about the values, goals and presuppositions that accompany the technical drivers of nanomedicine.

12.2 Generic and Cross-Cutting Ethical Issues

The first generic question often asked is whether applying nanotechnologies to medicine raises any 'new' ethical issues. The EC European Group on Ethics observed that, notwithstanding the revolution it promised, nanomedicine did not raise issues in biomedical ethics that had not been encountered and considered before [8]. While there was no dramatic new question for the European Commission legislative bodies to face, it did not mean that there are no ethical issues to consider. Nanotechnologies tend to be 'enabling technologies' which provide novel means to existing medical objectives, like rapid point-of-care diagnostics or targeting drug delivery to diseased cells. As is often the case, such new technologies may raise old problems in new ways, or amplify existing issues, or shed new light on them. As the Royal Society report noted, the important thing is to address them, whether new or not [10].

A second generic question relates to implicit values in nanomedicine. The 'enabling' concept does not represent the full story. Nanomedical innovations are not just neutral devices or tools, with no ethical significance. They are considered to be 'value-laden'. This means that the diagnostics, devices and drugs which are being enabled by nanotechnologies all, to some extent, embody certain values, visions and tacit assumptions about future medicine and healthcare. The innovators could be said to be co-producing values and artefacts at the same time. This is indeed inevitably the case for early stage innovation. But it is the task of ethical reflection, social analysis and public engagement to check that the values of technical experts are ones which cohere with ethical views in the wider society [21, 22]. Often, underlying changes of value are difficult to see at the time. The NanoMedRound Table report drew attention to the perspective that nanotechnologies are enabling technologies within a wider progressive reshaping of medicine through technologies in general, whose values and goals need to be duly examined [6].

It is important to consider how these changes relate to our understandings of the human person, of society, and of the role and limits of medicine. By nature, nanomedicine seeks to measure and intervene in the body at the most reduced scales. In doing so, it brings together disciplines with different contexts and concepts – physical and materials sciences, engineering, biotechnology, neurosciences, informatics and medicine. Applying an inherently reductionist focus from the analytical sciences to complex systems in the human body may create some conceptual tensions to the broader practices of medicine. The focus on biological functions at the smallest levels needs also to be duly related to our wider understandings of human beings, derived from culture, religion and ethics, and to concepts like human dignity, personhood, divine image, autonomy, and so on. What is the moral status of human beings, considered in relation to normal adult life, to the earliest (embryonic and/or foetal) and last weeks and months of life, and to people being subjects in medical research, or of novel or experimental medical treatments?

To take one example, at the end of life, no guarantees can be made of life quality into extreme old age, yet breakthroughs inspired or enabled by nanomedicine may mean that many more people will in future survive into a final, frail stage of life. How should we handle the ethical decisions arising from longer average life spans? Is human dignity only associated with a certain quality of life? Or is it something which is intrinsic to a human being, wholly independent of one's state of bodily health or capability? What is it that makes life worth living – is it only the possession of certain faculties, or is it something fundamental in being human, regardless of our frailty? These are old questions, but ones which nanomedicine may amplify in particular ways.

Again, as technologies develop at the brain-machine interface, these renew long-standing discussions about the relationship between one's identity, the mind/brain and the body. In pursuit of nanomedical solutions, how far should we develop devices which promote direct brain-machine interactions, or apply external or internal controls to the brain? Is our human responsibility or autonomy modified if we have a neurotransplant? Should technologies devised and permitted in a strict medical context then be applied without limit to non-medical interventions?

The Nano-2-Life review noted that, whilst welcoming the many new possibilities to treat disease and alleviate suffering, medicine should not be reduced to engineering solutions. In applying techniques derived from nanosciences, it is important not to lose sight of the wider values of medicine and health care which see the patient as more than mal-operating functions. Materials scientists and bioengineers may make very good devices to help, but they may not be the best doctors to treat the whole person, or to help people face the point when a condition is beyond even the best human ingenuity to treat [7].

A third type of cross-cutting ethical issues relates to social justice. In a world of much inequity, how far should we promote advanced nanomedical technologies, if the likelihood is that they will favour only a few who can afford them, or only those countries with the relevant innovation and regulatory infrastructure? For optimal cases, nano-based therapies may be cheap, or may achieve a net saving in long term healthcare costs. But for some applications, the additional expense of sophisticated therapies may place further financial strain on health care systems. These in turn pose further dilemmas for those responsible for apportioning stretched resources. If nano- and other technologies find medical treatments for hitherto intractable problems, medicine is likely to become constrained less by untreatable conditions, and more by the lack of resources to treat everyone. The situation is more acute, when considered in wider global terms. So much could be done to alleviate suffering on a very wide scale by much more basic health provisions than nanotechnology. There is thus also an opportunity cost, both in ethics and resources, in concentrating on developing high-tech medicine. Do we have the appropriate balance?

12.3   Diagnostics, Information and Predictive Medicine

One of the most far-reaching impacts of nanomedicine is likely to be in the area of diagnostics. It is foreseeable that nano-enabled 'lab-on-a-chip' devices may be able to perform a rapid genome analysis on a simple blood sample, for example. If the equipment could become affordable, what has hitherto been an expensive specialist analysis could become routinely used in a family doctor's surgery. Similarly, nanoscale methods and devices could act as 'biomarkers' to monitor chemical changes in the body's metabolism that would be early indicators of developing a disease conditions, long before the physiological appearance of the normal symptoms. Such rapid point-of-care diagnostics could greatly extend the scope of information available to doctors and patients about health status, both in the range of conditions, and the range of people.

This is seen as part of a major shift from the evidential medicine based on observable symptoms, to a predictive, pre-symptomatic, 'information based' medicine, in which the early indications of an incipient disease state could be picked up early, perhaps years before the symptoms were observed. The hope is that it should then be possible to address the condition long before it takes serious hold in the body. This may delay or reduce the onset, or even prevent it altogether.

A goal of this pre-symptomatic approach to medicine is to move beyond addressing clinical symptoms, or families known to be at risk of a particular disease, to be able to pick up indications in people who would consider themselves healthy. A core assumption in such knowledge-based medicine is that information is taken to be a universal good. In many cases having more specific and relevant information about the condition of a patient will indeed be of much benefit. But is this always the case? When examined more closely, the value of the information is highly context-dependent.

Three questions arise:

(a) How useful is information about our present and future health status?
(b) What preventive interventions are justified within the body, based on what level of diagnostic and especially pre-symptomatic information, and on what levels of probability?
(c) To whom should my health status information be known, other than myself?

Consider four types of preliminary indication of a medical condition that might be obtained using nano-enabled diagnostics. The first two are situations where the knowledge gives a clear and unambiguous diagnosis.

1. As the doctor, if you have an indication, you know what to do, but you don't usually know early enough. For example in atherosclerosis, the first indication may be a heart attack 'out of the blue'. If doctors and patients only knew that the condition was developing, actions could be taken that might delay the condition, or make it much less serious, or in the best cases could prevent it from developing altogether. Here, there is a strong ethical case for pre-symptomatic information, for knowing in advance.

2. In the second case, the disease will certainly develop, but there is nothing that can be done to prevent it. It is relatively uncommon to know with such certainty, and usually it is on the basis of genetic information. Huntington's disease is the classic case, where a late-onset highly distressing, terminal degenerative disease can be detected by a simple genetic test. If the defect is found, the outcome is unavoidable. The offer to test members of families in which the disease is known to run leads to two typical reactions. Some choose to have the test – to remove the uncertainty and know one way or the other – to have prior warning of what will indeed happen, or to have the relief of knowing that they and their children will be free of the disease. Others choose not to have test, not wishing to know any earlier in their lives whether so devastating and unpreventable degradation is about to happen to them.

   The other two situations are probabilistic. The diagnostic information only indicates an increased *propensity* to developing a certain condition, but it is not certain that it will develop significantly. Perhaps it is not clear that the level is sufficient for the condition to take hold, or the test or biomarker shows a positive indication of only one of several factors all of which need to come into play before the disease really develops. Some of these other factors may be known, but others may as yet be unknown. What the doctor can tell a patient is that they have a greater than average *probability* of developing the particular condition, perhaps a lot higher in some cases, but not that they will necessarily get it. Again there are two types.

3. The condition might never develop, but there is nothing that can be done, if it did. This is a probabilistic version of case 2. The worth of having the information is even more problematic.

4. In case 1, the doctor or patient can do something about it, but the condition may never actually develop. This presents a real challenge as to what to do. It will depend on the nature of what can be done – the degree of invasiveness to one's body, the restriction or disruption to one's daily life, activities and expectations, the risks involved. If the actions were a simply change of diet or getting more exercise, this might not pose much of a problem. If a much more invasive and profound intervention is involved, like a mastectomy or prostate removal, or if the procedure or treatment itself carries a significant risk, the patient is left with a dilemma. Moreover, if the indication now entails starting to take pharmaceutical drugs for the rest of one's life, there are likely to be side-effects of the drugs when taken on a long term basis.

These examples are given to illustrate the complex range of contexts into which the information provided by nano-diagnostics would be received by doctors, patients and their families and carers. Pre-symptomatic information may be very beneficial in many situations, but not in all. In case 1, the information may well be life-saving, whereas in case 2 it may be an advanced warning of one's death. In cases 3 and 4, the information is only probabilistic. It is not a foregone conclusion that such information is necessarily a benefit. The predicted condition may still never happen. People are likely to vary in their attitude. For some, the knowledge of the probability would represent prudent foreknowledge and some actions they could take, just in case. But for others it would just be more stress to one's life, and they would rather get on with living and face the situation if it arises.

Advanced knowledge of a future disease or condition also carries the personal problem of admitting to oneself that one is now an 'ill' person. Does a person in their 30's, who feels perfectly healthy and otherwise in the prime of life, now want to start lifelong preventive medicine, for a condition that might normally only appear in their 70's? And if it is not certain that it would develop, at what percent probability and what degree of seriousness of outcome, does one judge that it is indeed worth beginning such a course of action?

None of these are new issues in medicine. But what is different is their *scope*. Hitherto such questions were typically faced by families which carried a serious genetic disease, for which a test existed, but they were not often experienced in the wider population. The new situation that nanomedical diagnostics is likely to open up is that the range of testable medical conditions, and the availability of testing within the population, will be very much wider. The sorts of dilemmas indicated in the examples above are likely to become much more commonplace.

In considering the tests enabled by nanodiagnostics, careful consideration needs to be given to who wants the information, and under what circumstances? A recurring theme in literature, from Greek myths, through Macbeth to Harry Potter, is that having advanced information is something humans do not typically handle well. A considerable re-education may be needed of what people might in future expect from a visit to the doctor. A point-of-care genome analysis may tell her what antibiotic to prescribe for my persistent cough, but when does she tell me that my genome also shows, say, that I have my higher than average risk of colon cancer?

In such cases, at what point should people be told, and what should they be told? In general it is a doctor's duty to tell the patient material information for their health. But how does the doctor allow for the fact that I might prefer not to know, given that even to reveal to me that there *is* information is likely to prompt me to want to know what it's about, or else to worry about what it might be? Important factors will include such things as explaining to the patient about the extent and expectations from a test, and discussing how far they wish a preliminary result to be investigated further. Regulatory bodies have basic principles and guidelines on matters such as consent and confidentiality, see for example in [23], but these may need to be kept under review in the light of advances in nanodiagnostics in the next few years.

There is also an increased relevance to some long-standing ethical questions: to whom the information should be available, other than the patient and the doctor, and who has the right to interpret its implications outside my immediate health context? Insurance companies, employers, the police, or state databases may each consider they have legitimate claims to my data, under certain circumstances. There were serious concerns in the past over insurance obligations. Insurance companies feared people having tests and then taking out large insurance protections based on the result. Some people in families at risk did not wish to have a genetic test, however, not for fear of finding they have a susceptible gene, but for fear of what an insurance company might do with the result. In the UK there is an ongoing moratorium protecting people who undergo genetic tests from having to disclose the information to insurers, except in the limited cases [24]. The availability of tests has not grown as much as anticipated [25], but nanodiagnostics may change this picture in future.

A further issue is sheer volume of information which may become available, and how it could be handed and interpreted. Already the internet has made far more medical information available to patients, both good and bad, about diseases and treatments. How well can either doctors or patients cope with all the new specific information that may emerge about my health and its implications from nanomedicine? This is likely to change the balance of the doctor-patient relationship. Some people would no doubt welcome being able to take greater charge of their health in a more informed way, but others will prefer to leave most of it to the professionals. This is also dependent on the type of healthcare system and culture which one is in. The British National Health Service and the private healthcare systems of the USA represent two very different situations, for example.

Another practical implication is that a greatly increased degree of personal counselling is likely to be needed, to help families respond to the dilemmas which the additional knowledge may bring. Significant contact time between professional and patient may be needed, first to understand, to let the implications sink in, and then to begin to decide amongst possible options. Experience from genetic counselling suggests that this may require several meetings over a relatively short period of time. Counselling services for rare genetic conditions have needed relatively modest resources. If the dilemmas of pre-symptomatic medicine become commonplace, much greater emphasis will inevitably need to be put on counselling.

This may seem a long way from nanoscale biomarkers as means of monitoring the condition of certain key health parameters. But if the technologies are as successful as expected then, the impacts of success need also to be considered wisely. The complexities discussed above indicate that it would be naive to embrace nanodiagnostics as part of the 'knowledge economy', without assessing what knowledge is beneficial and what is not, and under what circumstances it should be given, when, and to whom. Information, as such, is blind to human circumstances. It is up to humans not to be driven only by the logic of data, but to take account also of wider values and considerations. To reveal what would normally be hidden can certainly have important advantages for some areas of medicine, but it may on occasions disrupt more natural patterns of human knowledge. The religions and literature of many cultures suggest that in a person's life there may a proper time to know, and a proper time not to know, about a future event, like a terminal disease.

12.4   Remote or Personal Monitoring of Health Status

A development closely related to diagnostics is to combine in a nanoscale device both the means to monitor a health parameter in a patient, and a way to transmit the information to another location. Such 'smart' nano-scale implants in the body may allow someone to go home sooner after an operation, if the healthcare professionals at the hospital can continue to follow the patient's recovery remotely. If key bio-markers fall or rise beyond prescribed warning levels, indicating that something is going wrong, this can act as an alert for action to be taken. This might be a local treatment, or bringing the patient back into hospital.

This concept is has considerable attractions. It would both reduce the time the patient is away from home and loved ones, and also free up much needed bed space in the hospital more quickly. In more remote and rural areas, like the Scottish islands, the ability to monitor a patient remotely could be of great benefit if he is 3 h drive away from a main hospital, provided sufficient local infrastructure was also on hand to respond at need. This could be extended to very elderly or chronically ill people, or those with a known susceptibility, like heart or stroke patients at risk of a recurrence. Again, changes in critical parameters could forewarn of the need to take preventative measures.

One downside is that this represents a degree of surveillance by a third party. One's whereabouts, and to some degree, something of one's private activities will be known and followed. Even with benevolent intentions, and restricted to particular professionals, family members and carers, some people may not welcome a sense of their privacy being 'snooped on'. On the other hand, it can provide a sense of security that someone is on hand to help if something starts to go wrong. A weekly video or phone link to a nurse to go through the week's readings can become a welcome reassurance, and a point of regular contact, for people living alone with a long term medical condition.

A second implication is the degree of responsibility that is shifted to the home, the patient and the carer. This may be something welcomed. If one has an elderly parent who is living on their own, it could ease some of the stress to know that particular health functions were being monitored. If I am the patient, it might be reassuring for me to keep a regular check on my critical measurements, which become part of the way of life of living with my condition. On the other hand, it can be a considerable additional stress to keep taking or checking a measurement, day in day out, and interpreting what it means if things are not going well.

Looking further into the future, if monitoring health parameters by implants or particles becomes commonplace, there may be pressure to use them for other than medical reasons. Elite sports and military use are examples where monitoring is already done, which are accepted within their special contexts. One could imagine equivalent arguments being put forward for certain occupations, like a pilot or bus driver where many other lives are in one's hands. Technically, it would be a relatively short step to a more general surveillance of performance in a work context, or for an insurer to want to monitor one's risk levels. But this would cross a significant ethical line where the primary beneficiary of people's private health information is no longer the individual him/herself, but various third parties. This may represent another clash between the value of human persons and a merely functional logic to use the capacities which some new nanotechnology may enable. At this point I would argue we lose the precedence of human values at our peril.

12.5 Implants and Targeted Delivery

The flag ship concept of nanomedicine is the notion of targeted drug delivery. Typically a pharmaceutical product is encapsulated in a nanoscale carrier, to which has been attached an array of ligands variously designed to carry the particle through hostile media in the body, to recognise specific diseased cells of the body, or toxins or other malicious entities, and, on encountering them, to release the active ingredient to do its job. It should do this without affecting healthy cells nearby, or other organs of the body. In particular, it is intended to overcome the problem of the introducing chemicals systemically. At present, in order to have a sufficient concentration of the active chemical in the affected organs, it has to be introduced into the whole body, and chemicals powerful enough to attach a cancer cell, say, may impact harmfully in many unintended elsewhere in the body, causing significant side effects. A growing number of targeted pharmaceuticals are in use, and many more are likely to follow.

There is a strong ethical case in favour of nanomedical methods which have a reasonable prospect of addressing the problems of systemic drug delivery. The primary ethical questions are about risk, long term implants, and overclaiming. There are issues of risk in relation to the nanocarrier and its behaviour and ultimate fate in the body. Because of their size, nanoparticles have the potential to pass through barriers of the body and end up in strange places. While the understanding of nanoparticle risks remains relatively poorly developed, a precautionary approach is appropriate. Given the present uncertainties of the technique, targeted delivery is better focused on the more serious or intractable medical conditions, until a body of substantial experience has been accumulated of these therapies in practice. Ongoing and long term monitoring needs to go hand in hand with more fundamental risk studies on different materials and formulations. Carriers that the body will naturally degrade or 'functionalised' particles have been seen as more favourable than pure inorganic materials which would remain largely inert.

This leads to a more general question about implanting nanoscale devices into the human body. This might be to monitor the progression of disease, to deliver therapies in situ, to provide scaffolds for replacing damaged or failing tissues, or to provide external monitoring of our health. How far should we make nano-technically enabled interventions in the body? Should this remain something done exceptionally, under particular conditions, as with macro interventions like hip replacements and stents? Or should we expect this eventually become the normal pattern, in widespread use? The technology may be beguiling. What other important factors need to be taken into consideration, and what takes precedence if conflicts between values arise?

The last point in this section is a tendency for some promoters of targeted delivery to overclaim about their products. For example, video simulations create an almost military scene of unmanned capsules zooming through blood vessels, seeking out their targets, and delivering their payload by precision impact on the affected cells. The military analogy is perhaps unfortunate, but the medical equivalents of collateral damage and wrong targeting of supposedly precision bombing, are relevant issues.

Targeted drug delivery should be a step change improvement compared with system delivery of a pharmaceutical, but risks of unintended consequences remain, say if the intended therapeutic molecule hits the wrong target, or if it has side effects which perhaps the researchers did not look for.

The tendency to exaggerate can sometimes present an ethical problem in itself. Hopes for the benefits of for high tech therapies are raised amongst vulnerable patients, or hard pressed policy makers. At earlier stages in innovation, the prospects claimed by the researcher or a company about a 'breakthrough' under idealised research conditions, with a view to attracting further funding, can raise misleading expectations. In practice, applications will only be realised after a critical review of their feasibility, clinical reliability, economics, and safety, and unfortunately many fail at one hurdle or another. Once approved for clinical use, a technique may work well in some patients but prove less effective in others. Such is the nature of innovation. A degree of modesty is therefore called for, both out of respect for the vulnerability of the human patients and recognition of the finite understanding of the method. At this nano-medical interface, the emphasis needs to be that of the doctor treating a human person, rather than the impersonal logic of technique, however good it may be.

12.6   Theranostics

A special case of implants is the so-called 'theranostic' device (*thera*peutic – diag-*nostic*) which would combine a measurement and monitoring role with some kind of therapeutic delivery. The delivery is activated in response to a critical change in level of a parameter which has just been measured. The attraction is not needing to wait before activating the therapeutic response, perhaps to maintain some function like blood sugar levels within a tolerated range, if these were about to drift outside the range. The advantage is that the remedial action is rapid and does not depend on the patient or perhaps a nurse to have to step in, notice the change and activate the response.

It depends, however, on having established a close numerical correlation between what is measured and the degree of therapeutic response. This has to have been established in advance with considerable precision and reliability in order to programme the device accordingly. One problem is applicability and reliability. Can one indeed produce an algorithm so robust, or so flexible, that it applies infallibly to all patients who would present with this condition? It may be possible to tune the device initially for the particular patient, but will the settings continue to be valid as the patient's metabolism changes, with different activities, at different times of day, times in a woman's monthly cycle, etc. To the extent that the device responds automatically, there must be a very high resilience to misleading data. How reliable in engineering terms is the equipment, for example as materials degrade with age, or pump flows become restricted? If modifications became required from time to time, would this be possible?

Conceptually, theranostics relies on the assumption that a necessarily limited set of measurements in a device is sufficient on its own to represent accurately a complex physiological change and to deliver the correct response without human intervention. Such an assumption requires a great deal of trust in the reliability of the design concept, the programming and the engineering. To produce a 'black box' of sufficient flexibility and reliability would seem to be a tall order. Indeed, some companies in this area are reluctant to make devices too automated, in recognition of factors like the variability in patients.

In the UK public consultation on potential research fields in nanomedicine, theranostics received a lower priority than several other nanomedical applications. People expressed concern that scientist might make such a device to be *too* smart. In envisaging theranostic devices implanted in their bodies, people felt they or a doctor should keep some control over the implant and its operation. A degree of human judgement was necessary rather than depending on algorithms and programming. It is another case where human values are needed to modulate engineering logic.

12.7 Regenerative Medicine

One of the most intriguing prospects of nanotechnology is to be able to construct material objects 'atom-by-atom' to any shape or form desired. While its more exaggerated claims have been rightly criticised, one potentially useful application is in regenerative medicine. The isolation of human embryonic stem cells and induced pluripotent cells have opened up many new possibilities to replace lost or damaged cells in vital organs of the body. One aspect is the possibility to regrow nerve cells, for example in spinal cord following an accident, or in the retina in certain cases of blindness. This requires an appropriate tissue scaffold to be grown starting at the nanoscale and building upwards. A number of nanomaterials are being investigated.

While the basic ethical rationale is very good, such techniques run into the serious ethical problems, if they entail the use of cells originally derived via the destruction of human embryos. It is not the place here to rehearse the arguments for and against human embryonic stem cells, but, suffice to say, in some countries and for some individual patients, such technologies would be ruled out unless they could be achieved based on non-embryonic sources of the cells. Fortunately some of the best prospects lie in encouraging the body's own stem cells to regrow, so the ethical problem may be avoidable, but it is important to aware of the issue. Other potential ethical questions would arise if it becomes possible to regrow brain tissue, and in attempts to construct organs by this method outside the body.

12.8 Risk and Uncertainty

The risks associated with the use of nanoparticles and nanoscale implants in the human body have been much discussed. These include the transport of particles to unintended parts of the body, side-effects on the body's metabolism, and the long term use of implants. Given that there will always be risk associated with these types of interventions, there is an ethical dimension to how one sets a tolerable level risk, and against what criteria. Once scientific data become available, the numerical probabilities and consequences remain as numbers on a page, until it is decided what levels constitute acceptable or unacceptable risks. These are ultimately ethical judgements. To make such judgements will require not only much good research but also much careful engagement with different publics, patients' groups and their carers.

Risk may be calculated with a good deal of reliability in areas of established engineering experience. Nanotechnologies, however, typically go beyond well trodden paths. A second dimension is thus how one handles the inevitable uncertainty associated with novel procedures involving tiny devices in the body. How precautionary should we be? From fields such as genetically modified food, two version of the precaution principle emerged – hard and soft. Hard precaution is the inclination not to proceed if a significant case for a risk of harm can be made. Soft precaution argues that one should proceed unless there is reasonable evidence that there is a risk of significant harm, but which at this point cannot be sufficiently evaluated. In principle both are resolvable by further evidence, one way or the other. In practice, however, some uncertainties are likely to remain intractable, or would take a very long time to assess. In the meantime, patients are longing for treatments to their conditions.

The key question will be at what point it is considered that enough is now known to proceed, or it is concluded that the intended process would entail unacceptable risk. Some general principles are that, in areas of uncertainty, the initial focus should be on applications with a high degree of medical benefit, the more serious diseases, on situations where there is a degree of reversibility or recovery from adverse effects, and where there is an ability to track the fate of nanoparticles or implants. But a strict ethical principle of 'do no harm' may not be achievable. This brings us to a final and fundamental point about the way we handle risk.

Since risk is inherent to the human condition, we should resist undue demands for certainty and safety, or a culture of blame for techniques which fail. There is a profound difference between negligence, given what you knew but did not act on, and not knowing something which no one had reason to know at the time. Hindsight can be very destructive in this respect. The question is, set against all the other risks of human living and the particular condition of the patient, how much or little risk is tolerable? And, having decided, everyone involved should recognise that no one can guarantee success.

12.9 Human Enhancement

One cannot end a chapter on the ethics of nanomedicine without briefly considering the issue known as human enhancement. All the examples considered so far address using nanotechnologies for explicitly medical goals. A topic of increasing debate in the last 10 years has been whether we should use these, or other technological methods, to enhance the human body beyond its present capacities. Should nanotechnology only 'make humans better', in the sense of treating diseases and injuries, or should we use it to 'make better humans', by using technology to improve the basic specification of the human body and brain directly?

The first significant study, the 2002 US report NBIC report on 'converging technologies for improving human performances', was optimistic about this latter prospect [26]. A European expert group urged submitting enhancement goals to wider social scrutiny, if our humanity is not to be redefined by a techno-logic driven primarily by technical and economic feasibility [27]. More recent reports are from the UK academies [28] and the Dutch Rathenau Instituut [29] as well as an increasing academic literature and several European Commission studies, some broadly approving, and some more critical ([30–34]). It must be said that only a few technologies exist today. Most of what is discussed in these reports remains as future prospects whose practical feasibility is very uncertain, and some have criticised the field for indulging in too much speculative ethics [35]. The implications are sufficiently far reaching to take the question seriously, however, and we summarise some of the key issues.

The first is the basic ethical issue of whether we should seek to make serious changes to human body and its metabolism. This depends on one's view of the human being and of human technological intervention. Traditional presuppositions hold that there are moral or societal bounds which should act as a restrain on what may otherwise be feasible technically. These limits are drawn from the insights of the religious and cultural traditions, philosophy and theology, the arts and humanities, and the social sciences. Christian thinking for example grounds human nature in God's creation of human beings 'in God's image', although two recent studies considered that this did not mean that enhancements are necessarily prohibited, as such [36, 37]. Traditional views are challenged by transhumanist belief that humans are destined to go beyond our current biological limitations.

It may be helpful to think of this question in terms of three general views. One view sees the human body and its capacities as something 'given' which it is not to be majorly changed. The transhumanist philosophy regards the human body as evolutionary, in principle open to be changed without limit. An intermediate position recognises that humans could be changed in degree but not without limits, not change whatever we regard as our human nature. In summary these are: change nothing, change anything, or draw lines as to limit what may be changed.

The second ethical question is what is meant by the idea human enhancement, and whether enhancements really do enhance? The assumption is made that if I do something that improves some functional capability in my body or brain, it is an enhancement. But on what basis are we to judge whether actually it constitutes an enhancement, beyond a purely subjective view? A focus on improving human performance, for example, seems too limited a criterion, compared with more holistic concepts of the human person. The assumption that to be that little bit faster, stronger, smarter, more retentive, more musical, we are somehow happier and better as humans seems too vulnerable to things going wrong. A better question would be in what sense are we better as human beings by having a particular capacity enhanced? It might indeed be appealing to do certain things better than one could naturally, but would it make the difference between a good life and a poor one? Beyond a certain basic point of physical survival and necessity, what matters most to humans are their relationships. Wider issues such as love, friendship, creativity and spirituality seem to matter more than functional abilities.

A few examples serve to illustrate that enhancements may not be as straightforward as the term suggests at first sight. Suppose retinal implants would provide a true recovery of sight to some blind people, and this is extended to offer vision into the infrared region for the normal sighted. It is said that this would have considerable safety advantages for night driving, for example. On the other hand, would one use the new capacity to drive faster, and not more safely? And why not simply use some form of spectacles to achieve the same end? Secondly, cognitive enhancing drugs have become used by students concentrating for exams, but the value only exists as long as only a few have the advantage. If all students used it, there would be no competitive advantage, but no one would then dare stop using it. Thus all would become locked into a pointless 'enhancement', and one which would not reflect their true ability. A third example is in the field of sport. There are plenty of examples where the over-amplification of critical functions can be pursued out of proportion to the rest of the body, resulting in significant overall damage. The same harmful imbalances have been observed in genetic engineering of animals for faster growth rate, both by selective breeding and molecular intervention [38]. We may need to consider rather carefully before calling an intervention an enhancement.

A third issue is reliability, risk and regulation. Whereas medical devices and pharmaceuticals are subject to strict and complex testing and regulation, there is little or no regulation of enhancements. There is no comparable system to test and guarantee that an implant or a chemical enhancement both does 'what it says on the packet' and is safe and reliable, and is not a false product of the modern equivalent of the quack doctor. Recent experience of unscrupulous and invalidated stem cell treatments underlines the importance of having a system of regulation and validation.

Safety testing also raises a problem. Riskier medical procedures are justified only for the more serious medical conditions. For enhancement technologies there is no comparable balancing good of saving life or preventing serious suffering. This marks an important distinction between nanotechnologies to address medical conditions and those intended to enhance healthy human beings. In academic debate, some have criticised this distinction, because it assumes ideas of a 'human nature' and what is 'normal' to humans, which are merely human constructs of our times, but which do not have any ultimate grounding philosophically. As observed above, this depends somewhat on one's world view.

In contrast to this, one of the findings of a recent public engagement study was that people do seem to have an implicit sense of normalcy in human capacities, even if it would be hard to pin it down to any sort of definition [39]. In assessing a range of potential human enhancement applications, people often made a distinction: technologies to bring people who are ill or disabled in some way 'up' to the norm, were broadly accepted, but using the same technologies to take healthy people beyond the norm, were viewed sceptically or even objected to. Some special situations were perhaps acceptable, for example, for rescue workers in a disaster to take a drug to do without sleep. But to do this in everyday life was seen as abnormal.

To a first approximation the medical – enhancement distinction seems valid. There are indeed situations where the distinction is blurred, which should be considered on their own merits, rather than invalidating the distinction. As Holm has pointed out, we do not think yellow and red are not valid as colours just because they can blur to form orange [40]. Thus whereas nanomedicine would be generally favoured, enhancements were viewed much more critically.

A last main group of issues of human enhancement are its many social implications. Amongst these are three of particular note. The first is a general point that the various implications of enhancement technologies are too important to be treated just as matters of personal preference, but should be regulated at a societal level in most cases. The second is a deep concern that in practice human enhancement is promoted primarily by and for those who are already in high levels of economic and social advantage. Enhancements would inevitably be available mostly for the rich and priviledged, thereby enhancing their advantages still further. Advocates of enhancements point out that any new technology tends to create new winners and losers, and that one should not object to enhancement on that basis. Many things once considered luxuries are now cheap and widely available, like televisions, mobile phones and computers. But suppose enhancement technologies really did prove to be as good as some claim, this would give those who could afford to use them 'hard-wired' advantages. While those less well-off were waiting for the prices to come down, the rich would get ever further ahead, in what would become a 'nano-divide'. Some argue that to pursue personal human enhancements, without regard for those who miss out, people might be enhanced functionally but diminished in humanity [36].

The final social issue is whether, faced with the issues like poverty, poor health, climate change, and global food security, the idea of human enhancement is largely a distraction for the well-off, and which misses the point. It might be argued that what is wrong with the human condition is not a lack of strength, longevity, intelligence, beauty, athleticism, art, science or even education. It lies in deeper moral and spiritual shortcomings of humanity, individually and collectively, as the world's ongoing conflicts show. However much we 'enhanced' ourselves physically, these inherent human failings would remain because these would seem to lie beyond technical fixes.

12.10   Conclusions and Postscript

This chapter has considered a range of ethical and social issues associated with likely advances in nanotechnologies, primarily as applied to medicine, but also possible enhancements of the human body. The technical logic of nano-enabled medicine always needs moderating with ethical reflection to apply human values to achieve wise solutions. In many cases a good ethical case can be made for the considerable medical benefits, but enthusiasm for new technical solutions should not lose sight of the wider perspective of human values, and the long experience of the practice of medicine. Nanomedicine needs to maintain a human face. Human enhancement, on the other hand, does not have the ethical benefit of making an ill person better. It depends for its appeal on a more elusive idea of making 'better' humans. It remains to be seen whether it would actually offer significant improvements to the human condition that would outweigh social concerns, risks and practical problems. It also begs the wider question whether the deepest needs of the human condition cannot be met by technology.

Important social and conceptual changes are likely to accompany the application of nanomedicine, especially in pre-symptomatic diagnostic information giving advanced knowledge about our future health status. We think of ourselves as relatively well or relatively ill. But if in the long term, nanotechnologies might eventually make much hitherto unsuspected data about our bodies accessible to us, what now is a well person? If read-outs of genes, chemicals or other parameters will represent almost any body function, may we find that we are all to some extent 'ill', or at least probably ill?

In many cases, such knowledge will be welcome and valuable, and in some circumstances even be life saving. But a sense of proportion is also needed about 'knowledge-based medicine' and our health status. In his witty Victorian English tale, *Three Men in a Boat*, Jerome K. Jerome recounts going to the British Museum library to look up an ailment in a medical encyclopaedia. But out of curiosity he reads on and finds that he seems to show the symptoms for half of the diseases in the book. 'I went into that reading room a happy healthy man. I crawled out a decrepit wreck.' He went to his doctor with a full list of his supposed diseases, and is given a prescription. He goes to the pharmacy and discovers the prescription is for a daily diet of beef steak, a pint of beer, good exercise and early to bed ... 'and don't stuff your head with things you don't understand!' [41].

References

1. CLINAM (2014) European foundation for clinical nanomedicine. www.clinam.org. Accessed
2. Mnyusiwalla A, Daar AS, Singer PA (2003) Mind the gap: science and ethics in nanotechnology, nanotechnology. Nanotechnology 14:R9–R13
3. Bruce DM (2006a) Ethical and social issues in nanobiotechnologies. EMBO Rep 7:754–758
4. Bruce DM (2006b) Nano2Life ethics: a scoping paper on ethical and social issues in nanobiotechnologies. In: Ach JS, Siep L (eds) Nano-bio-ethics: ethical dimensions of nanobiotechnology. Lit Verlag, Münster, pp 63–83
5. Bennett DJ, Schuurbiers D (2005) Nanobiotechnology: responsible action on issues in society and ethics (NANOBIO-RAISE). In: Nanotech 2005: technical proceedings of the 2005 NSTI nanotechnology conference and trade show, vol 2, pp. 765–768. http://www.nsti.org/publications/Nanotech/2005/pdf/402.pdf. Accessed 17 Jan 2013
6. NanoMed Round Table (2011) NanoMed round table: A report on the nanomedicine environment, http://www.philosophie.tu-darmstadt.de/media/institut_fuer_philosophie/diesunddas/nordmann/nanomed.pdf. Accessed 17 and 18 Jan 2013
7. ObservatoryNANO (2010) ObservatoryNANO 2nd annual report on ethical and societal aspects of nanotechnology, http://www.observatorynano.eu/project/filesystem/files/NanobioethicsApril2010.pdf. 27 Sept 2014

8. DEEPEN (2009) Deepening ethical engagement and participation in emerging nanotechnologies. http://www.geography.dur.ac.uk/Projects/Default.aspx?alias=www.geography.dur.ac.uk/projects/deepen. Accessed 29 July 2013

9. EGE (2007) Ethical aspects of nanomedicine, Opinion 21 of the European group on ethics in science and new technologies to the European Union, 17 Jan 2007. European Commission, Brussels.   http://ec.europa.eu/bepa/european-group-ethics/docs/publications/opinion_21_nano_en.pdf. Accessed 18 Jan 2013

10. ETC Group (2003) The big down: Atomtech – technologies converging at the nano-scale. ETC Group, Winnipeg

11. The Royal Society, The Royal Academy of Engineering (2004) Nanoscience and nanotechnologies: opportunities and uncertainties. The Royal Society, London

12. European Commission (2005) Eurobarometer 63: public opinion in the European Union, European Commission, Brussels. http://ec.europa.eu/public_opinion/archives/eb_arch_en.htm. Accessed 29 July 2013

13. Gavelin K, Wilson R, Doubleday R (2007) Democratic nanotechnologies? The final report of the Nanotechnology Engagement group. Involve, London. http://www.involve.org.uk/neg. Accessed 18 Jan 2013

14. Grobe A, Rissanen M, Funda P, de Beer J and Jonas U (2012) Nanotechnologies from the consumers' point of view: what consumers know and what they would like to know. Stiftung Risiko Dialog, St Gallen. http://www.risiko-dialog.ch/images/RD-Media/PDF/Themen/Nanotechnologie/Konsumentenstudie_Nano_2011_final.pdf (in German). English summary, http://www.risiko-dialog.ch/images/RD-Media/PDF/Themen/Nanotechnologie/Consumerstudy_Nano_EN.pdf. Accessed 18 Jan 2013

15. Rathenau Instituut (2011) Governance of nanotechnology in the Netherlands – Informing and engaging in different social spheres. Rathenau Instituut, Den Haag, The Netherlands. http://www.rathenau.nl/en/publications/publication/governance-of-nanotechnology-in-the-netherlands-informing-and-engaging-in-different-social-spher.html. Accessed 18 Jan 2013

16. DEFRA (2011) Nanotechnologies stakeholder group, 2005–2010. UK Department for the Environment, Food and Rural affairs, London. http://archive.defra.gov.uk/environment/quality/nanotech/research.htm. Accessed 29 July 2013

17. Friends of the Earth (2006) Nanomaterials, sunscreens and cosmetics. http://nano.foe.org.au/sites/default/files/FoEA%20nano%20cosmetics%20report%202MB.pdf. Accessed 18 Jan 2013

18. Friends of the Earth (2011) Nano-silver: policy failure puts public health at risk. http://nano.foe.org.au/sites/default/files/Nano-silver_2011%20Aus%20v2%20web.pdf. Accessed 18 Jan 2013

19. Phillips L (2012) Nanotechnology: armed resistance. Nature 488:576–579, 30 Aug 2012, http://www.nature.com/news/nanotechnology-armed-resistance-1.11287. Accessed 18 Jan 2013

20. Bhattachary D, Stockley R, Hunter A (2008) Nanotechnology for healthcare, prepared for the Engineering and Physical Sciences Research Council, July 2008. Engineering and Physical Sciences Research Council, Swindon

21. Davies S, Macnaghten P, Kearnes M (eds) (2009) Reconfiguring responsibility: lessons for public policy (Part 1 of the report on Deepening debate on nanotechnology). Durham University, Durham

22. Ferrari A, Nordmann A (eds) (2009) Reconfiguring responsibility: lessons for nanoethics (Part 2 of the report on Deepening debate on nanotechnology). Durham University, Durham

23. General Medical Council (2008) Consent: patients and doctors making decisions together. UK General Medical Council, Manchester. http://www.gmc-uk.org/static/documents/content/Consent_-_English_0911.pdf. Accessed 18 Jan 2013

24. Frodshamon G (2012) Insurance will continue to be unaffected by predictive genetic tests, says Government. BioNews 663:2 July 2012. http://www.bionews.org.uk/page_154877.asp. Accessed 18 Jan 2013

25. DoH (2007) See for example, Genetics and Insurance Committee, Sixth report from January 2007 to December 2007. UK Department of Health, London. http://www.dh.gov.uk/prod_con-

sum_dh/groups/dh_digitalassets/@dh/@en/documents/digitalasset/dh_084686.pdf. Accessed 18 Jan 2013

26. Roco MC, Bainbridge WS (eds) (2002) Converging technologies for improving human performances, US National Science Foundation report. US National Science Foundation, Washington, DC

27. Nordmann A (ed) (2004) Converging technologies and the natural, social and cultural world, report of a European Commission expert group. European Commission, Brussels

28. Academy of Medical Sciences (2012) Joint academies project on human enhancement and the future of work. Academy of Medical Sciences, London. http://www.acmedsci.ac.uk/p47prid102.html. Accessed 18 Jan 2013

29. Rathenau Instituut (2012) Human enhancement. Rathenau Instituut, Den Haag, The Netherlands. http://www.rathenau.nl/en/publications/publication/human-enhancement-1.html. Accessed 18 Jan 2013

30. Savalescu J, Bostrom N (2009) Human enhancement. OUP, Oxford

31. ETHENTECH (2012) Ethics of enhancement technologies. EC FP7 ETHENTECH project: Final report to the European Commission, 2012, (in press)

32. Bruce DM (ed) (2007) Human enhancement: ethical reflections on emerging nanobio-technologies, Report of an expert working group of the EC NanoBio-Raise programme. http://www.edinethics.co.uk/nano/enhancement.html

33. Ferrari A, Coenen C, Grunwald A (2012) Visions and ethics in current discourse on human enhancement. Nanoethics 6:215–229

34. Nordmann A (2007) If and then: a critique of speculative nanoethics. NanoEthics 1:31–46

35. Meulen R, Savulescu J, Kahane G (eds) (2011) Enhancing human capacities, Wiley-Blackwell, Oxford. This collection was derived in part from the EC FP6 Science and society project ENHANCE – Enhancing human capacities: ethics, regulation and European policy. http://www.euprojekt.su.se/index.php/kb_83/io_1278/io.html. Accessed 18 Jan 2013

36. Conference of European Churches (2009) Human enhancement – A discussion document. Conference of European churches (CEC), Strasbourg. http://csc.ceceurope.org/fileadmin/filer/csc/Ethics_Biotechnology/Human_enhancement_final_March_10.pdf. Accessed 18 Jan 2013

37. Committee of European Catholic Bishops Conferences (2009) On the prospects for improvement of the human being (human enhancement) 25 May 2009, Committee of European catholic bishops conferences (COMECE). Brussels

38. Bruce D, Bruce A (eds) (1998) Engineering genesis. Earthscan, London, pp 113–114

39. Bruce DM, Walker P (2012) Democs games on human enhancement played to end June 2012 within the ETHENTECH Project. Submitted to the European Commission, (in press)

40. Holm S, McNamee M (2011) Physical enhancement: what baseline, whose judgement? In: Enhancing human capacities. Wiley-Blackwell, Oxford, pp 291–303

41. Jerome JK (1889) Three men in a boat. Penguin, London, pp 7–10